ESSAYS UPON HEREDITY AND KINDRED BIOLOGICAL PROBLEMS

BY

August Weismann

Oxford

1889

CONTENTS.

Translator. A. E. Shipley

Translator. Selmar Schönland

Abstracts of Professor Weismann's Essays on Heredity and Kindred

Problems, already Published in this Country.

I.   A short abstract in 'Nature,' Vol. XXXVII, pp. 541-542, by P. C. Mitchell.

II. A short abstract in 'Nature,' Vol. XXXVIII, pp. 156-157, by P. C. Mitchell.

III. A short article on the subject of this Essay in 'The Nineteenth Century' for May, 1885, by A. E. Shipley.

IV. Abstract in 'Nature,' Vol. XXXIII, pp. 154-157, by Professor Moseley.

V.   Abstract in 'Nature,' Vol. XXXIV, pp. 629-632, by Professor Moseley.

VI. Abstract in 'Nature,' Vol. XXXVI, pp. 607-609, by Professor Weismann.

VII, VIII. The Essays being of so recent a date no abstract has yet appeared in this country.

A criticism of Professor Weismann's theories will be found in 'The Physiology of Plants,' by Professor Vines, Lecture XXIII, pp. 660 et seqq.

THE DURATION OF LIFE.

preface

The following paper was read at the meeting of the Association of German Naturalists at Salzburg, on September 21st, 1881; and it is here printed in essentially the same form. A somewhat longer discussion of a few points has been now intercalated; these were necessarily omitted from the lecture itself for the sake of brevity, and are, therefore, not contained in the account printed in the Proceedings of the fifty-fourth meeting of the Association.

Further additions would not have been admissible without an essential change of form, and therefore I have not put into the text a note which ought otherwise to have been there, and which is now to be found in the Appendix, as Note 8. It fills up a gap which was left in the text, for the above-mentioned reason, by attempting to give an explanation of the normal death of cells of tissues—an explanation which is required if we are to maintain that unicellular organisms are so constituted as to be potentially immortal.

The other parts of the Appendix contain, partly further expansions, partly proofs of the views brought forward in the text, and above all a compilation of all the observations which are known to me upon the duration of life in several groups of animals. I am indebted to several eminent specialists for the communication of many data, which are among the most exact that I have been able to obtain. Thus Dr. Hagen of Cambridge (U.S.A.) was kind enough to send me an account of his observations upon insects of different orders: Mr. W. H. Edwards of West Virginia, and Dr. Speyer of Rhoden—their experience with butterflies. Dr. Adler of Schleswig sent me data upon the duration of life in *Cynipidae*, which have a special value, as they are accompanied by very exact observations upon the conditions of life in these animals; hence in this case we can directly examine the factors upon which, as I believe, the duration of life is chiefly based. Sir John Lubbock in England, and Dr. August Forel of Zürich, have had the kindness to send me an account of their observations upon ants, and S. Clessin of Ochsenfurth his researches upon our native land and fresh-water Mollusca.

In publishing these valuable communications, together with all facts which I have been able to collect from literature upon the subject of the duration of life, and the little which I have myself observed upon this subject, I hope to provide a stimulus for further observation in this field, which has been hitherto much neglected. The views which I have brought forward in this paper are based on a comparatively small number of facts, at least as far as the duration of life in various species is concerned. The larger the number of accurate data which are supplied, and the more exactly the duration of life and its conditions are ascertained, the more securely will it be possible to establish our views upon the causes which determine the duration of life.

A. W.

Naples, *Dec. 6, 1881.*

I.

THE DURATION OF LIFE.

With your permission, I will bring before you to-day some thoughts upon the subject of the duration of life. I can scarcely do better than begin with the simple but significant words of Johannes Müller: 'Organic bodies are perishable; while life maintains the appearance of immortality in the constant succession of similar individuals, the individuals themselves pass away.'

Omitting, for the time being, any discussion as to the precise accuracy of this statement, it is at any rate obvious that the life of an individual has its natural limit, at least among those animals and plants which are met with in every-day life. But it is equally obvious that the limits are very differently placed in the various species of animals and plants. These differences are so manifest that they have given rise to popular sayings. Thus Jacob Grimm mentions an old German saying, 'A wren lives three years, a dog three times as long as a wren, a horse three times as long as a dog, and a man three times as long as a horse, that is eighty-one years. A donkey attains three times the age of a man, a wild goose three times that of a donkey, a crow three times that of a wild goose, a deer three times that of a crow, and an oak three times the age of a deer.'

If this be true a deer would live 6000 years, and an oak nearly 20,000 years. The saying is certainly not founded upon exact observation, but it becomes true if looked upon as a general statement that the duration of life is very different in different organisms.

The question now arises as to the causes of these great differences. How is it that individuals are endowed with the power of living long in such very various degrees?

One is at first tempted to seek the answer by an appeal to the differences in morphological and chemical structure which separate species from one another. In fact all attempts to throw light upon the subject which have been made up to the present time lie in this direction.

All these explanations are nevertheless insufficient. In a certain sense it is true that the causes of the duration of life must be contained in the organism itself, and cannot be found in any of its external conditions or circumstances. But structure and chemical composition—in short the physiological constitution of the body in the ordinary sense of the words—are not the only factors which determine duration of life. This conclusion forces itself upon our attention as soon as the attempt is made to explain existing facts by these factors alone: there must be some other additional cause contained in the organism as an unknown and invisible part of its constitution, a cause which determines the duration of life.

The size of the organism must in the first place be taken into consideration. Of all organisms in the world, large trees have the longest lives. The Adansonias of the Cape Verd Islands are said to live for 6000 years. The largest animals also attain the greatest age. Thus there is no doubt that whales live for some hundreds of years. Elephants live 200 years, and it would not be difficult to construct a descending series of animals in which the duration of life diminishes in almost exact proportion to the decrease in the size of the body. Thus a horse lives forty years, a blackbird eighteen, a mouse six, and many insects only a few days or weeks.

If however the facts are examined a little more closely it will be observed that the great age (200 years) reached by an elephant is also attained by many smaller animals, such as the pike and carp. The horse lives forty years, but so does a cat or a toad; and a sea anemone has been known to live for over fifty years. The duration of life in a pig (about twenty years) is the same as that in a crayfish, although the latter does not nearly attain the hundredth part of the weight of a pig.

It is therefore evident that length of life cannot be determined by the size of the body alone. There is, however, some relation between these two attributes. A large animal lives longer than a small one because it is larger; it would not be able to become even comparatively large unless endowed with a comparatively long duration of life.

Apart from all other reasons, no one could imagine that the gigantic body of an elephant could be built up like that of a mouse in three weeks, or in a single day like that of the larva of certain flies. The gestation of an elephant lasts for nearly two years, and maturity is only reached after a lapse of about twenty-four years.

Furthermore, to ensure the preservation of the species, a longer time is required by a large animal than by a small one, when both have reached maturity. Thus Leuckart and later Herbert Spencer have pointed out that the absorbing surface of an animal only increases as the square of its length, while its size increases as the cube; and it therefore follows that the larger an animal becomes, the greater will be the difficulty experienced in assimilating any nourishment over and above that which it requires for its own needs, and therefore the more slowly will it reproduce itself.

But although it may be stated generally that the duration of the period of growth and length of life are longest in the largest animals, it is nevertheless impossible to maintain that there is any fixed relation between the two; and Flourens was mistaken when he considered that the length of life was always equivalent to five times the duration of the period of growth. Such a conclusion might be accepted in the case of man if we set his period of growth at twenty years and his length of life at a hundred; but it cannot be accepted for the majority of other Mammalia. Thus the horse lives from forty to fifty years, and the latter age is at least as frequently reached among horses as a hundred years among men; but the horse becomes mature in four years, and the length of its life is thus ten or twelve times as long as its period of growth.

The second factor which influences the duration of life is purely physiological: it is the rate at which the animal lives, the rapidity with which assimilation and the other vital processes take place. Upon this point Lotze remarks in his Microcosmus—'Active and restless mobility destroys the organized body: the swift-footed animals hunted by man, as also dogs, and even apes, are inferior in length of life to man and the larger beasts of prey, which satisfy their needs by a few vigorous efforts.' 'The inertness of the Amphibia is, on the other hand, accompanied by relatively great length of life.'

There is certainly some truth in these observations, and yet it would be a great mistake to assume that activity necessarily implies a short life. The most active birds have very long lives, as will be shown later on: they live as long as and sometimes longer than the majority of Amphibia which reach the same size. The organism must not be looked upon as a heap of combustible material, which is completely reduced to ashes in a certain time the length of which is determined by size, and by the rate at which it burns; but it should be rather compared to a fire, to which fresh fuel can be continually added, and which, whether it burns quickly or slowly, can be kept burning as long as necessity demands.

The connection between activity and shortness of life cannot be explained by supposing that a more rapid consumption of the body occurs, but it is explicable because the increased rate at which the vital processes take place permit the more rapid achievement of the aim and purpose of life, viz. the attainment of maturity and the reproduction of the species.

When I speak of the aim and purpose of life, I am only using figures of speech, and I do not mean to imply that nature is in any way working consciously.

When I was speaking of the relation between duration of life and the size of the body, I might have added another factor which also exerts some influence, viz. the complexity of the structure. Two organisms of the same size, but belonging to different grades of organization, will require different periods of time for their development. Certain animals of a very lowly organization, such as the Rhizopoda, may attain a diameter of ·5 mm. and may thus become larger than many insects' eggs. Yet under favourable circumstances an Amoeba can divide into two animals in ten minutes, while no insect's egg can develope into the young animal in a less period than twenty-four hours. Time is required for the development of the immense number of cells which must in the latter case arise from the single egg-cell.

Hence we may say that the peculiar constitution of an animal does in part determine the length of time which must elapse before reproduction begins. The period before reproduction is however only part of the whole life of an animal, which of course extends over the total period during which the animal exists.

Hitherto it has always been assumed that the duration of this total period is solely determined by the constitution of the animal's body. But the assumption is erroneous. The strength of the spring which drives the wheel of life does not solely depend upon the size of the wheel itself or upon the material of which it is made; and, leaving the metaphor, duration of life is not exclusively determined by the size of the animal, the complexity of its structure, and the rate of its metabolism. The facts are plainly and clearly opposed to such a supposition.

How, for instance, can we explain from this point of view the fact that the queen-ant and the workers live for many years, while the males live for a few weeks at most? The sexes are not distinguished by any great difference in size or complexity of body, or in the rate of metabolism. In all these three particulars they must be looked upon as precisely the same, and yet there is this immense difference between the lengths of their lives.

I shall return later on to this and other similar cases, and for the present I assume it to be proved that physiological considerations alone cannot determine the duration of life. It is not these which alone determine the strength of the spring which moves the machinery of life; we know that springs of different strengths may be fixed in machines of the same kind and quality. This metaphor is however imperfect, because we cannot imagine the existence of any special force in an organism which determines the duration of its life; but it is nevertheless useful because it emphasises the fact that the duration of life is forced upon the organism by causes outside itself, just as the spring is fixed in its place by forces outside the machine, and not only fixed in its place, but chosen of a certain strength so that it will run down after a certain time.

To put it briefly, I consider that duration of life is really dependent upon adaptation to external conditions, that its length, whether longer or shorter, is governed by the needs of the species, and that it is determined by precisely the same mechanical process of regulation as that by which the structure and functions of an organism are adapted to its environment.

Assuming for the moment that these conclusions are valid, let us ask how the duration of life of any given species can have been determined by their means. In the first place, in regulating duration of life, the advantage to the species, and not to the individual, is alone of any importance. This must be obvious to any one who has once thoroughly thought out the process of natural selection. It is of no importance to the species whether the individual lives longer or shorter, but it is of importance that the individual should be enabled to do its work towards the maintenance of the species. This work is reproduction, or the formation of a sufficient number of new individuals to compensate the species for those which die. As soon as the individual has performed its share in this work of compensation, it ceases to be of any value to the species, it has fulfilled its duty and may die. But the individual may be of advantage to the species for a longer period if it not only produces offspring, but tends them for a longer or shorter time, either by protecting, feeding, or instructing them. This last duty is not only undertaken by man, but also by animals, although to a smaller extent; for instance, birds teach their young to fly, and so on.

We should therefore expect to find that, as a rule, life does not greatly outlast the period of reproduction except in those species which tend their young; and as a matter of fact we find that this is the case.

All mammals and birds outlive the period of reproduction, but this never occurs among insects except in those species which tend their young. Furthermore, the life of all the lower animals ceases also with the end of the reproductive period, as far as we can judge.

Duration of life is not however determined in this way, but only the point at which its termination occurs relatively to the cessation of reproduction. The duration itself depends first upon the length of time which is required for the animal to reach maturity—that is, the duration of its youth, and, secondly, upon the length of the period of fertility—that is the time which is necessary for the individual to produce a sufficient number of descendants to ensure the perpetuation of the species. It is precisely this latter point which is determined by external conditions.

There is no species of animal which is not exposed to destruction through various accidental agencies—by hunger or cold, by drought or flood, by epidemics, or by enemies, whether beasts of prey or parasites. We also know that these causes of death are only apparently accidental, or at least that they can only be called accidental as far as a single individual is concerned. As a matter of fact a far greater number of individuals perish through the operation of these agencies than by natural death. There are thousands of species of which the existence depends upon the destruction of other species; as, for example, the various kinds of fish which feed on the countless minute Crustacea inhabiting our lakes.

It is easy to see that an individual is, *ceteris paribus*, more exposed to accidental death when the natural term of its life becomes longer; and therefore the longer the time required by an individual for the production of a sufficient number of descendants to ensure the existence of the species, the greater will be the number of individuals which perish accidentally before they have fulfilled this important duty. Hence it follows, first, that the number of descendants produced by any individual must be greater as the duration of its reproductive period becomes longer; and, secondly, the surprising result that nature does not tend to secure the longest possible life to the adult individual, but, on the contrary, tends to shorten the period of reproductive activity as far as possible, and with this the duration of life; but these conclusions only refer to the animal and not to the vegetable world.

All this sounds very paradoxical, but the facts show that it is true. At first sight numerous instances of remarkably long life seem to refute the argument, but the contradictions are only apparent and disappear on closer investigation.

Birds as a rule live to a surprisingly great age. Even the smallest of our native singing birds lives for ten years, while the nightingale and blackbird live from twelve to eighteen years. A pair of eider ducks were observed to make their nest in the same place for twenty years, and it is believed that these birds sometimes reach the age of nearly one hundred years. A cuckoo, which was recognised by a peculiar note in its call, was heard in the same forest for thirty-two consecutive years. Birds of prey, and birds which live in marshy districts, become much older, for they outlive more than one generation of men.

Schinz mentions a bearded vulture which was seen sitting on a rock upon a glacier near Grindelwald, and the oldest men in Grindelwald had, when boys, seen the same bird sitting on the same rock. A white-headed vulture in the Schönbrunn Zoological Gardens had been in captivity for 118 years, and many examples are known of eagles and falcons reaching an age of over 100 years. Finally, we must not forget Humboldt's Atur parrot from the Orinoco, concerning which the Indians said that it could not be understood because it spoke the language of an extinct tribe.

It is therefore necessary to ask how far we can show that such long lives are really the shortest which are possible under the circumstances.

Two factors must here be taken into consideration; first, that the young of birds are greatly exposed to destructive agencies; and, secondly, that the structure of a bird is adapted for flight and therefore excludes the possibility of any great degree of fertility.

Many birds, like the stormy petrel, the diver, guillemot, and other sea-birds, lay only a single egg, and breed (as is usually the case with birds) only once a year. Others, such as birds of prey, pigeons, and humming-birds, lay two eggs, and it is only those which fly badly, such as jungle fowls and pheasants, which produce a number of eggs (about twenty), and the young of these very species are especially exposed to those dangers which more or less affect the offspring of all birds. Even the eggs of our most powerful native bird of prey, the golden eagle, which all animals fear, and of which the eyrie, perched on a rocky height, is beyond the reach of any enemies, are very frequently destroyed by late frosts or snow in spring, and, at the end of the year in winter, the young birds encounter the fiercest of foes, viz. hunger. In the majority of birds, the egg, as soon as it is laid, becomes exposed to the attacks of enemies; martens and weasels, cats and owls, buzzards and crows are all on the look out for it. At a later period the same enemies destroy numbers of the helpless young, and in winter many succumb in the struggle against cold and hunger, or to the numerous dangers which attend migration over land and sea, dangers which decimate the young birds.

It is impossible directly to ascertain the exact number which are thus destroyed; but we can arrive at an estimate by an indirect method. If we agree with Darwin and Wallace in believing that in most species a certain degree of constancy is maintained in the number of individuals of successive generations, and that therefore the number of individuals within the same area remains tolerably uniform for a certain period of time; it follows that, if we know the fertility and the average duration of life of a species, we can calculate the number of those which perish before reaching maturity. Unfortunately the average length of life is hardly known with certainty in the case of any species of bird. Let us however assume, for the sake of argument, that the individuals of a certain species live for ten years, and that they lay twenty eggs in each year; then of the 200 eggs which are laid during the ten years, which constitute the lifetime of an individual, 198 must be destroyed, and only two will reach maturity, if the number of individuals in the species is to remain constant. Or to take a concrete example; let us fix the duration of life in the golden eagle at 60 years, and its period of immaturity (of which the length is not exactly known) at ten years, and let us assume that it lays two eggs a year;—then a pair will produce 100 eggs in 50 years, and of these only two will develope into adult birds; and thus on an average a pair of eagles will only succeed in bringing a pair of young to maturity once in fifty years. And so far from being an exaggeration, this calculation rather under-estimates the proportion of mortality among the young; it is sufficient however to enforce the fact that the number of young destroyed must reach in birds a very high figure as compared with the number of those which survive [See Note 1].

If this argument holds, and at the same time the fertility from physical and other grounds cannot be increased, it follows that a relatively long life is the only means by which the maintenance of the species of birds can be secured. Hence a great length of life is proved to be an absolute necessity for birds.

I have already mentioned that these animals demonstrate most clearly that physiological considerations do not by any means suffice to explain the duration of life. Although all vital processes take place with greater rapidity and the temperature of the blood is higher in birds than in mammals, yet the former greatly surpass the latter in length of life. Only in the largest Mammalia,—the whales and the elephants—is the duration of life equal to or perhaps greater than that of the longest lived birds. If we compare the relative weights of these animals, the Mammalia are everywhere at a disadvantage. Even such large animals as the horse and bear only attain an age of fifty years at the outside; the lion lives about thirty-five years, the wild boar twenty-five, the sheep fifteen, the fox fourteen, the hare ten, the squirrel and the mouse six years [See Note 2]; but the golden eagle, though it does not weigh more than from 9-12 pounds, and is thus intermediate as regards weight between the hare and the fox, attains nevertheless an age which is ten times as long. The explanation of this difference is to be found first in the much greater fertility of the smaller Mammalia, such as the rabbit or mouse, and secondly in the much lower mortality among the young of the larger Mammalia. The minimum duration of life necessary for the maintenance of the species is therefore much lower than it is among birds. Even here, however, we are not yet in possession of exact statistics indicating the number of young destroyed; but it is obvious that Mammalia possess over birds a great advantage in their intra-uterine development. In Mammalia the destruction of young only begins after birth, while in birds it begins during the development of the embryo. This distinction is in fact carried even further, for many mammals protect their young against enemies for a long time after birth.

It is unnecessary to go further into the details of these cases, or to consider whether and to what extent every class of the animal kingdom conforms to these principles. Thus to consider all or even most of the classes of the animal kingdom would be quite impossible at the present time, because our knowledge of the duration of life among animals is very incomplete. Biological problems have for a long time excited less interest than morphological ones. There is nothing or almost nothing to be found in existing zoological text books upon the duration of life in animals; and even monographs upon single classes, such as the Amphibia, reptiles, or even birds, contain very little on this subject. When we come to the lower animals, knowledge on this point is almost entirely wanting. I have not been able to find a single reference to the age in Echinodermata, and very little about that of worms, Crustacea, and Coelenterata [See Note 4]. The length of life in many molluscan species is very well known, because the age can be determined by markings on the shell [See Note 5]. But even in this group, any exact knowledge, such as would be available for our purpose, is still wanting concerning such necessary points as the degree of fertility, the relation to other animals, and many other factors.

Data the most exact in all respects are found among the insects [See Note 3], and to this class I will for a short time direct your special attention. We will first consider the duration of larval life. This varies very greatly, and chiefly depends upon the nature of the food, and the ease or difficulty with which it can be procured. The larvae of bees reach the pupal stage in five to six days; but it is well known that they are fed with substances of high nutritive value (honey and pollen), and that they require no great effort to obtain the food, which lies heaped up around them. The larval life in many *Ichneumonidae* is but little longer, being passed in a parasitic condition within other insects; abundance of accessible food is thus supplied by the tissues and juices of the host. Again, the larvae of the blow-fly become pupae in eight to ten days, although they move actively in boring their way under the skin and into the tissues of the dead animals upon which they live. The life of the leaf-eating caterpillars of butterflies and moths lasts for six weeks or longer, corresponding to the lower nutritive value of their food and the greater expenditure of muscular energy in obtaining it. Those caterpillars which live upon wood, such as *Cossus ligniperda*, have a larval life of two to three years, and the same is true of hymenopterous insects with similar habits, such as *Sirex*.

Furthermore, predaceous larvae require a long period for attaining their full size, for they can only obtain their prey at rare intervals and by the expenditure of considerable energy. Thus among the dragon-flies larval life lasts for a year, and among many may-flies even two or three years.

All these results can be easily understood from well-known physiological principles, and they indicate that the length of larval life is very elastic, and can be extended as circumstances demand; for otherwise carnivorous and wood-eating larvae could not have survived in the phyletic development of insects. Now it would be a great mistake to suppose that there is any reciprocal relation between duration of life in the larva and in the mature insect, or imago; or, to put it differently, to suppose that the total duration of life is the same in insects of the same size and activity, so that the time which is spent in the larval state is, as it were, deducted from the life of the imago, and *vice versa*. That this cannot be the case is shown by the fact already alluded to, that among bees and ants larval life is of the same length in males and females, while there is a difference of some years between the lengths of their lives as imagos.

The life of the imago is generally very short, and not only ends with the close of the period of reproduction, as was mentioned above, but this latter period is also itself extremely short [See Note 3].

The larva of the cockchafer devours the roots of plants for a period of four years, but the mature insect with its more complex structure endures for a comparatively short time; for the beetle itself dies in about a month after completing its metamorphosis. And this is by no means an extreme case. Most butterflies have an even shorter life, and among the moths there are many species (as in the *Psychidae*) which only live for a few days, while others again, which reproduce by the parthenogenetic method, only live for twenty-four hours. The shortest life is found in the imagos of certain may-flies, which only live four to five hours. They emerge from the pupa-case towards the evening, and as soon as their wings have hardened, they begin to fly, and pair with one another. Then they hover over the water; their eggs are extruded all at once, and death follows almost immediately.

The short life of the imago in insects is easily explained by the principles set forth above. Insects belong to the number of those animals which, even in their mature state, are very liable to be destroyed by others which are dependent upon them for food; but they are at the same time among the most fertile of animals, and often produce an astonishing number of eggs in a very short time. And no better arrangement for the maintenance of the species under such circumstances can be imagined than that supplied by diminishing the duration of life, and simultaneously increasing the rapidity of reproduction.

This general tendency is developed to very different degrees according to conditions peculiar to each species. The shortening of the period of reproduction, and the duration of life to the greatest extent which is possible, depends upon a number of co-operating circumstances, which it is impossible to enumerate completely. Even the manner in which the eggs are laid may have an important effect. If the larva of the may-fly lived upon some rare and widely distributed food-plant instead of at the bottom of streams, the imagos would be compelled to live longer, for they would be obliged—like many moths and butterflies—to lay their eggs singly or in small clusters, over a large area. This would require both time and strength, and they could not retain the rudimentary mouth which they now possess, for they would have to feed in order to acquire sufficient strength for long flights; and—whether they were carnivorous like dragon-flies, or honey-eating like butterflies—their feeding would itself cause a further expenditure of both time and strength, which would necessitate a still further increase in the duration of life. And as a matter of fact we find that dragon-flies and swift-flying hawk-moths often live for six or eight weeks and sometimes longer.

We must also remember that in many species the eggs are not mature immediately after the close of the pupal stage, but that they only gradually ripen during the life of the imago, and frequently, as in many beetles and butterflies, do not ripen simultaneously, but only a certain number at a time. This depends, first, upon the amount of reserve nutriment accumulated in the body of the insect during larval life; secondly, upon various but entirely different circumstances, such as the power of flight. Insects which fly swiftly and are continually on the wing, like hawk-moths and dragon-flies, cannot be burdened with a very large number of ripe eggs. In these cases the gradual ripening of the eggs becomes necessary, and involves an increase in the duration of life. In Lepidoptera, we see how the power of flight diminishes step by step as soon as other circumstances permit, and simultaneously how the eggs ripen more and more rapidly, while the length of life becomes shorter, until a minimum is reached. Only two stages in the process of transformation can be mentioned here.

The strongest flyers—the hawk-moths and butterflies—must be looked upon as the most specialised and highest types among the Lepidoptera. Not only do they possess organs for flight in their most perfect form, but also organs for feeding—the characteristic spiral proboscis or 'tongue.'

There are certain moths (among the Bombyces) of which the males fly as well as the hawk-moths, while the females are unable to use their large wings for flight, because the body is too heavily weighted by a mass of eggs, all of which reach maturity at the same time. Such species, as for instance *Aglia tau*, are unable to distribute their eggs over a wide area, but are obliged to lay them all in a single spot. They can however do this without harm to the species, because their caterpillars live upon forest trees, which provide abundant food for a larger number of larvae than can be produced by the eggs of a single female. The eggs of *Aglia tau* are deposited directly after pairing, and shortly afterwards the insect dies at the foot of the tree among the moss-covered roots of which it has passed the winter in the pupal state. The female moth seldom lives for more than three or four days; but the males which fly swiftly in the forests, seeking for the less abundant females, live for a much longer period, certainly from eight to fourteen days[2].

The females of the *Psychidae* also deposit all their eggs in one place. The grasses and lichens upon which their caterpillars live grow close at hand upon the surface of the earth and stones, and hence the female moth does not leave the ground, and generally does not even quit the pupa-case, within which it lays its eggs; as soon as this duty is finished, it dies. In relation to these habits the wings and mouth of the female are rudimentary, while the male possesses perfectly developed wings.

The causes which have regulated the length of life in these cases are obvious enough, yet still more striking illustrations are to be found among insects which live in colonies.

The duration of life varies with the sex in bees, wasps, ants, and termites: the females have a long life, the males a short one; and there can be no doubt that the explanation of this fact is to be found in adaptation to external conditions of life.

The queen-bee—the only perfect female in the hive—lives two to three years, and often as long as five years, while the male bees or drones only live four to five months. Sir John Lubbock has succeeded in keeping female and working ants alive for seven years—a great age for insects[3],—while the males only lived a few weeks.

These last examples become readily intelligible when we remember that the males neither collect food nor help in building the hive. Their value to the colony ceases with the nuptial flight, and from the point of view of utility it is easy to understand why their lives should be so short [See Note 7 and Note 9]. But the case is very different with the female. The longest period of reproduction possible, when accompanied by very great fertility, is, as a rule, advantageous for the maintenance of the species. It cannot however be attained in most insects, for the capability of living long would be injurious if all individuals fell a prey to their enemies before they had completed the full period of life. Here it is otherwise: when the queen-bee returns from her nuptial flight, she remains within the hive until her death, and never leaves it. There she is almost completely secure from enemies and from dangers of all kinds; thousands of workers armed with stings protect, feed, and warm her; and in short there is every chance of her living through the full period of a life of normal length. And the case is entirely similar with the female ant. In neither of these insects is there any reason why the advantages which follow from a lengthened period of reproductive activity should be abandoned [See Note 6].

That an increase in the length of life has actually taken place in such cases seems to be indicated by the fact that both sexes of the saw-flies—the probable ancestors of bees and ants—have but a short life. On the other hand, the may-flies afford an undoubted instance of the shortening of life. Only in certain species is life as short as I have indicated above; in the majority it lasts for one or more days. The extreme cases, with a life of only a few hours, form the end of a line of development tending in the direction of a shortened life. This is made clear by the fact that one of these may-flies (*Palingenia*) does not even leave its pupa-skin, but reproduces in the so-called sub-imago stage.

It is therefore obvious that the duration of life is extremely variable, and not only depends upon physiological considerations, but also upon the external conditions of life. With every change in the structure of a species, and with the acquisition of new habits, the length of its life may, and in most cases must, be altered.

In answering the question as to the means by which the lengthening or shortening of life is brought about, our first appeal must be to the process of natural selection. Duration of life, like every other characteristic of an organism, is subject to individual fluctuations. From our experience with the human species we know that long life is hereditary. As soon as the long-lived individuals in a species obtain some advantage in the struggle for existence, they will gradually become dominant, and those with the shortest lives will be exterminated.

So far everything is quite simple; but hitherto we have only considered the external mechanism, and we must now further inquire as to the concomitant internal means by which such processes are rendered possible.

This brings us face to face with one of the most difficult problems in the whole range of physiology,—the question of the origin of death. As soon as we thoroughly understand the circumstances upon which normal death depends in general, we shall be able to make a further inquiry as to the circumstances which influence its earlier or later appearance, as well as to any functional changes in the organism which may produce such a result.

The changes in the organism which result in normal death,—senility so-called,—have been most accurately studied among men. We know that with advancing age certain alterations take place in the tissues, by which their functional activity is diminished; that these changes gradually increase, and finally either lead to direct or so-called normal death, or produce indirect death by rendering the organism incapable of resisting injuries due to external influences. These senile changes have been so well described from the time of Burdach and Bichat to that of Kussmaul, and are so well known, that I need not enter into further details here.

In answer to an inquiry as to the causes which induce these changes in the tissues, I can only suggest that the cells which form the vital constituents of tissues are worn out by prolonged use and activity. It is conceivable that the cells might be thus worn out in two ways; either the cells of a tissue remain the same throughout life, or else they are being continually replaced by younger generations of cells, which are themselves cast off in their turn.

In the present state of our knowledge the former alternative can hardly be maintained. Millions of blood corpuscles are continually dying and being replaced by new ones. On both the internal and external surfaces of the body countless epithelial cells are being incessantly removed, while new ones arise in their place; the activity of many and probably of all glands is accompanied by a change in their cells, for their secretions consist partly of detached and partly of dissolved cells; it is stated that even the cells of bone, connective tissue, and muscle undergo the same changes, and nervous tissue alone remains, in which it is doubtful whether such a renewal of cells takes place. And yet as regards even this tissue, certain facts are known which indicate a normal, though probably a slow renewal of the histological elements. I believe that one might reasonably defend the statement,—in fact, it has already found advocates,—that the vital processes of the higher (i.e. multicellular) animals are accompanied by a renewal of the morphological elements in most tissues.

This statement leads us to seek the origin of death, not in the waste of single cells, but in the limitation of their powers of reproduction. Death takes place because a worn-out tissue cannot for ever renew itself, and because a capacity for increase by means of cell-division is not everlasting, but finite [See Note 8]. This does not however imply that the immediate cause of death lies in the imperfect renewal of cells, for death would in all cases occur long before the reproductive power of the cells had been completely exhausted. Functional disturbances will appear as soon as the rate at which the worn-out cells are renewed becomes slow and insufficient.

But it must not be forgotten that death is not always preceded by senility, or a period of old age. For instance, in many of the lower animals death immediately follows the most important deed of the organism, viz. reproduction. Many Lepidoptera, all may-flies, and many other insects die of exhaustion immediately after depositing their eggs. Men have been known to die from the shock of a strong passion. Sulla is said to have died as the result of rage, whilst Leo X succumbed to an excess of joy. Here the psychical shock caused too intense an excitement of the nervous system. In the same manner the exercise of intense effort may also produce a similarly fatal excitement in the above-mentioned insects. At any rate it is certain that when, for some reason, this effort is not made, the insect lives for a somewhat longer period.

It is clear that in such animals as insects we can only speak figuratively of normal death, if we mean by this an end which is not due to accident. In these animals an accidental end is the rule, and is therefore, strictly speaking, normal [See Note 9].

Assuming the truth of the above-mentioned hypothesis as to the causes of normal death, it follows that the number of cell-generations which can proceed from the egg-cell is fixed for every species, at least within certain limits; and this number of cell-generations, if attained, corresponds to the maximum duration of life in the individuals of the species concerned. Shortening of life in any species must depend upon a decrease in the number of successive cell-generations, while conversely, the lengthening of life depends upon an increase in the number of cell-generations over those which were previously possible.

Such changes actually take place in plants. When an annual plant becomes perennial, the change—one in every way possible—can only happen by the production of new shoots, i. e. by an increase in the number of cell-generations. The process is not so obvious in animals, because in them the formation of young cells does not lead to the production of new and visible parts, for the new material is merely deposited in the place of that which is worn out and disappears. Among plants, on the other hand, the old material persists, its cells become lignified, and it is built over by new cells which assume the functions of life.

It is certainly true that the question as to the necessity of death in general does not seem much clearer from this point of view than from the purely physiological one. This is because we do not know why a cell must divide 10,000 or 100,000 times and then suddenly stop. It must be admitted that we can see no reason why the power of cell-multiplication should not be unlimited, and why the organism should not therefore be endowed with everlasting life. In the same manner, from a physiological point of view, we might admit that we can see no reason why the functions of the organism should ever cease.

It is only from the point of view of utility that we can understand the necessity of death. The same arguments which were employed to explain the necessity for as short a life as possible, will with but slight modification serve to explain the common necessity of death[4].

Let us imagine that one of the higher animals became immortal; it then becomes perfectly obvious that it would cease to be of value to the species to which it belonged. Suppose that such an immortal individual could escape all fatal accidents, through infinite time,—a supposition which is of course hardly conceivable. The individual would nevertheless be unable to avoid, from time to time, slight injuries to one or another part of its body. The injured parts could not regain their former integrity, and thus the longer the individual lived, the more defective and crippled it would become, and the less perfectly would it fulfil the purpose of its species. Individuals are injured by the operation of external forces, and for this reason alone it is necessary that new and perfect individuals should continually arise and take their place, and this necessity would remain even if the individuals possessed the power of living eternally.

From this follows, on the one hand, the necessity of reproduction, and, on the other, the utility of death. Worn-out individuals are not only valueless to the species, but they are even harmful, for they take the place of those which are sound. Hence by the operation of natural selection, the life of our hypothetically immortal individual would be shortened by the amount which was useless to the species. It would be reduced to a length which would afford the most favourable conditions for the existence of as large a number as possible of vigorous individuals, at the same time.

If by these considerations death is shown to be a beneficial occurrence, it by no means follows that it is to be solely accounted for on grounds of utility. Death might also depend upon causes which lie in the nature of life itself. The floating of ice upon water seems to us to be a useful arrangement, although the fact that it does float depends upon its molecular structure and not upon the fact that its doing so is of any advantage to us. In like manner the necessity of death has been hitherto explained as due to causes which are inherent in organic nature, and not to the fact that it may be advantageous.

I do not however believe in the validity of this explanation; I consider that death is not a primary necessity, but that it has been secondarily acquired as an adaptation. I believe that life is endowed with a fixed duration, not because it is contrary to its nature to be unlimited, but because the unlimited existence of individuals would be a luxury without any corresponding advantage. The above-mentioned hypothesis upon the origin and necessity of death leads me to believe that the organism did not finally cease to renew the worn-out cell material because the nature of the cells did not permit them to multiply indefinitely, but because the power of multiplying indefinitely was lost when it ceased to be of use.

I consider that this view, if not exactly proved, can at any rate be rendered extremely probable.

It is useless to object that man (or any of the higher animals) dies from the physical necessity of his nature, just as the specific gravity of ice results from its physical nature. I am quite ready to admit that this is the case. John Hunter, supported by his experiments on *anabiosis*, hoped to prolong the life of man indefinitely by alternate freezing and thawing; and the Veronese Colonel Aless. Guaguino made his contemporaries believe that a race of men existed in Russia, of which the individuals died regularly every year on the 27th of November, and returned to life on the 24th of the following April. There cannot however be the least doubt, that the higher organisms, as they are now constructed, contain within themselves the germs of death. The question however arises as to how this has come to pass; and I reply that death is to be looked upon as an occurrence which is advantageous to the species as a concession to the outer conditions of life, and not as an absolute necessity, essentially inherent in life itself.

Death, that is the end of life, is by no means, as is usually assumed, an attribute of all organisms. An immense number of low organisms do not die, although they are easily destroyed, being killed by heat, poisons, &c. As long, however, as those conditions which are necessary for their life are fulfilled, they continue to live, and they thus carry the potentiality of unending life in themselves. I am speaking not only of the Amoebae and the low unicellular Algae, but also of far more highly organized unicellular animals, such as the Infusoria.

The process of fission in the Amoeba has been recently much discussed, and I am well aware that the life of the individual is generally believed to come to an end with the division which gives rise to two new individuals, as if death and reproduction were the same thing. But this process cannot be truly called death. Where is the dead body? what is it that dies? Nothing dies; the body of the animal only divides into two similar parts, possessing the same constitution. Each of these parts is exactly like its parent, lives in the same manner, and finally also divides into two halves. As far as these organisms are concerned, death can only be spoken of in the most figurative sense.

There are no grounds for the assumption that the two halves of an Amoeba are differently constituted internally, so that after a time one of them will die while the other continues to live. Such an idea is disproved by a recently discovered fact. It has been noticed in *Euglypha* (one of the Foraminifera) and in other low animals of the same group, that when division is almost complete, and the two halves are only connected by a short strand, the protoplasm of both parts begins to circulate, and for some time passes backwards and forwards between the two halves. A complete mingling of the whole substance of the animal and a resulting identity in the constitution of each half is thus brought about before the final separation [See Note 10].

The objection might perhaps be raised that, if the parent animal does not exactly die, it nevertheless disappears as an individual. I cannot however let this pass unless it is also maintained that the man of to-day is no longer the same individual as the boy of twenty years ago. In the growth of man, neither structure nor the components of structure remain precisely the same; the material is continually changing. If we can imagine an Amoeba endowed with self-consciousness, it might think before dividing 'I will give birth to a daughter,' and I have no doubt that each half would regard the other as the daughter, and would consider itself to be the original parent. We cannot however appeal to this criterion of personality in the Amoeba, but there is nevertheless a criterion which seems to me to decide the matter: I refer to the continuity of life in the same form.

Now if numerous organisms, endowed with the potentiality of never-ending life, have real existence, the question arises as to whether the fact can be understood from the point of view of utility. If death has been shown to be a necessary adaptation for the higher organisms, why should it not be so for the lower also? Are they not decimated by enemies? are they not often imperfect? are they not worn out by contact with the external world? Although they are certainly destroyed by other animals, there is nothing comparable to that deterioration of the body which takes place in the higher organisms. Unicellular animals are too simply constructed for this to be possible. If an infusorian is injured by the loss of some part of its body, it may often recover its former integrity, but if the injury is too great it dies. The alternative is always perfect integrity or complete destruction.

We may now leave this part of the subject, for it is obvious that normal death, that is to say, death which arises from internal causes, is an impossibility among these lower organisms. In those species at any rate in which fission is accompanied by a circulation of the protoplasm of the parent, the two halves must possess the same qualities. Since one of them is endowed with a potentiality for unending life, and must be so endowed if the species is to persist, it is clear that the other exactly similar half must be endowed with equal potentiality.

Let us now consider how it happened that the multicellular animals and plants, which arose from unicellular forms of life, came to lose this power of living for ever.

The answer to this question is closely bound up with the principle of division of labour which appeared among multicellular organisms at a very early stage, and which has gradually led to the production of greater and greater complexity in their structure.

The first multicellular organism was probably a cluster of similar cells, but these units soon lost their original homogeneity. As the result of mere relative position, some of the cells were especially fitted to provide for the nutrition of the colony, while others undertook the work of reproduction. Hence the single group would come to be divided into two groups of cells, which may be called somatic and reproductive—the cells of the body as opposed to those which are concerned with reproduction. This differentiation was not at first absolute, and indeed it is not always so to-day. Among the lower Metazoa, such as the polypes, the capacity for reproduction still exists to such a degree in the somatic cells, that a small number of them are able to give rise to a new organism,—in fact new individuals are normally produced by means of so-called buds. Furthermore, it is well known that many of the higher animals have retained considerable powers of regeneration; the salamander can replace its lost tail or foot, and the snail can reproduce its horns, eyes, etc.

As the complexity of the Metazoan body increased, the two groups of cells became more sharply separated from each other. Very soon the somatic cells surpassed the reproductive in number, and during this increase they became more and more broken up by the principle of the division of labour into sharply separated systems of tissues. As these changes took place, the power of reproducing large parts of the organism was lost, while the power of reproducing the whole individual became concentrated in the reproductive cells alone.

But it does not therefore follow that the somatic cells were compelled to lose the power of unlimited cell-production, although in accordance with the law of heredity, they could only give rise to cells which resembled themselves, and belonged to the same differentiated histological system. But as the fact of normal death seems to teach us that they have lost even this power, the causes of the loss must be sought outside the organism, that is to say, in the external conditions of life; and we have already seen that death can be very well explained as a secondarily acquired adaptation. The reproductive cells cannot lose the capacity for unlimited reproduction, or the species to which they belong would suffer extinction. But the somatic cells have lost this power to a gradually increasing extent, so that at length they became restricted to a fixed, though perhaps very large number of cell-generations. This restriction, which implies the continual influx of new individuals, has been explained above as a result of the impossibility of entirely protecting the individual from accidents, and from the deterioration which follows them. Normal death could not take place among unicellular organisms, because the individual and the reproductive cell are one and the same: on the other hand, normal death is possible, and as we see, has made its appearance, among multicellular organisms in which the somatic and reproductive cells are distinct.

I have endeavoured to explain death as the result of restriction in the powers of reproduction possessed by the somatic cells, and I have suggested that such restriction may conceivably follow from a limitation in the number of cell-generations possible for the cells of each organ and tissue. I am unable to indicate the molecular and chemical properties of the cell upon which the duration of its power of reproduction depends: to ask this is to demand an explanation of the nature of heredity—a problem the solution of which may still occupy many generations of scientists. At present we can hardly venture to propose any explanation of the real nature of heredity.

But the question must be answered as to whether the kind and degree of reproductive power resides in the nature of the cell itself, or in any way depends upon the quality of its nutriment.

Virchow, in his 'Cellular Pathology,' has remarked that the cells are not only nourished, but that they actively supply themselves with food. If therefore the internal condition of the cell decides whether it shall accept or reject the nutriment which is offered, it becomes conceivable that all cells may possess the power of refusing to absorb nutriment, and therefore of ceasing to undergo further division.

Modern embryology affords us many proofs, in the segmentation of the ovum, and in the subsequent developmental changes, that the causes of the different forms of reproductive activity witnessed in cells lie in the essential nature of the cells themselves. Why does the segmentation of one half of certain eggs proceed twice as rapidly as that of the other half? why do the cells of the ectoderm divide so much more quickly than those of the endoderm? Why does not only the rate, but also the number of cells produced (so far as we can follow them) always remain the same? Why does the multiplication of cells in every part of the blastoderm take place with the exact amount of energy and rapidity necessary to produce the various elevations, folds, invaginations, etc., in which the different organs and tissues have their origin, and from which finally the organism itself arises? There can be no doubt that the causes of all these phenomena lie within the cells themselves; that in the ovum and the cells which are immediately derived from it, there exists a tendency towards a certain determined (I might almost say specific) mode and energy of cell-multiplication. And why should we regard this inherited tendency as confined to the building up of the embryo? why should it not also exist in the young, and later in the mature animal? The phenomena of heredity which make their appearance even in old age afford us proofs that a tendency towards a certain mode of cell-multiplication continues to regulate the growth of the organism during the whole of its life.

The above-mentioned considerations show us that the degree of reproductive activity present in the tissues is regulated by internal causes while the natural death of an organism is the termination—the hereditary limitation—of the process of cell-division, which began in the segmentation of the ovum.

Allow me to suggest a further consideration which may be compared with the former. The organism is not only limited in time, but also in space: it not only lives for a limited period, but it can only attain a limited size. Many animals grow to their full size long before their natural end: and although many fishes, reptiles, and lower animals are said to grow during the whole of their life, we do not mean by this that they possess the power of unlimited growth any more than that of unlimited life. There is everywhere a maximum size, which, as far as our experience goes, is never surpassed. The mosquito never reaches the size of an elephant, nor the elephant that of a whale.

Upon what does this depend? Is there any external obstacle to growth? Or is the limitation entirely imposed from within?

Perhaps you may answer, that there is an established relation between the increase of surface and mass, and it cannot be denied that these relations do largely determine the size of the body. A beetle could never reach the size of an elephant, because, constituted as it is, it would be incapable of existence if it attained such dimensions. But nevertheless the relations between surface and mass do not form the only reason why any given individual does not exceed the average size of its species. Each individual does not strive to grow to the largest possible size, until the absorption from its digestive area becomes insufficient for its mass; but it ceases to grow because its cells cannot be sufficiently nourished in consequence of its increased size. The giants which occasionally appear in the human species prove that the plan upon which man is constructed can also be carried out on a scale which is far larger than the normal one. If the size of the body chiefly depends upon amount of nutriment, it would be possible to make giants and dwarfs at will. But we know, on the contrary, that the size of the body is hereditary in families to a very marked extent; in fact so much so that the size of an individual depends chiefly upon heredity, and not upon amount of food.

These observations point to the conclusion that the size of the individual is in reality pre-determined, and that it is potentially contained in the egg from which the individual develops.

We know further that the growth of the individual depends chiefly upon the multiplication of cells and only to a slight extent upon the growth of single cells. It is therefore clear that a limit of growth is imposed by a limitation in the processes by which cells are increased, both as regards the number of cells produced and the rate at which they are formed. How could we otherwise explain the fact that an animal ceases to grow long before it has reached the physiologically attainable maximum of its species, without at the same time suffering any loss of vital energy?

In many cases at least, the most important duty of an organism, viz. reproduction, follows upon the attainment of full size—a fact which induced Johannes Müller to reject the prevailing hypothesis which explained the death of animals as due to 'the influences of the inorganic environment, which gradually wear away the life of the individual.' He argued that, if this were the case, 'the organic energy of an individual would steadily decrease from the beginning,' while the facts indicate that this is not so[5].

If it is further asked why the egg should give rise to a fixed number of cell-generations, although perhaps a number which varies widely within certain limits, we may now refer to the operation of natural selection upon the relation of surface to mass, and upon other physiological necessities which are peculiar to the species. Because a certain size is the most favourable for a certain plan of organization, the process of natural selection determined that such a size should be within certain variable limits, characteristic of each species. This size is then transmitted from generation to generation, for when once established as normal for the species, the most favourable size is potentially present in the reproductive cell from which each individual is developed.

If this conclusion holds, and I believe that no essential objection can be raised against it, then we have in the limitation in space a process which is exactly analogous to the limitation in time, which we have already considered. The latter limitation—the duration of life—also depends upon the multiplication of cells, the rapid increase of which first gave rise to the characteristic form of the mature body, and then continued at a slower rate. In the mature animal, cell-reproduction still goes on, but it no longer exceeds the waste; for some time it just compensates for loss, and then begins to decline. The waste is not compensated for, the tissues perform their functions incompletely, and thus the way for death is prepared, until its final appearance by one of the three great *Atria mortis*.

I admit that facts are still wanting upon which to base this hypothesis. It is a pure supposition that senile changes are due to a deficient reproduction of cells: at the same time this supposition gains in probability when we are enabled to reduce the limitations of the organism in both time and space to one and the same principle. It cannot however be asserted under any circumstances that it is a pure supposition that the ovum possesses a capacity for cell-multiplication which is limited both as to numbers produced and rate of production. The fact that each species maintains an average size is a sufficient proof of the truth of this conclusion.

Hitherto I have only spoken of animals and have hardly mentioned plants. I should not have been able to consider them at all, had it not happened that a work of Hildebrand's [See Note 12] has recently appeared, which has, for the first time, provided us with exact observations on the duration of plant-life.

The chief results obtained by this author agree very well with the view which I have brought before you to-day. Hildebrand shows that the duration of life in plants also is by no means completely fixed, and that it may be very considerably altered through the agency of the external conditions of life. He shows that, in course of time, and under changed conditions of life, an annual plant may become perennial, or *vice versa*. The external factors which influence the duration of life are here however essentially different, as indeed we expect them to be, when we remember the very different conditions under which the animal and vegetable kingdoms exist. During the life of animals the destruction of mature individuals plays a most important part, but the existence of the mature plant is fairly well secured; their chief period of destruction is during youth, and this fact has a direct influence upon the degree of fertility, but not upon the duration of life. Climatic considerations, especially the periodical changes of summer and winter, or wet and dry seasons, are here of greater importance.

It must then be admitted that the dependence of the duration of life upon the external conditions of existence is alike common to plants and animals. In both kingdoms the high multicellular forms with well-differentiated organs contain the germs of death, while the low unicellular organisms are potentially immortal. Furthermore, an undying succession of reproductive cells is possessed by all the higher forms, although this may be but poor consolation to the conscious individual which perishes. Johannes Müller is therefore right, when in the sentence quoted at the beginning of my lecture, he speaks of an 'appearance of immortality' which passes from each individual into that which succeeds it. That which remains over, that which persists, is not the individual itself,—not the complex aggregate of cells which is conscious of itself,—but an individuality which is outside its consciousness, and of a low order,—an individuality which is made up of a single cell, which arises from the conscious individual. I might here conclude, but I wish first, in a few words, to protect myself against a possible misunderstanding.

I have repeatedly spoken of immortality, first of the unicellular organism, and secondly of the reproductive cell. By this word I have merely intended to imply a duration of time which appears to be endless to our human faculties. I have no wish to enter into the question of the cosmic or telluric origin of life on the earth. An answer to this question will at once decide whether the power of reproduction possessed by these cells is in reality eternal or only immensely prolonged, for that which is without beginning is, and must be, without end.

The supposition of a cosmic origin of life can only assist us if by its means we can altogether dispense with any theory of spontaneous generation. The mere shifting of the origin of life to some other far-off world cannot in any way help us. A truly cosmic origin in its widest significance will rigidly limit us to the statement—*omne vivum e vivo*—to the idea that life can only arise from life, and has always so arisen,—to the conclusion that organic beings are eternal like matter itself.

Experience cannot help us to decide this question; we do not know whether spontaneous generation was the commencement of life on the earth, nor have we any direct evidence for the idea that the process of development of the living world carries the end within itself, or for the converse idea that the end can only be brought about by means of some external force.

I admit that spontaneous generation, in spite of all vain efforts to demonstrate it, remains for me a logical necessity. We cannot regard organic and inorganic matter as independent of each other and both eternal, for organic matter is continually passing, without residuum, into the inorganic. If the eternal and indestructible are alone without beginning, then the non-eternal and destructible must have had a beginning. But the organic world is certainly not eternal and indestructible in that absolute sense in which we apply these terms to matter itself. We can, indeed, kill all organic beings and thus render them inorganic at will. But these changes are not the same as those which we induce in a piece of chalk by pouring sulphuric acid upon it; in this ease we only change the form, and the inorganic matter remains. But when we pour sulphuric acid upon a worm, or when we burn an oak tree, these organisms are not changed into some other animal and tree, but they disappear entirely as organized beings and are resolved into inorganic elements. But that which can be completely resolved into inorganic matter must have also arisen from it, and must owe its ultimate foundation to it. The organic might be considered eternal if we could only destroy its form, but not its nature.

It therefore follows that the organic world must once have arisen, and further that it will at some time come to an end. Hence we must speak of the eternal duration of unicellular organisms and of reproductive cells in the Metazoa and Metaphyta in that particular sense which signifies, when measured by our standards, an immensely long time.

Yet who can maintain that he has discovered the right answer to this important question? And even though the discovery were made, can any one believe that by its means the problem of life would be solved? If it were established that spontaneous generation did actually occur, a new question at once arises as to the conditions under which the occurrence became possible. How can we conceive that dead inorganic matter could have come together in such a manner as to form living protoplasm, that wonderful and complex substance which absorbs foreign material and changes it into its own substance, in other words grows and multiplies?

And so, in discussing this question of life and death, we come at last—as in all provinces of human research—upon problems which appear to us to be, at least for the present, insoluble. In fact it is the quest after perfected truth, not its possession, that falls to our lot, that gladdens us, fills up the measure of our life, nay! hallows it.

APPENDIX.

Note 1. The Duration of Life among Birds.

There is less exact knowledge upon this subject than we might expect, considering the existing number of ornithologists and ornithological societies with their numerous publications. It has neither been possible nor necessary for my purpose to look up all the widely-scattered references which are to be found upon the subject. Many of these are doubtless unknown to me; for we are still in want of a compilation of accurately determined observations in this department of zoology. I print the few facts which I have been able to collect, as a slight contribution towards such a compilation.

Small singing birds live from eight to eighteen years: the nightingale, in captivity, eight years, but longer according to some writers: the blackbird, in captivity, twelve years, but both these birds live longer in the natural state. A 'half-bred nightingale built its nest for nine consecutive years in the same garden' (Naumann, 'Vögel Deutschlands,' p. 76).

Canary birds in captivity attain an age of twelve to fifteen years (l. c., p. 76).

Ravens have lived for almost a hundred years in captivity (l. c., Bd. I. p. 125).

Magpies in captivity live twenty years, and, 'without doubt,' much longer in the natural state (l. c., p. 346).

Parrots 'in captivity have reached upwards of a hundred years' (l. c., p. 125).

A single instance of the cuckoo (alluded to in the text) is mentioned by Naumann as reaching the age of thirty-two years (l. c., p. 76).

Fowls live ten to twenty years, the golden pheasant fifteen years, the turkey sixteen years, and the pigeon ten years (Oken, 'Naturgeschichte, Vögel,' p. 387).

A golden eagle which 'died at Vienna in the year 1719, had been captured 104 years previously' (Brehm, 'Leben der Vögel,' p. 72).

A falcon (species not mentioned) is said to have attained an age of 162 years (Knauer, 'Der Naturhistoriker,' Vienna, 1880).

A white-headed vulture which was taken in 1706 died in the Zoological Gardens at Vienna (Schönbrunn) in 1824, thus living 118 years in captivity (l. c.).

The example of the bearded vulture, mentioned in the text, is quoted from Schinz's 'Vögel der Schweiz,' p. 196.

The wild goose must live for upwards of 100 years, according to Naumann (l. c., p. 127). The proof of this is not, however, forthcoming. A wild goose which had been wounded reached its eighteenth year in captivity.

Swans are said to have lived 300 years(?), (Naumann, l. c., p. 127).

It is evident that observations upon the duration of life in wild birds can only rarely be made, and that they are usually the result of chance and cannot be verified. It is on this account all the more to be desired that every ascertained fact should be collected.

If the long life of birds has been correctly interpreted as compensation for their feeble fertility and for the great mortality of their young, it will be possible to estimate the length of life in a species, without direct observation, if we only know its fertility and the percentage of individuals destroyed. This percentage can, however, at best, be known only as an average. If we consider, for example, the enormous number of sea birds which breed in summer on the rocks and cliffs of the northern seas, and if we remember that the majority of these birds lay but one, or at most two eggs yearly, and that their young are exposed to very many destructive agencies, we are forced to the conclusion that they must possess a very long life, so that the breeding period may be many times repeated. Their number does not diminish. Year after year countless numbers of these birds cover the rocks, from summit to sea line; millions of them rest there, and rise in the air like a thick cloud whenever they are disturbed. Even in those localities which are every year visited by man in order to effect their capture, the number does not appear to decrease, unless the birds are disturbed and are therefore prompted to seek other breeding-places. From the small island of St.

Kilda, off Scotland, 20,000 young gannets (*Sula*) and an immense number of eggs are annually collected; and although this bird only lays a single egg yearly and takes four years to attain maturity, the numbers do not diminish[6]. 30,000 sea-gulls' eggs and 20,000 terns' eggs are yearly exported from the breeding-places on the island of Sylt, but in this case it appears that a systematic disturbance of the birds is avoided by the collectors, and no decrease in their numbers has yet taken place[7]. The destruction of northern birds is not only caused by man, but also by various predaceous mammals and birds. Indeed the dense mass of birds which throng the cliffs is a cause of destruction to many of the young and to the eggs, which are pushed over the edge of the rocks. According to Brehm the foot of these cliffs is 'always covered with blood and the dead bodies of fledglings.'

Such birds must attain a great age or they would have been exterminated long ago: the minimum duration of life necessary for the maintenance of the species must in their case be a very high one.

Note 2. The Duration of Life among Mammals.

The statements upon this subject in the text are taken from many sources; from Giebel's 'Säugethiere,' from Oken's 'Naturgeschichte,' from Brehm's 'Illustrirtem Thierleben,' and from an essay of Knauer in the 'Naturhistoriker,' Vienna, 1880.

Note 3. The Duration of Life among Mature Insects.

A short statement of the best established facts which I have been able to find is given below. I have omitted the lengthening of imaginal life which is due to hybernation in certain species. In almost all orders of insects there are certain species which emerge from the pupa in the autumn, but which first reproduce in the following spring. The time spent in the torpid condition during winter cannot of course be reckoned with the active life of the species, for its vital activity is either entirely suspended for a time by freezing (*Anabiosis*: Preyer[8]), or it is at any rate never more than a *vita minima*, with a reduction of assimilation to its lowest point.

The following account does not make any claim to contain all or even most of the facts scattered through the enormous mass of entomological literature, and much less all that is privately known by individual entomologists. It must therefore be looked upon as merely a first attempt, a nucleus, around which the principal facts can be gradually collected. It is unnecessary to give any special information as to the duration of larval life, for numerous and exact observations upon this part of the subject are contained in all entomological works.

I. Orthoptera.

Gryllotalpa. The eggs are laid in June or July, and the young are hatched in from two to three weeks; they live through the winter, and become sexually mature in the following May or June. 'When the female has deposited her eggs, her body collapses, and afterwards she does not survive much longer than a month.' 'According as the females are younger or older, they live a longer or shorter life, and hence some females are even found in the autumn' (Rösel, 'Insektenbelustigungen,' Bd. II. p. 92). Rösel believes that the female watches the eggs until they are hatched, and this explains the fact that she outlives the process of oviposition by about a month. It is not stated whether the males die at an earlier period.

Gryllus campestris becomes sexually mature in May, and sings from June till October, 'when they all die' (Oken, 'Naturgeschichte,' Bd. II. Abth. iii. p. 1527). It is hardly probable that any single individual lives for the whole summer; probably, as in the case of *Gryllotalpa*, the end of the life of those individuals which first become mature, overlaps the beginning of the life of others which reach maturity at a later date.

Locusta viridissima and *L. verrucivora* are mature at the end of August; they lay their eggs in the earth during the first half of September and then die. It is probable that the females do not live for more than four weeks in the mature state. It is not known whether the males of this or other species of locusts live for a shorter period.

I have found *Locusta cantans* in plenty, from the beginning of September to the end of the month. In captivity they die after depositing their eggs: the males are probably more short-lived, for towards the middle and end of September they are much less plentiful than the females.

Acridium migratorium 'dies after the eggs are laid' (Oken, 'Naturgeschichte').

The male *Termes* probably live for a short time only, although exact observations upon the point are wanting. The females 'seem sometimes to live four or five years,' as I gather from a letter from Dr. Hagen, of Cambridge, Mass., U.S.A.

Ephemeridae. Rösel, speaking of *Ephemera vulgata* ('Insektenbelustigungen,' Bd. II. der Wasserinsekten, 2te Klasse, p. 60 et seq.), says:—'Their flight commences at sunset, and comes to an end before midnight, when the dew begins to fall.' 'The pairing generally takes place at night and lasts but a short time. As soon as the insects have shed their last skin, in the afternoon or evening, they fly about in thousands, and pair almost immediately; but by the next day they are all dead. They continue to emerge for many days, so that when yesterday's swarm is dead, to-day a new swarm is seen emerging from the water towards the evening.' 'They not only drop their eggs in the water, but wherever they may happen to be,—on trees, bushes, or the earth. Birds, trout and other fish lie in wait for them.'

Dr. Hagen writes to me—'It is only in certain species that life is so short. The female *Palingenia* does not live long enough to complete the last moult of the sub-imago. I believe that a female imago has never been seen. The male imago, often half in its sub-imago skin, fertilizes the female sub-imago and immediately the contents of both ovaries are extruded, and the insect dies. It is quite possible that the eggs pass out by rupturing the abdominal segments.'

Libellula. All dragon-flies live in the imago condition for some weeks; at first they are not capable of reproduction, but after a few days they pair.

Lepisma saccharina. An individual lived for two years in a pill-box, without any food except perhaps a little *Lycopodium* dust[9].

II. Neuroptera.

Phryganids 'live in the imago stage for at least a week and probably longer, apparently without taking food' (letter from Dr. Hagen).

According to the latest researches *Phrygane grandis*[10] never contains food in its alimentary canal, but only air, although it contains the latter in such quantities that the anterior end of the chylific ventricle is dilated by it.

III. Strepsiptera.

The larva requires for its development a rather shorter time than that which is necessary for the grub of the bee into the body of which it has bored. The pupa stage lasts eight to ten days. The male, which flies about in a most impetuous manner, lives only two to three hours, while the female lives for some days. Possibly the pairing does not take place until the female is two to three days old. The viviparous female seems to produce young only once in a lifetime, and then dies: it is at present uncertain whether she also produces young parthenogenetically (cf. Siebold, 'Ueber Paedogenesis der Strepsipteren,' Zeitschr. f. Wissensch. Zool., Band. XX, 1870).

IV. Hemiptera.

Aphis. Bonnet ('Observations sur les Pucerons,' Paris, 1745) had a parthenogenetic female of *Aphis euonymi* in his possession for thirty-one days, from its birth, during which time it brought forth ninety-five larvae. Gleichen kept a parthenogenetic female of *Aphis mali* fifteen to twenty-three days.

Aphis foliorum ulmi. The mother of a colony which leaves the egg in May is 2‴ long at the end of July: it therefore lives for at least two and a half months (De Geer, 'Abhandlungen zur Geschichte der Insekten,' 1783, III. p. 53).

Phylloxera vastatrix. The males are merely ephemeral sexual organisms, they have no proboscis and no alimentary canal, and die immediately after fertilizing the female.

Pemphigus terebinthi. The male as well as the female sexual individuals are wingless and without a proboscis; they cannot take food and consequently live but a short time,—far shorter than the parthenogenetic females of the same species (Derbès, 'Note sur les aphides du pistachier térébinthe,' Ann. des sci. nat., Tom. XVII, 1872).

Cicada. In spite of the numerous and laborious descriptions of the Cicadas which have appeared during the last two centuries, I can only find precise statements as to the duration of life in the mature insect in a single species. P. Kalm, writing upon the North American *Cicada septemdecim*, which sometimes appears in countless numbers, states that 'six weeks after (such a swarm had been first seen) they had all disappeared.' Hildreth puts the life of the female at from twenty to twenty-five days. This agrees with the fact that the Cicada lays many hundred eggs (Hildreth states a thousand); sixteen to twenty at a time being inserted into a hole which is bored in wood, so that the female takes some time to lay her eggs (Oken, 'Naturgeschichte,' 2ter Bd. 3te Abth. p. 1588 et seq.).

Acanthia lectularia. No observations have been made upon the bed bug from which the normal length of its life can be ascertained, but many statements tend to show that it is exceedingly long-lived, and this is advantageous for a parasite of which the food (and consequently growth and reproduction) is extremely precarious. They can endure starvation for an astonishingly long period, and can survive the most intense cold. Leunis ('Zoologie,' p. 659) mentions the case of a female which was shut up in a box and forgotten: after six months' starvation it was found not only alive but surrounded by a circle of lively young ones. Göze found bugs in the hangings of an old bed which had not been used for six years: 'they appeared white like paper.' I have myself observed a similar case, in which the starving animals were quite transparent. De Geer placed some bugs in an unheated room in the cold winter of 1772, when the thermometer fell to -33°C: they passed the whole winter in a state of torpidity, but revived in the following May. (De Geer, Bd. III. p. 165, and Oken, 'Naturgeschichte,' 2ter Bd. 3te Abth. p. 1613.)

V. Diptera.

Pulex irritans. Oken says of the flea ('Naturgeschichte,' Bd. II. Abth. 2, p. 759) that 'death follows the deposition of the eggs in the course of two or three days, even if the opportunity of sucking blood is given them.' The length of time which intervenes between the emergence from the cocoon and fertilization or the deposition of eggs is not stated.

Sarcophaga carnaria. The female fly dies ten to twelve hours after the birth of the viviparous larvae; the time intervening between the exit from the cocoon and the birth of the young is not given (Oken, quoting Réaumur, 'Mém. p. s. à l'hist. Insectes,' Paris, 1740-48, IV).

Musca domestica. In the summer the common house-fly begins to lay eggs eight days after leaving the cocoon: she then lays several times. (See Gleichen, 'Geschichte der gemeinen Stubenfliege,' Nuremberg, 1764.)

Eristalis tenax. The larva of this large fly lives in liquid manure, and has been described and figured by Réaumur as the rat-tailed larva. I kept a female which had just emerged from the cocoon, from August 30th till October 4th, in a large gauze-covered glass vessel. The insect soon learnt to move freely about in its prison, without attempting to escape; it flew round in circles, with a characteristic buzzing sound, and obtained abundant nourishment from a solution of sugar, provided for it. From September 12th it ceased to fly about, except when frightened, when it would fly a little way off. I thought that it was about to die, but matters took an unexpected turn, and on the 26th of September it laid a large packet of eggs, and again on the 29th of the same month another packet of similar size. The flight of the animal had been probably impeded by the weight of the mass of ripe eggs in its body. The deposition of eggs was probably considerably retarded in this case, because fertilization had not taken place. The fly died on the 4th of October, having thus lived for thirty-five days. Unfortunately, I have been unable to make any experiments as to the duration of life in the female when males are also present.

VI. Lepidoptera.

I am especially indebted to Mr. W. H. Edwards[11], of Coalburgh, W. Virginia, and to Dr. Speyer, of Rhoden, for valuable letters relating to this order.

The latter writes, speaking of the duration of life in imagos generally:—'It is, to my mind, improbable that any butterfly can live as an imago for a twelvemonth. Specimens which have lived through the winter are only rarely seen in August, even when the summer is late. A worn specimen of *Vanessa cardui* has, for instance, been found at this time' ('Entomolog. Nachrichten,' 1881, p. 146).

In answer to my question as to whether the fact that certain Lepidoptera take no solid or liquid food, and are, in fact, without a functional mouth, may be considered as evidence for an adaptation of the length of life to the rapid deposition of eggs, Dr. Speyer replies:—'The wingless females of the *Psychidae* do not seem to possess a mouth, at any rate I cannot find one in *Psyche unicolor* (*graminella*). They do not leave the case during life, and certainly do not drink water. The same is true of the wingless female of *Heterogynis*, and of *Orgyia ericae*, and probably of all the females of the genus *Orgyia*; and as far as I can judge from cabinet specimens, it is probably true of the males of *Heterogynis* and *Psyche*. I have never seen the day-flying *Saturnidae*, *Bombycidae*, and other Lepidoptera with a rudimentary proboscis, settle in damp places, or suck any moist substance, and I doubt if they would ever do this. The sucking apparatus is probably deficient.'

In answer to my question as to whether the males of any species of butterfly or moth are known to pass a life of different length from that of the female, Dr. Speyer stated that he knew of no observations on this point.

The following are the only instances of well-established direct observations upon single individuals, in my possession[12]:—

Pieris napi, var. *bryoniae* [male] and [female], captured on the wing: lived in confinement ten days, and were then killed.

Vanessa prorsa lived at most ten days in confinement.

Vanessa urticae lived ten to thirteen days in confinement.

Papilio ajax. According to a letter from Mr. W. H. Edwards, the female, when she leaves the pupa, contains unripe eggs in her body, and lives for about six weeks—calculating from the first appearance of this butterfly to the disappearance of the same generation[13]. The males live longer, and continue to fly when very worn and exhausted. A worn female is very seldom seen;—'I believe the female does not live long after laying her eggs, but this takes some days, and probably two weeks.'

Lycaena violacea. According to Mr. Edwards, the first brood of this species lives three to four weeks at the most.

Smerinthus tiliae. A female, which had just emerged from the pupa, was caught on June 24th; on the 29th pairing took place; on the 1st of July she laid about eighty eggs, and died the following day. She lived nine days, taking no food during this period, and she only survived the deposition of eggs by a single day.

Macroglossa stellatarum. A female, captured on the wing and already fertilized, lived in confinement from June 28th to July 4th. During this time she laid about eighty eggs, at intervals and singly; she then disappeared, and must have died, although the body could not be found among the grass at the bottom of the cage in which she was confined.

Saturnia pyri. A pair which quitted the cocoons on the 24th or 25th of April, remained in coitu from the 26th until May 2nd—six or seven days; the female then laid a number of eggs, and died.

Psyche graminella. The fertilized female lives some days, and the unfertilized female over a week (Speyer).

Solenobia triquetrella. 'The parthenogenetic form (I refer to the one which I have shown to be parthenogenetic in Oken's 'Isis,' 1846, p. 30) lays a mass of eggs in the abandoned case, soon after emergence. The oviposition causes her body to shrivel up, and some hours afterwards she dies. The non-parthenogenetic female of the same species remains for many days, waiting to be fertilized; if this does not occur, she lives over a week.' 'The parthenogenetic female lives for hardly a day, and the same is true of the parthenogenetic females of another species of *Solenobia*' (*S. inconspicuella*?). Letter from Dr. Speyer.

Psyche calcella, O. The males live a very short time; 'those which leave the cocoon in the evening are found dead on the following morning, with their wings fallen off, at the bottom of their cage.' Dr. Speyer.

Eupithecia, sp. (*Geometridae*), 'when well-fed, live for three to four weeks in confinement; the males fertilize the females frequently, and the latter continue to lay eggs when they are very feeble, and are incapable of creeping or flying.' Dr. Speyer.

The conclusions and speculations in the text seem to be sufficiently supported from this short series of observations. There remains, as we see, much to be done in this field, and it would well repay a lepidopterist to undertake some exact observations upon the length of life in different butterflies and moths, with reference to the conditions of life—the mode of egg-laying, the degeneracy of the wings, and of the external mouth-parts or the closure of the mouth itself. It would be well to ascertain whether such closure does really take place, as it undoubtedly does in certain plant-lice.

VII. Coleoptera.

Melolontha vulgaris. Cockchafers, which I kept in an airy cage with fresh food and abundant moisture, did not in any case live longer than thirty-nine days. One female only, out of a total number of forty-nine, lived for this period; a second lived thirty-six days, a third thirty-five, and a fourth and fifth twenty-four days; all the rest died earlier. Of the males, only one lived as long as twenty-nine days. These periods are less by some days than the true maximum duration of life, for the beetles were captured in the field, and had lived for at least a day; but the difference cannot be great, when we remember that out of forty-nine beetles, only three females lived thirty-five to thirty-nine days, and only one male twenty-nine days. Those that died earlier had probably lived for some considerable time before being caught.

Exact experiments with pupae which have survived the winter would show whether the female really lives for ten days more than the male, or whether the results of my experiment were merely accidental. I may add that coitus frequently took place during the period of captivity. One pair, observed in this condition on the 17th, separated in the evening; they paired again on the morning of the 18th, and separated in the middle of the day. Coitus took place between another pair on the 22nd, and again on the 26th.

I watched the gradual approach of death in many individuals: some days before it ensued, the insects became sluggish, ceased to fly and to eat, and only crept a little way off when disturbed: they then fell to the ground and remained motionless, apparently dead, but moved their legs when irritated, and sometimes automatically. Death came on gradually and imperceptibly; from time to time there was a slow movement of the legs, and at last, after some hours, all signs of life ceased.

In one case only I found bacteria present in great numbers in the blood and tissues; in the other individuals which had recently died, the only noticeable change was the unusual dryness of the tissues.

Carabus auratus. An experiment with an individual, caught on May 27th, gave the length of life at fourteen days; this is probably below the average, since the beetles are found, in the wild state, from the end of May until the beginning of July.

Lucanus cervus. Captured individuals, kept in confinement, and fed on a solution of sugar, never lived longer than fourteen days, and as a rule not so long. The beetles appear in June and July, and certainly cannot live much over a month. As is the case with many beetles appearing during certain months, the length of the individual life is shorter than the period over which they are found. Accurate information, especially as to any difference between the lengths of life in the sexes, is not obtainable.

Isolated accounts of remarkably long lives among beetles are to be found scattered throughout the literature of the subject. Dr. Hagen, of Cambridge, Mass., has been kind enough to draw my attention to these, and to send me some observations of his own.

Cerambyx heros. One individual lived in confinement from August until the following year[14].

Saperda carcharias. An individual lived from the 5th of July until the 24th of July of the next year[15].

Buprestis splendens. A living individual was removed from a desk which had stood in a London counting-house for thirty years; from the condition of the wood it was evident that the larva had been in it before the desk was made[16].

Blaps mortisaga. One individual lived three months, and two others three years.

Blaps fatidica. One individual which was left in a box and forgotten, was found alive when the box was opened six years afterwards.

Blaps obtusa. One lived a year and a half in confinement.

Eleodes grandis and *E. dentipes.* Eight of these beetles from California were kept in confinement and without food for two years by Dr. Gissler, of Brooklyn; they were then sent to Dr. Hagen who kept them another year.

Goliathus cacicus. One individual lived in a hot-house for five months.

In addition to these cases, Dr. Hagen writes to me: 'Among the beetles which live for more than a year,—*Blaps, Pasimachus, (Carabidae)*—and among ants, almost thirty per cent. are found with the cuticle worn out and cracked, and the powerful mandibles so greatly worn down that species were formerly founded upon this point. The mandibles are sometimes worn down to the hypodermis.'

From the data before me I am inclined to believe that in certain beetles the normal length of life extends over some years, and this is especially the case with the *Blapidae*. It seems probable that in these cases another factor is present,—a *vita minima*, or apparent death, a sinking of the vital processes to a minimum in consequence of starvation, which we might call the hunger sleep, after the analogy of winter sleep. The winter sleep is usually ascribed to cold alone, and some insects certainly become so torpid that they appear to be dead when the temperature is low. But cold does not affect all insects in this way. Among bees, for example, the activity of the insects diminishes to a marked extent at the beginning of winter, but if the temperature continues to fall, they become active again, run about, and as the bee-keepers say, 'try to warm themselves by exercise'; by this means they keep some life in them. If the frost is very severe, they die. In the tropics the period of hibernation for many animals coincides with the time of maximum heat and drought. This shows that the organism can be brought into the condition of a *vita minima* in various ways, and it would not be at all remarkable if such a state were induced in certain insects by hunger. Exact experiments however are the only means by which such a suggestion can be tested, and I have already commenced a series of experiments. The fact that certain beetles live without food for many years (even six) can hardly be explained on any other supposition, for these insects consume a fair amount of food under normal conditions, and it is inconceivable that they could live for years without food, if the metabolism were carried on with its usual energy.

A very striking example, showing that longevity may be induced by the lengthening of the period of reproductive activity, is communicated to me by Dr. Adler in the following note: 'Three years ago I accidentally noticed that ovoviviparous development takes place in *Chrysomela varians*,— a fact which I afterwards discovered had been already described by another entomologist.

'The egg passes through all the developmental stages in the ovary; when these are completed the egg is laid, and a minute or two afterwards the larva breaks through the egg-shell. In each division of the ovary the eggs undergo development one at a time; it therefore follows that they are laid at considerable intervals, so that a long life becomes necessary in order to ensure the development of a sufficiently long series of eggs. Hence it comes about that the females live a full year. Among other species of *Chrysomela* two generations succeed each other in a year, and the duration of life in the individual varies from a few months to half a year.'

VIII. Hymenoptera.

Cynipidae. I have been unable to find any accurate accounts of the duration of life in the imagos of saw-flies or ichneumons; but on the other hand I owe to the kindness of Dr. Adler, an excellent observer of the *Cynipidae*, the precise accounts of that family which are in my possession. I asked Dr. Adler the general question as to whether there was any variation in the duration of life among the *Cynipidae* corresponding to the conditions under which the deposition of eggs took place; whether those species which lay many eggs, or of which the oviposition is laborious and protracted, lived longer than those species which lay relatively few eggs, or easily and quickly find the suitable places in which to deposit them.

Dr. Adler fully confirmed my suppositions and supported them by the following statements:—

'The summer generation of *Neuroterus* (*Spathegaster*) has the shortest life of all *Cynipidae*. Whether captured or reared from the galls I have only kept them alive on an average for three to four days. In this generation the work of oviposition requires the shortest time and the least expenditure of energy, for the eggs are simply laid on the surface of a leaf. The number of eggs in the ovary is also smaller than that of other species, averaging about 200. This form of *Cynips* can easily lay 100 eggs a day.

'The summer generation of *Dryophanta* (*Spathegaster Taschenbergi, verrucosus*, etc.) lives somewhat longer; I have kept them in confinement for six to eight days. The oviposition requires a considerable expenditure of time and strength, for the ovipositor has to pierce the rather tough mid-rib or vein of a leaf. The number of eggs in the ovary averages 300 to 400.

'The summer generation of *Andricus*, which belongs to the extensive genus *Aphilotrix*, have also a long life. I have kept the smaller *Andricus* (such as *A. nudus, A. cirratus, A. noduli*) alive for a week, and the larger (*A. inflator, A. curvator, A. ramuli*) for two weeks. The smaller species pierce the young buds when quite soft, but the larger ones bore through the fully grown buds protected by tough scales. The ovary of the former contains 400 to 500 eggs, that of the latter over 600.

'The agamic winter generations live much longer. The species of *Neuroterus* have the shortest life; they live for two weeks at the outside; on the other hand, species of *Aphilotrix* live quite four weeks, and *Dryophanta* and *Biorhiza* even longer. I have kept *Dryophanta scutellaris* alive for three months. The number of eggs in these agamic *Cynipidae* is much larger: *Dryophanta* and *Aphilotrix* contain 1200 and *Neuroterus* about 1000.'

It is evidently, therefore, a general rule that the duration of life is directly proportional to the number of eggs and to the time and energy expended in oviposition. It must of course be understood that, here as in all other instances, these are not the only factors which determine the duration of life, but many other factors, at present unknown, may be in combination with them and assist in producing the result. For example, it is very probable that the time of year at which the imagos appear exerts some indirect influence. The long-lived *Biorhiza* emerges from the gall in the middle of winter, and at once begins to deposit eggs in the oak buds. Although the insect is not sensitive to low temperature, for I have myself seen oviposition proceeding when the thermometer stood at 5° R., yet very severe frost would certainly lead to interruption and would cause the insect to shelter itself among dead leaves on the ground. Such interruptions may be of long duration and frequently repeated, so that the remarkably long life of this species may perhaps be looked upon as an adaptation to its winter life.

Ants. Lasius flavus lays its eggs in the autumn, and the young larvae pass the winter in the nest. The males and females leave the cocoons in June, and pair during July and August. The males fly out of the nest with the females, but they do not return to it; 'they die shortly after pairing.' It is also believed that the females do not return to the nest, but found new colonies; this point is however one of the most uncertain in the natural history of ants. On the other hand it is quite certain that the female may live for years within the nest, continuing to lay fertilized eggs. Old females are sometimes found in the colony, with their jaws worn down to the hypodermis.

Breeding experiments confirm these statements. P. Huber[17] and Christ have already put the life of the female at three to four years, and Sir John Lubbock, who has been lately occupied with the natural history of ants, was able to keep a female worker of *Formica sanguinea* alive for five years; and he has been kind enough to write and inform me that two females of *Formica fusca*, which he captured in a wood together with ten workers, in December 1874, are still alive (July 1881), so that these insects live as imagos for six and a half years or more[18].

On the other hand, Sir John Lubbock never succeeded in keeping the males 'alive longer than a few weeks.' Both the older and more recent observers agree in stating that female ants, like queen bees, are always protected as completely as possible from injury and danger. Dr. A. Forel, whose thorough knowledge of Swiss ants is well known, writes to me,—'The female ants are only once fertilized, and are then tended by the workers, being cleaned and fed in the middle of the nest: one often finds them with only three legs, and with their chitinous armour greatly worn. They never leave the centre of the nest, and their only duty is to lay eggs.'

With regard to the workers, Forel believes that their constitution would enable them to live as long as the females (as the experiments of Lubbock also indicate), and the fact that in the wild state they generally die sooner than the females is 'certainly connected with the fact that they are exposed to far greater dangers.' The same relation seems also to obtain among bees, but with them it has not been shown that in confinement the workers live as long as the queens.

Bees. According to von Berlepsch[19] the queen may as an exception live for five years, but as a rule survives only two or three years. The workers always seem to live for a much shorter period, generally less than a year. Direct experiments upon isolated or confined bees, or upon marked individuals in the wild state, do not prove this, but the statistics obtained by bee-keepers confirm the above. Every winter the numbers in a hive diminish from 12,000-20,000 to 2000-3000. The queen lays the largest number of eggs in the spring, and the workers which die before the winter are replaced by those which emerge in the summer, autumn or during a mild winter. The queen lays eggs at such a variable rate throughout the year that the above-mentioned inequality in numbers is explained. The workers do not often live for more than six to seven months, and at the time of their greatest labour, (May to July), only three months. An attempt to calculate the length of life of the workers and drones by taking stock at the end of summer, gives six months for the former and four months for the latter[20].

The drones do not as a rule live so long as four months, for they meet with a violent death before the end of this period. The well-known slaughter of the drones is not, according to the latest observations, brought about directly by means of the stings of the workers, but by these latter driving away the useless drones from the food so that they perish of starvation.

Wasps. It is interesting that among these near relations of the bees, the life of the female should be much shorter, corresponding to the much lower degree of specialization found in the colonies. The females of *Polistes gallica* and of *Vespa* not only lay eggs but take part in building the cells and in collecting food; they are therefore obliged to use all parts of the body more actively and especially the wings, and are exposed to greater danger from enemies.

It is well known from Leuckart's observations, that the so-called 'workers' of *Polistes gallica* and *Bombus* are not arrested females like the workers of a bee-hive, but are females which although certainly smaller, are in every way capable of being fertilized and of reproduction. Von Siebold has nevertheless proved that they are not fertilized, but reproduce parthenogenetically.

The fertilized female which survives the winter, commences to found a colony at the beginning of May: the larvæ, which hatch from the first eggs, which are about fifteen in number, become pupæ at the beginning of June, and the imagos appear towards the end of the same month. These are all small 'workers,' and they perform such good service in tending the second brood, that the latter attain the size of the female which founded the colony; only differing from her in the perfect condition of their wings, for by this time her wings are greatly worn away.

The males appear at the beginning of July; their spermatozoa are mature in August, and pairing then takes place with certain 'special females which require fertilization' which have in the meantime emerged from their cocoons. These are the females which live through the winter and found new colonies in the following spring. The old females of the previous winter die, and do not live beyond the summer at the beginning of which they founded colonies. At the first appearance of frost, the young fertilized females seek out winter quarters; the males which never survive the winter, do not take this course, but perish in October. The parthenogenetic females, which remain in the nest during the nuptial flight, also perish.

The males of *Polistes gallica* do not live longer than three months—from July to the beginning of October; the parthenogenetic females live a fortnight longer at the outside—from the middle of June to October, but the later generations have a shorter life. The sexual females alone live for about a year, including the winter sleep.

A similar course of events takes place in the genus *Vespa*. In both these genera the possibility of reproduction is not restricted to a single female in the nest, but is shared by a number of females. In the genus *Apis* alone is the division of labour complete, so that only a single female (the queen) is at any one time capable of reproduction, a power which differentiates it from the sterile workers.

Note 4. The Duration of Life of the Lower Marine Animals.

I have only met with one definite statement in the literature of this part of the subject. It concerns a sea anemone,—which is a solitary and not a colonial form. The English zoologist Dalyell, in August, 1828, removed an *Actinia mesembryanthemum* from the sea and placed it in an aquarium[21]. It was a very fine individual, although it had not quite attained the largest size; and it must have been at least seven years old, as proved by comparison with other individuals reared from the egg. In the year 1848, it was about thirty years old, and in the twenty years during which it had been in captivity it had produced 334 young Actiniae. Prof. Dohrn, of Naples, tells me that this Actinia is still living to-day, and is shown as a curiosity to those who visit the Botanical Gardens in Edinburgh. It is now (1882) at least sixty-one years old[22].

Note 5. The Duration of Life in
Indigenous Terrestrial and Fresh-water Mollusca.

I am indebted to Herr Clessin—the celebrated student of our mollusca—for some valuable notes upon our indigenous snails and bivalves (*Lamellibranchiata*). I could not incorporate them in the text, for a number of necessary details as to the conditions of life are at present entirely unknown, or are at least only known in a very fragmentary manner. No statistics as to the amount of destruction suffered by the young are available, and even the number of eggs produced annually is only known for a few species. I nevertheless include Herr Clessin's very interesting communications, as a commencement to the life statistics of the Mollusca.

(1) '*Vitrinae* are annual; the old animals die in the spring, after having produced the spawn from which the young develope. These continue to grow until the following spring.'

(2) 'The *Succineae* are mostly biennial; *Succinea putris* probably triennial. Fertilization takes place from June till the beginning of August, and the young develope until the autumn. *Succinea Pfeifferi* and *S. elegans* live through the winter, and the fact is proved by very distinct annual markings. Reproduction takes place in July and August of the following year, and they die in the autumn. They continue to grow until their death.'

(3) 'The shells of our native species of *Pupa*, *Clausilia*, and *Bulimus* (with the exception of *Bulimus detritus*) show but faint annual markings. They can hardly require more than two years for their complete development. The great number of living individuals with full-sized shells belonging to these genera, as compared with the number which possess smaller shells, makes it probable that these animals live in the mature condition longer than our other *Helicidae*. I have always found full-sized shells present in at least two-thirds of the individuals of these genera characterized by much-coiled shells—a proportion which I have never seen among our larger

Helicidae. Nevertheless direct observations as to the length of life in the mature condition are still wanting.'

(4) 'The *Helicidae* live from two to four years; *Helix sericea, H. hispida*, two to three years; *H. hortensis, H. nemoralis, H. arbustorum*, as a rule three years; *H. pomatia* four years. Fertilization is not in these species strictly confined to any one time of year, but in the case of old animals takes place in the spring, as soon as the winter sleep is over; while in the two-year-old animals it also happens later in the summer.'

(5) 'The *Hyalineae* are mostly biennial: they seldom live three years, and even in the largest species such an age is probably exceptional. The smallest *Hyalineae* and *Helicidae* live at most two years. The length of life is dependent upon the time at which the parents are fertilized, for this decides whether the young begin to shift for themselves early in the summer or later in the autumn, and so whether the first year's growth is large or small.'

(6) 'The species of *Limnaeus, Planorbis*, and *Ancylus* live two to three years, that is they take two to three years to attain the full size. *L. auricularis* is mostly biennial, *L. palustris* and *L. pereger* two to three years: I have found that the latter, in the mountains at Oberstorf in the Bavarian Alps, may exceptionally attain the age of four years, that is, it may possess three clearly defined annual markings, whilst the specimens from the plain never showed more than two.'

(7) 'The *Paludinidae* attain an age of three or four years.'

(8) 'The smaller bivalves, *Pisidium* and *Cyclas*, do not often live for more than two years: the larger *Najadae*, on the other hand, often live for more than ten years, and indeed they are not full grown until they possess ten to fourteen annual markings. It is possible that habitat may have great influence upon the length of life in this order.'

'*Unio* and *Anodonta* become sexually mature in the third to the fifth year.'

As far as I am aware but few statements exist upon the length of life in marine mollusca, and these are for the most part very inexact. The giant bivalve *Tridacna gigas* must attain an age of 60 to 100 years[23]. All *Cephalopods* live for at least over a year, and most of them well over ten years; and the giant forms, sometimes mistaken for 'sea-serpents,' must require many decades in which to attain such a remarkable size. L. Agassiz has determined the length of life in a large sea snail, *Natica heros*, by sorting a great number of individuals according to their sizes: he places it at 30 years[24].

I am glad to be able to communicate an observation made at the Zoological Station at Naples upon the length of life in *Ascidians*. The beautiful white *Cionea intestinalis* has settled in great numbers in an aquarium at the Station, and Professor Dohrn tells me that it produces three generations annually, and that each individual lives for about five months, and then reproduces itself and dies. External conditions accounting for this early death have not been discovered.

It is known that the freshwater *Polyzoa* are annual, but it is not known whether the first individuals produced from a colony in the spring, live for the whole summer. The length of life is also unknown in single individuals of any marine Polyzoon.

Clessin's accurate statements upon the freshwater Mollusca, previously quoted, show that a surprisingly short length of life is the general rule. Only those forms of which the large size requires that many years shall elapse before the attainment of sexual maturity, live ten years or over (*Unio, Anodonta*); indeed, our largest native snail (*Helix pomatia*) only lives for four years, and many small species only one year, or two years if the former time is insufficient to render them sexually mature. These facts seem to indicate, as I think, that these molluscs are exposed to great destruction in the adult state, indeed to a greater extent than when they are young, or, at any rate, to an equal extent. The facts appear to be the reverse of those found among birds. The fertility is enormous; a single mussel contains several hundred thousand eggs; the destruction of young as compared with the number of eggs produced is distinctly smaller than in birds, therefore a much shorter duration of the life of each mature individual is rendered possible, and further becomes advantageous because the mature individuals are exposed to severe destruction.

However it can only be vaguely suggested that this is the case, for positive proofs are entirely absent. Perhaps the destruction of single mature individuals does not play so important a part as the destruction of their generative organs. The ravages of parasitic animals (*Trematodes*) in the internal organs of snails and bivalves are well known to zoologists. The ovaries of the latter are often entirely filled with parasites, and such animals are then incapable of reproduction.

Besides, molluscs have many enemies, which destroy them both on land and in water. In the water,—fish, frogs, newts, ducks and other water-fowl, and on land many birds, the hedgehog, toads, etc., largely depend upon them for food.

If the principles developed in this essay apply to the freshwater Mollusca, we must then infer that snails which maintain the mature condition—the capability of reproduction—for one year, are in this state more exposed to destruction from the attacks of enemies than those species which remain sexually mature for two or three years, or that the latter suffer from a greater proportional loss of eggs and young.

Note 6. Unequal Length of Life in the two Sexes.

This inequality is frequently found among insects. The males of the remarkable little parasites infesting bees, the *Strepsiptera*, only live for two to three hours in the mature condition, while the wingless, maggot-like, female lives eight days: in this case, therefore, the female lives sixty-four times as long as the male. The explanation of these relations is obvious; a long life for the male would be useless to the species, while the relatively long life of the female is a necessity for the species, inasmuch as she is viviparous, and must nourish her young until their birth.

Again, the male of *Phylloxera vastatrix* lives for a much shorter period than the female, and is devoid of proboscis and stomach, and takes no food: it fertilizes the female as soon as the last skin has been shed and then dies.

Insects are not the only animals among which we find inequality in the length of life of the two sexes. Very little attention has been hitherto directed to this matter, and we therefore possess little or no accurate information as to the duration of life in the sexes, but in some cases we can draw inferences either from anatomical structure or from the mode of development. Thus, male *Rotifers* never possess mouth, stomach, or intestine, they cannot take food, and without doubt live much shorter lives than the females, which are provided with a complete alimentary canal. Again, the dwarf males of many parasitic *Copepods*—low Crustacea—and the 'complementary males' of *Cirrhipedes* (or barnacles) are devoid of stomach, and must live for a much shorter time than the females; and the male *Entoniscidae* (a family of which the species are endo-parasitic in the larger Crustacea), although they can feed, die after fertilizing the females; while the latter then take to a parasitic life, produce eggs, and continue to live for some time. It is supposed that the dwarf male of *Bonellia viridis* does not live so long by several years as the hundred times larger female, and it too has no mouth to its alimentary canal. These examples might be further increased by reference to zoological literature.

In most cases the female lives longer than the male, and this needs no special explanation; but the converse relation is conceivable, when, for instance, the females are much rarer than the males, and the latter lose much time in seeking them. The above-mentioned case of *Aglia tau* probably belongs to this category.

We cannot always decide conclusively whether the life of one sex has been lengthened or that of the other shortened; both these changes must have taken place in different cases. There is no doubt that a lengthening of life in the female has arisen in the bees and ants, for both sexes of the saw-flies, which are believed to be the ancestors of bees, only live for a few weeks. But among the *Strepsiptera* the shorter life of the male must have been secondarily acquired, since we only rarely meet with such an extreme case in insects.

Note 7. Bees.

It has not been experimentally determined whether the workers, which are usually killed after some months, would live as long as the queen, if they were artificially protected from danger in the hive; but I think that this is probable, because it is the case among ants, and because the peculiarity of longevity must be latent in the egg. As is well known, the egg which gives rise to the queen is identical with that which produces a worker, and differences in the nutrition alone decide whether a queen or a worker shall be formed. It is therefore probable that the duration of life in queen and worker is potentially the same.

Note 8. Death of the Cells in higher Organisms.

The opinion has been often expressed that the inevitable appearance of normal 'death' is dependent on the wearing out of the tissues in consequence of their functional activity. Bertin says, referring to animal life[25]:—'L'observation des faits y attache l'idée d'une terminaison fatale, bien que la raison ne découvre nullement les motifs de cette nécessité. Chez les êtres qui font partie du règne animal l'exercice même de la rénovation moléculaire finit par user le principe qui l'entretient sans doute parceque le travail d'échange ne s'accomplissant pas avec une perfection mathématique, il s'établit dans la figure, comme dans la substance de l'être vivant, une déviation insensible, et que l'accumulation des écarts finit par amener un type chimique ou morphologique incompatible avec la persistance de ce travail.'

Here the replacement of the used-up elements of tissue by new ones is not taken into account, but an attempt is made to show that the functions of the whole organism necessarily cause it to waste away. But the question at once arises, whether such a result does not depend upon the fact that the single histological elements,—the cells,—are worn out by the exercise of function. Bertin admits this to be the case, and this idea of the importance of changes in the cells themselves is everywhere gaining ground. But although we must admit that the histological elements do, as a matter of fact, wear out, in multicellular animals, this would not prove that, nor explain why, such changes must follow from the nature of the cell and the vital processes which take place within it. Such an admission would merely suggest the question:—how is it that the cells in the tissues of higher animals are worn out by their function, while cells which exist in the form of free and independent organisms possess the power of living for ever? Why should not the cells of any tissue, of which the equilibrium is momentarily disturbed by metabolism, be again restored, so that the same cells continue to perform their functions for ever:—why cannot they live without their properties suffering alteration? I have not sufficiently touched upon this point in the text, and as it is obviously important it demands further consideration.

In the first place, I think we may conclude with certainty from the unending duration of unicellular organisms, that such wearing out of tissue cells is a secondary adaptation, that the death of the cell, like general death, has arisen with the complex, higher organisms. Waste does not depend upon the intrinsic nature of the cells, as the primitive organisms prove to us, but it has appeared as an adaptation of the cells to the new conditions by which they are surrounded when they come into combination, and thus form the cell-republic of the metazoan body. The replacement of cells in the tissues must be more advantageous for the functions of the whole organism than the unlimited activity of the same cells, inasmuch as the power of single cells would be much increased by this means. In certain cases, these advantages are obvious, as for example in many glands of which the secretions are made up of cast-off cells. Such cells must die and be separated from the organism, or the secretion would come to an end. In many cases, however, the facts are obscure, and await physiological investigation. But in the meantime we may draw some conclusions from the effects of growth, which are necessarily bound up with a certain rate of production of new cells. In the process of growth a certain degree of choice between the old cells which have performed their functions up to any particular time, and the new ones which have appeared between them, is as it were left to the organism.

The organism may thus, figuratively speaking, venture to demand from the various specific cells of tissues a greater amount of work than they are able to bear, during the normal length of their life, and with the normal amount of their strength. The advantages gained by the whole organism might more than compensate for the disadvantages which follow from the disappearance of single cells. The glandular secretions which are composed of cell-detritus, prove that the cells of a complex organism may acquire functions which result in the loosening of their connexion with the living cell-community of the body, and their final separation from it. And the same facts hold with the blood corpuscles, for the exercise of their function results in ultimate dissolution. Hence it is not only conceivable, but in every way probable, that many other functions in the higher organisms involve the death of the cells which perform them, not because the living cell is necessarily worn out and finally killed by the exercise of any ordinary vital process, but because the specific functions in the economy of the cell community which such cells undertake to perform, involve the death of the cells themselves. But the fact that such functions have appeared,—involving as they do the sacrifice of a great number of cells,—entirely depends upon the replacement of the old by newly formed cells, that is by the process of reproduction in cells[26].

We cannot *a priori* dispute the possibility of the existence of tissues in which the cells are not worn out by the performance of function, but such an occurrence appears to be improbable when we recollect that the cells of all tissues owe their constitution to a very far-reaching process of division of labour, which leaves them comparatively one-sided, and involves the loss of many properties of the unicellular, self-sufficient organism. At any rate we only know of potential immortality in the cells which constitute independent unicellular organisms, and the nature of these is such that they are continually undergoing a complete process of reformation.

If we did not find any replacement of cells in the higher organism, we should be induced to look upon death itself as the direct result of the division of labour among the cells, and to conclude that the specific cells of tissues have lost, as a consequence of the one-sided development of their activities, the power of unending life, which belongs to all independent primitive cells. We should argue that they could only perform their functions for a certain time, and would then die, and with them the organism whose life is dependent upon their activity. The longer they are occupied with the performance of special functions, the less completely do they carry out the phenomena of life, and hence they lead to the appearance of retrogressive changes. But the replacement of cells is certain in many tissues (in glands, blood, etc.), so that we can never seek a satisfactory explanation in the train of reasoning indicated above, but we must assume the existence of limits to the replacement of cells. In my opinion, we can find an explanation of this in the general relations of the single individual to its species, and to the whole of the external conditions of life; and this is the explanation which I have suggested and have attempted to work out in the text.

Note 9. Death by Sudden Shock.

The most remarkable example of this kind of death known to me, is that of the male bees. It has been long known that the drone perishes while pairing, and it was usually believed that the queen bites it to death. Later observations have however shown that this is not the case, but that the male suddenly dies during copulation, and that the queen afterwards bites through the male intromittent organ, in order to free herself from the dead body. In this case death is obviously due to sudden excitement, for when the latter is artificially induced, death immediately follows. Von Berlepsch made some very interesting observations on this point, 'If one catches a drone by the wings, during the nuptial flight, and holds it free in the air without touching any other part, the penis is protruded and the animal instantly dies, becoming motionless as though killed by a shock. The same thing happens if one gently stimulates the dorsal surface of the drone on a similar occasion. The male is in such an excited and irritable condition that the slightest muscular movement or disturbance causes the penis to be protruded[27].' In this case death is caused by the so-called nervous shock. The humble-bees are not similarly constituted, for the male does not die after fertilizing the female, 'but withdraws its penis and flies away.' But the death of male bees, during pairing, must not be regarded as normal death. Experiment has shown that these insects can live for more than four months[28]. They do not, as a matter of fact, generally live so long; for—although the workers do not, as was formerly believed, kill them after the fertilization of the queen, by direct means—they prevent them from eating the honey and drive them from the hive, so that they die of hunger[29].

We must also look upon death which immediately, or very quickly, follows upon the deposition of eggs as death by sudden shock. The females of certain species of *Psychidae*, when they reproduce sexually, may remain alive for more than a week waiting for a male: after fertilization, however, they lay their eggs and die, while the parthenogenetic females of the same species lay their eggs and die immediately after leaving the cocoon; so that while the former live for many days, the latter do not last for more than twenty-four hours. 'The parthenogenetic form of *Solenobia triquetrella*, soon after emergence, lays all her eggs together in the empty case, becomes much shrunken, and dies in a few hours.' (Letter from Dr. Speyer, Rhoden.)

Note 10. Intermingling during the Fission of Unicellular Organisms[30].

Fission is quite symmetrical in *Amoebae*, so that it is impossible to recognise mother and daughter in the two resulting organisms. But in *Euglypha* and allied forms the existence of a shell introduces a distinguishing mark by which it is possible to discriminate between the products of fission; so that the offspring can be differentiated from the parent. The parent organism, before division, builds the parts of the shell for the daughter form. These parts are arranged on the surface of that part of the protoplasm, external to the old shell, which will be subsequently separated as the daughter-cell. On this part the spicules are arranged and unite to form the new shell. The division of the nucleus takes place after that of the protoplasm, so that the daughter-cell is for some time without a nucleus. Although we can in this species recognise the daughter-cell for some time after separation from the parent by the greater transparency of its younger shell, it is nevertheless impossible to admit that the characteristics of the two animals are in any way different, for just before the separation of the two individuals a circulation of the protoplasm through both shells takes place after the manner described in the text, and there is therefore a complete intermingling of the substance of the two bodies.

The difference between the products is even greater after transverse fission of the *Infusoria*, for a new anus must be formed at the anterior part and a new mouth posteriorly. It is not known whether any circulation of the protoplasm takes place, as in *Euglypha*. But even if this does not occur, there is no reason for believing that the two products of division possess a different duration of life.

The process of fission in the *Diatomaceae* seems to me to be theoretically important, because here, as in the previously-mentioned *Monothalamia* (*Euglypha*, etc.), the new silicious skeleton is built up within the primary organism, but not, as in *Euglypha*, for the new individual only, but for both parent and daughter-cell alike[31]. If we compare the diatom shell to a box, then the two halves of the old shell would form two lids, one for each of the products of fission, while a new box is built up afresh for each of them. In this case there is an absolute equality between the products of fission, so far as the shell is concerned.

Note 11. Regeneration.

A number of experiments have been recently undertaken, in connection with a prize thesis at Würzburg, in order to test the powers of regeneration possessed by various animals. In all essential respects the results confirm the statements of the older observers, such as Spallanzani. Carrière has also proved that snails can regenerate not only their horns and eyes, but also part of the head when it has been cut off, although he has shown that Spallanzani's old statement that they can regenerate the whole head, including the nervous system, is erroneous[32].

Note 12. The Duration of Life in Plants.

The title of the work on this subject mentioned in the Text is 'Die Lebensdauer und Vegetationsweise der Pflanzen, ihre Ursache und ihre Entwicklung,' F. Hildebrand, Engler's botanische Jahrbücher, Bd. II. 1. und 2. Heft, Leipzig, 1881.

Note 13.

[Many interesting facts and conclusions upon the subject of this essay will be found in a volume by Professor E. Ray Lankester, 'On comparative Longevity in Man and the lower Animals,' Macmillan and Co., 1870.—E. B. P.]

Footnotes for the Appendix to Essay I.

1. Humboldt's 'Ansichten der Natur.'

2. This estimate is derived from observation of the time during which these insects are to be seen upon the wing. Direct observations upon the duration of life in this species are unknown to me.

3. [Sir John Lubbock has now kept a queen ant alive for nearly 15 years. See note 2 {note 18 below} on p. 51.—E. B. P.]

4. [After reading these proofs Dr. A. R. Wallace kindly sent me an unpublished note upon the production of death by means of natural selection, written by him some time between 1865 and 1870. The note contains some ideas on the subject, which were jotted down for further elaboration, and were then forgotten until recalled by the argument of this Essay. The note is of great interest in relation to Dr. Weismann's suggestions, and with Dr. Wallace's permission I print it in full below.

'The Action of Natural Selection in Producing Old Age, Decay, and Death.

'Supposing organisms ever existed that had not the power of natural reproduction, then since the absorptive surface would only increase as the square of the dimensions while the bulk to be nourished and renewed would increase as the cube, there must soon arrive a limit of growth.

Now if such an organism did not produce its like, accidental destruction would put an end to the species. Any organism therefore that, by accidental or spontaneous fission, could become two organisms, and thus multiply itself indefinitely without increasing in size beyond the limits most favourable for nourishment and existence, could not be thus exterminated: since the individual only could be accidentally destroyed,—the race would survive. But if individuals did not die they would soon multiply inordinately and would interfere with each other's healthy existence. Food would become scarce, and hence the larger individuals would probably decompose or diminish in size. The deficiency of nourishment would lead to parts of the organism not being renewed; they would become fixed, and liable to more or less slow decomposition as dead parts within a living body. The smaller organisms would have a better chance of finding food, the larger ones less chance. That one which gave off several small portions to form each a new organism would have a better chance of leaving descendants like itself than one which divided equally or gave off a large part of itself. Hence it would happen that those which gave off very small portions would probably soon after cease to maintain their own existence while they would leave a numerous offspring. This state of things would be in any case for the advantage of the race, and would therefore, by natural selection, soon become established as the regular course of things, and thus we have the origin of *old age*, *decay*, and *death*; for it is evident that when one or more individuals have provided a sufficient number of successors they themselves, as consumers of nourishment in a constantly increasing degree, are an injury to those successors. Natural selection therefore weeds them out, and in many cases favours such races as die almost immediately after they have left successors. Many moths and other insects are in this condition, living only to propagate their kind and then immediately dying, some not even taking any food in the perfect and reproductive state.'—E. B. P.]

5. Johannes Müller, 'Physiologie,' Bd. I. p. 31, Berlin, 1840.

6. Oken, 'Naturgeschichte,' Stuttgart, 1837, Bd. IV. Abth. 1.

7. Brehm, 'Leben der Vögel,' p. 278.

8. 'Naturwissenschaftliche Thatsachen und Probleme,' Populäre Vorträge, Berlin, 1880; *vide* Appendix.

9. 'Entomolog. Mag.,' vol. i. p. 527, 1833.

10. Imhof, 'Beiträge zur Anatomie der *Perla maxima*,' Inaug. Diss., Aarau, 1881.

11. Mr. Edwards has meanwhile published these communications in full; cf. 'On the length of life of Butterflies,' Canadian Entomologist, 1881, p. 205.

12. When no authority is given, the observations are my own.

13. In the paper quoted above, Edwards, after weighing all the evidence, reduces the length of life from three to four weeks.

14. 'Entomolog. Mag.,' vol. i. p. 527, 1823.

15. Ibid.

16. Ibid.

17. 'Recherches sur les mœurs des Fourmis indigènes,' Genève, 1810.

18. These two female ants were still alive on the 25th of September following Sir John Lubbock's letter, so that they live at least seven years. Cf. 'Observations on Ants, Bees, and Wasps,' Part VIII. p. 385; Linn. Soc. Journ. Zool., vol. xv. 1881.

[Sir John Lubbock has kindly given me further information upon the duration of life of these two queen ants. Since the receipt of his letter, the facts have been published in the Journal of the Linnean Society (Zoology), vol. xx. p. 133. I quote in full the passage which refers to these ants:—

'Longevity.—It may be remembered that my nests have enabled me to keep ants under observation for long periods, and that I have identified workers of *Lasius niger* and *Formica fusca* which were at least seven years old, and two queens of *Formica fusca* which have lived with me ever since December 1874. One of these queens, after ailing for some days, died on the 30th July, 1887. She must then have been more than thirteen years old. I was at first afraid that the other one might be affected by the death of her companion. She lived, however, until the 8th August, 1888, when she must have been nearly fifteen years old, and is therefore by far the oldest insect on record.

'Moreover, what is very extraordinary, she continued to lay fertile eggs. This remarkable fact is most interesting from a physiological point of view. Fertilization took place in 1874 at the latest. There has been no male in the nest since then, and, moreover, it is, I believe, well established that queen ants and queen bees are fertilized once for all. Hence the spermatozoa of 1874 must have retained their life and energy for thirteen years, a fact, I believe, unparalleled in physiology.'

'I had another queen of *Formica fusca* which lived to be thirteen years old, and I have now a queen of *Lasius niger* which is more than nine years old, and still lays fertile eggs, which produce female ants.'

Both the above-mentioned queens may have been considerably older, for it is impossible to estimate their age at the time of capture. It is only certain (as Sir John Lubbock informs me in his letter) that they must have been at least nine months old (when captured), as the eggs of *F. fusca* are laid in March or early in April.' The queens became gradually 'somewhat lethargic and stiff in their movements (before their death), but there was no loss of any limb nor any abrasion.' This last observation seems to indicate that queen ants may live for a much longer period in the wild state, for it is stated above that the chitin is often greatly worn, and some of the limbs lost (see pp. 48, 51, and 52).—E. B. P.]

19. A. von Berlepsch, 'Die Biene und ihre Zucht,' etc., 3rd ed.; Mannheim, 1872.

20. E. Bevan, 'Ueber die Honigbiene und die Länge ihres Lebens;' abstract in Oken's 'Isis,' 1844, p. 506.

21. Dalyell, 'Rare and Remarkable Animals of Scotland,' vol. ii. p. 203; London, 1848.

22. [Mr. J. S. Haldane has kindly obtained details of the death of the sea anemone referred to by the author. It died, by a natural death, on August 4, 1887, after having appeared to become gradually weaker for some months previous to this date. It had lived ever since 1828 in the same small glass jar in which it was placed by Sir John Dalyell. It must have been at least 66 years old when it died.—E.B.P.]

23. Bronn, 'Klassen und Ordnungen des Thierreichs,' Bd. III. p. 466; Leipzig.

24. Bronn, l. c.

25. Cf. the article 'Mort' in the 'Encyclop. Scienc. Méd.' vol. M. p. 520.

26. Roux, in his work 'Der Kampf der Theile im Organismus,' Jena 1881, has attempted to explain the manner in which division of labour has arisen among the cells of the higher organisms, and to render intelligible the mechanical processes by which the purposeful adaptations of the organism have arisen.

27. von Berlepsch, 'Die Biene und ihre Zucht,' etc.

28. Oken, 'Isis,' 1844, p. 506.

29. von Berlepsch, l. c., p. 165.

30. Cf. August Gruber, 'Der Theilungsvorgang bei Euglypha alveolata,' and 'Die Theilung der monothalamen Rhizopoden,' Z. f. W. Z., Bd. XXXV. and XXXVI., p. 104, 1881.

31. Cf. Victor Hensen, 'Physiologie d. Zeugung,' p. 152.

32. Cf. J. Carrière, 'Ueber Regeneration bei Landpulmonaten,' Tagebl. der 52. Versammlg. deutsch. Naturf. pp. 225-226.

II.

ON HEREDITY.

1883.

ON HEREDITY.

PREFACE.

The following essay was my inaugural lecture as Pro-Rector of the University of Freiburg, and was delivered publicly in the hall of the University, on June 21, 1883; it first appeared in print in the following August. Only a few copies of the first edition were available for the public, and it is therefore now reprinted as a second edition, which only differs from the first in a few not unimportant improvements and additions.

The title which I have chosen requires some explanation. I do not propose to treat of the whole problem of heredity, but only of a certain aspect of it—the transmission of acquired characters which has been hitherto assumed to occur. In taking this course I may say that it was impossible to avoid going back to the foundation of all the phenomena of heredity, and to determine the substance with which they must be connected. In my opinion this can only be the substance of the germ-cells; and this substance transfers its hereditary tendencies from generation to generation, at first unchanged, and always uninfluenced in any corresponding manner, by that which happens during the life of the individual which bears it. If these views, which are indicated rather than elaborated in this paper, be correct, all our ideas upon the transformation of species require thorough modification, for the whole principle of evolution by means of exercise (use and disuse), as proposed by Lamarck, and accepted in some cases by Darwin, entirely collapses.

The nature of the present paper—which is a lecture and not an elaborate treatise—necessitates that only suggestions and not an exhaustive treatment of the subject could be given. I have also abstained from giving further details in the form of an appendix, chiefly because I could hardly have attempted to complete a treatment of the whole range of the subject, and I hope to refer again to these questions in the future, when new experiments and observations have been made.

I am very glad to see that such an important authority as Pflüger[33] has in the meantime come to the same opinion, from an entirely different direction—an opinion which forms the foundation of the views here brought forward, namely, that heredity depends upon the continuity of the molecular substance of the germ from generation to generation.

A. W.

II.

ON HEREDITY.

With your permission I wish to bring before you to-day my views on a problem of general biological interest—the problem of heredity.

Heredity is the process which renders possible that persistence of organic beings throughout successive generations, which is generally thought to be so well understood and to need no special explanation. Nevertheless our minds cannot fail to be much perplexed by the multiplicity of its manifestations, and to be greatly puzzled as to its real nature. A celebrated German physiologist says[34], 'Although many hands have at all times endeavoured to break the seal which hides the theory of heredity from our view, the results achieved have been but small; and we are in a certain degree justified in looking with little hope upon new efforts undertaken in this direction. We must nevertheless endeavour from time to time to ascertain how far we have advanced towards a complete explanation.'

Such a course is in every way advisable, for we are not dealing with phenomena which from their very nature are incomprehensible by man. The great complexity of the subject has alone rendered it hitherto insuperable, but in the province of heredity we certainly have not reached the limits of attainable knowledge.

From this point of view heredity bears some resemblance to certain anatomical and physiological problems, e. g. the structure and function of the human brain. Its structure—with so many millions of nerve-fibres and nerve-cells—is of such extraordinary complexity that we might well despair of ever completely understanding it. Each fibre is nevertheless distinct in itself, while its connection with the nearest nerve-cell can be frequently traced, and the function of many groups of cell elements is already known. But it would seem to be impossible to unravel the excessively complex network into which the cells and fibres are knit together; and hence to arrive at the function of each single element appears to be also beyond our reach. We have not however commenced to untie this Gordian knot without some hope of success, for who can say how far human perseverance may be able to penetrate into the mechanism of the brain, and to reveal a connected structure and a common principle in its countless elements? But surely this work will be most materially assisted by the simultaneous investigation of the structure and function of the nervous system in the lower forms of life—in the polypes and jelly-fish, worms and Crustacea. In the same way we should not abandon the hope of arriving at a satisfactory knowledge of the processes of heredity, if we consider the simplest processes of the lower animals as well as the more complex processes met with in the higher forms.

The word heredity in its common acceptation, means that property of an organism by which its peculiar nature is transmitted to its descendants. From an eagle's egg an eagle of the same species developes; and not only are the characteristics of the species transmitted to the following generation, but even the individual peculiarities. The offspring resemble their parents among animals as well as among men.

On what does this common property of all organisms depend?

Häckel was probably the first to describe reproduction as 'an overgrowth of the individual,' and he attempted to explain heredity as a simple continuity of growth. This definition might be considered as a play upon words, but it is more than this; and such an interpretation rightly applied, points to the only path which, in my opinion, can lead to the comprehension of heredity.

Unicellular organisms, such as Rhizopoda and Infusoria, increase by means of fission. Each individual grows to a certain size, and then divides into two parts, which are exactly alike in size and structure, so that it is impossible to decide whether one of them is younger or older than the other. Hence in a certain sense these organisms possess immortality: they can, it is true, be destroyed, but, if protected from a violent death, they would live on indefinitely, and would only from time to time reduce the size of their overgrown bodies by division. Each individual of any such unicellular species living on the earth to-day is far older than mankind, and is almost as old as life itself.

From these unicellular organisms we can to a certain extent understand why the offspring, being in fact a part of its parents, must therefore resemble the latter. The question as to why the part should resemble the whole leads us to a new problem, that of assimilation, which also awaits solution. It is, at any rate, an undoubted fact that the organism possesses the power of taking up certain foreign substances, viz. food, and of converting them into the substance of its own body.

Among these unicellular organisms, heredity depends upon the continuity of the individual during the continual increase of its body by means of assimilation.

But how is it with the multicellular organisms which do not reproduce by means of simple division, and in which the whole body of the parent does not pass over into the offspring?

In such animals sexual reproduction is the chief means of multiplication. In no case has it always been completely wanting, and in the majority of cases it is the only kind of reproduction.

In these animals the power of reproduction is connected with certain cells which, as germ-cells, may be contrasted with those which form the rest of the body; for the former have a totally different rôle to play; they are without significance for the life of the individual[35], and yet they alone possess the power of preserving the species. Each of them can, under certain conditions, develope into a complete organism of the same species as the parent, with every individual peculiarity of the latter reproduced more or less completely. How can such hereditary transmission of the characters of the parent take place? how can a single reproductive cell reproduce the whole body in all its details?

Such a question could be easily answered if we were only concerned with the continuity of the substance of the reproductive cells from one generation to another; for this can be demonstrated in some cases, and is very probable in all. In certain insects the development of the egg into the embryo, that is the segmentation of the egg, begins with the separation of a few small cells from the main body of the egg. These are the reproductive cells, and at a later period they are taken into the interior of the animal and form its reproductive organs. Again, in certain small freshwater Crustacea (*Daphnidae*) the future reproductive cells become distinct at a very early period, although not quite at the beginning of segmentation, i. e. when the egg has divided into not more than thirty segments. Here also the cells which are separated early form the reproductive organs of the animal. The separation of the reproductive cells from those of the body takes place at a still later period, viz. at the close of segmentation, in *Sagitta*—a pelagic free-swimming form. In Vertebrata they do not become distinct from the other cells of the body until the embryo is completely formed. Thus, as their development shows, a marked antithesis exists between the substance of the undying reproductive cells and that of the perishable body-cells. We cannot explain this fact except by the supposition that each reproductive cell potentially contains two kinds of substance, which at a variable time after the commencement of embryonic development, separate from one another, and finally produce two sharply contrasted groups of cells.

It is evidently unimportant, as regards the question of heredity, whether this separation takes place early or late, inasmuch as the molecular constitution of the reproductive substance is determined before the beginning of development. In order to understand the growth and multiplication of cells, it must be conceded that all protoplasmic molecules possess the power of growing, that is of assimilating food, and of increasing by means of division. In the same manner the molecules of the reproductive protoplasm, when well nourished, grow and increase without altering their peculiar nature, and without modifying the hereditary tendencies derived from the parents. It is therefore quite conceivable that the reproductive cells might separate from the somatic cells much later than in the examples mentioned above, without changing the hereditary tendencies of which they are the bearers. There may be in fact cases in which such separation does not take place until after the animal is completely formed, and others, as I believe that I have shown[36], in which it first arises one or more generations later, viz. in the buds produced by the parent. Here also there is no ground for the belief that the hereditary tendencies of the reproductive molecules are in any way changed by the length of time which elapses before their separation from the somatic molecules. And this theoretical deduction is confirmed by observation, for from the egg of a Medusa, produced by the budding of a Polype, a Polype, in the first instance, and not a Medusa arises. Here the molecules of the reproductive substance first formed part of the Polype, and later, part of the Medusa bud, and, although they separated from the somatic cells in the bud, they nevertheless always retain the tendency to develope into a Polype.

We thus find that the reproduction of multicellular organisms is essentially similar to the corresponding process in unicellular forms; for it consists in the continual division of the reproductive cell; the only difference being that in the former case the reproductive cell does not form the whole individual, for the latter is composed of the millions of somatic cells by which the reproductive cell is surrounded. The question, 'How can a single reproductive cell contain the germ of a complete and highly complex individual?' must therefore be re-stated more precisely in the following form, 'How can the substance of the reproductive cells potentially contain the somatic substance with all its characteristic properties?'

The problem which this question suggests, becomes clearer when we employ it for the explanation of a definite instance, such as the origin of multicellular from unicellular animals. There can be no doubt that the former have originated from the latter, and that the physiological principle upon which such an origin depended, is the principle of division of labour. In the course of the phyletic development of the organized world, it must have happened that certain unicellular individuals did not separate from one another immediately after division, but lived together, at first as equivalent elements, each of which retained all the animal functions, including that of reproduction. The *Magosphaera planula* of Häckel proves that such perfectly homogeneous cell-colonies exist[37], even at the present day. Division of labour would produce a differentiation of the single cells in such a colony: thus certain cells would be set apart for obtaining food and for locomotion, while certain other cells would be exclusively reproductive. In this way colonies consisting of somatic and of reproductive cells must have arisen, and among these for the first time death appeared. For in each case the somatic cells must have perished after a certain time, while the reproductive cells alone retained the immortality inherited from the Protozoa. We must now ask how it becomes possible that one kind of cell in such a colony, can produce the other kind by division? Before the differentiation of the colony each cell always produced others similar to itself. How can the cells, after the nature of one part of the colony is changed, have undergone such changes in *their* nature that they can now produce more than one kind of cell?

Two theories can be brought forward to solve this problem. We may turn to the old and long since abandoned *nisus formativus*, or adapting the name to modern times, to a phyletic force of development which causes the organism to change from time to time. This *vis a tergo* or teleological force compels the organism to undergo new transformations without any reference to the external conditions of life. This theory throws no light upon the numerous adaptations which are met with in every organism; and it possesses no value as a scientific explanation.

Another supposition is that the primary reproductive cells are influenced by the secondary cells of the colony, which, by their adaptability to the external conditions of life, have become somatic cells: that the latter give off minute particles which entering into the former, cause such changes in their nature that at the next succeeding cell-division they are compelled to break up into dissimilar parts.

At first sight this hypothesis seems to be quite reasonable. It is not only conceivable that particles might proceed from the somatic to the reproductive cells, but the very nutrition of the latter at the expense of the former is a demonstration that such a passage actually takes place. But a closer examination reveals immense difficulties. In the first place, the molecules of the body devoured are never simply added to those of the feeding individual without undergoing any change, but as far as we know, they are really assimilated[138], that is, converted into the molecules of the latter. We cannot therefore gain much by assuming that a number of molecules can pass from the growing somatic cells into the growing reproductive cells, and can be deposited unchanged in the latter, so that, at their next division, the molecules are separated to become the somatic cells of the following generation. How can such a process be conceivable, when the colony becomes more complex, when the number of somatic cells becomes so large that they surround the reproductive cells with many layers, and when at the same time by an increasing division of labour a great number of different tissues and cells are produced, all of which must originate *de novo* from a single reproductive cell? Each of these various elements must, *ex hypothesi*, give up certain molecules to the reproductive cells; hence those which are in immediate contact with the latter would obviously possess an advantage over those which are more remote. If then any somatic cell must send the same number of molecules to each reproductive cell[39], we are compelled to suspend all known physical and physiological conceptions, and must make the entirely gratuitous assumption of an affinity on the part of the molecules for the reproductive cells. Even if we admit the existence of this affinity, its origin and means of control remain perfectly unintelligible if we suppose that it has arisen from differentiation of the complete colony. An unknown controlling force must be added to this mysterious arrangement, in order to marshal the molecules which enter the reproductive cell in such a manner that their arrangement corresponds with the order in which they must emerge as cells at a later period. In short, we become lost in unfounded hypotheses.

It is well known that Darwin has attempted to explain the phenomena of heredity by means of a hypothesis which corresponds to a considerable extent with that just described. If we substitute gemmules for molecules we have the fundamental idea of Darwin's provisional hypothesis of pangenesis. Particles of an excessively minute size are continually given off from all the cells of the body; these particles collect in the reproductive cells, and hence any change arising in the organism, at any time during its life, is represented in the reproductive cell[40]. Darwin believed that he had by this means rendered the transmission of acquired characters intelligible, a conception which he held to be necessary in order to explain the development of species. He himself pointed out that the hypothesis was merely provisional, and that it was only an expression of immediate, and by no means satisfactory knowledge of these phenomena.

It is always dangerous to invoke some entirely new force in order to understand phenomena which cannot be readily explained by the forces which are already known.

I believe that an explanation can in this case be reached by an appeal to known forces, if we suppose that characters acquired (in the true sense of the term) by the parent cannot appear in the course of the development of the offspring, but that all the characters exhibited by the latter are due to primary changes in the germ.

This supposition can obviously be made with regard to the above-mentioned colony with its constituent elements differentiated into somatic and reproductive cells. It is conceivable that the differentiation of the somatic cells was not primarily caused by a change in their own structure, but that it was prepared for by changes in the molecular structure of the reproductive cell from which the colony arose.

The generally received idea assumes that changes in the external conditions can, in connection with natural selection, call forth persistent changes in an organism; and if this view be accepted it must be as true of all Metazoa as it is of unicellular or of homogeneous multicellular organisms. Supposing that the hypothetical colonies, which were at first entirely made up of similar cells, were to gain some advantages, if in the course of development, the molecules of the reproductive cells, from which each colony arose became distributed irregularly in the resulting organism, there would be a tendency towards the perpetuation of such a change, wherever it appeared as the result of individual variability. As a result of this change the colony would no longer remain homogeneous, and its cells would become dissimilar from the first, because of the altered arrangement of the molecules in the reproductive cells. Nothing prevents us from assuming that, at the same time, the nature of a part of the molecule may undergo still further change, for the molecules are by nature complex, and may split up or combine together.

If then the reproductive cells have undergone such changes that they can produce a heterogeneous colony as the result of continual division, it follows that succeeding generations must behave in exactly the same manner, for each of them is developed from a portion of the reproductive cell from which the previous generation arose, and consists of the same reproductive substance as the latter.

From this point of view the exact manner in which we imagine the subsequent differentiation of the colony to be potentially present in the reproductive cell, becomes a matter of comparatively small importance. It may consist in a different molecular arrangement, or in some change of chemical constitution, or it may be due to both these causes combined. The essential point is that the differentiation was originally due to some change in the reproductive cells, just as this change itself produces all the differentiations which appear in the ontogeny of all species at the present day. No one doubts that the reason why this or that form of segmentation takes place, or why this or that species finally appears, is to be found in the ultimate structure of the reproductive cells. And, as a matter of fact, molecular differentiation and grouping, whether present from the beginning or first appearing in the course of development, plays a rôle which can be almost directly observed in certain species. The first segmentation furrow divides the egg of such species into an opaque and a clear half, or, as is often the case among Medusae, into a granular outer layer and a clear central part, corresponding respectively with the ectoderm and endoderm which are formed at a later period. Such early differentiations are only the visible proofs of certain highly complex molecular rearrangements in the cells, and the fact appears to indicate that we cannot be far wrong in maintaining that differentiations which appear in the course of ontogeny depend upon the chemical and physical constitution of the molecules in the reproductive cell.

At the first appearance of the earliest Metazoa alluded to above, only two kinds of cells, somatic and reproductive, arose from the segmentation of the reproductive cell. The reproductive cells thus formed must have possessed exactly the same molecular structure as the mother reproductive cell, and would therefore pass through precisely the same developmental changes. We can easily imagine that all the succeeding stages in the development of the Metazoa have been due to the same causes which were efficient at the earliest period. Variations in the molecular structure of the reproductive cells would continue to appear, and these would be increased and rendered permanent by means of natural selection, when their results, in the alteration of certain cells in the body, were advantageous to the species. The only condition necessary for the transmission of such changes is that a part of the reproductive substance (the germ-plasm) should always remain unchanged during segmentation and the subsequent building up of the body, or in other words, that such unchanged substance should pass into the organism, and after the lapse of a variable period, should reappear as the reproductive cells. Only in this way can we render to some extent intelligible the transmission of those changes which have arisen in the phylogeny of the species; only thus can we imagine the manner in which the first somatic cells gradually developed in numbers and in complexity.

It is only by supposing that these changes arose from molecular alterations in the reproductive cell that we can understand how the reproductive cells of the next generation can originate the same changes in the cells which are developed from them; and it is impossible to imagine any way in which the transmission of changes, produced by the direct action of external forces upon the somatic cells, can be brought about[41].

The difficulty or the impossibility of rendering the transmission of acquired characters intelligible by an appeal to any known force has been often felt, but no one has hitherto attempted to cast doubts upon the very existence of such a form of heredity.

There are two reasons for this: first, observations have been recorded which appear to prove the existence of such transmission; and secondly, it has seemed impossible to do without the supposition of the transmission of acquired characters, because it has always played such an important part in the explanation of the transformation of species.

It is perfectly right to defer an explanation, and to hesitate before we declare a supposed phenomenon to be impossible, because we are unable to refer it to any of the known forces. No one can believe that we are acquainted with all the forces of nature. But, on the other hand, we must use the greatest caution in dealing with unknown forces; and clear and indubitable facts must be brought forward to prove that the supposed phenomena have a real existence, and that their acceptance is unavoidable.

It has never been proved that acquired characters are transmitted, and it has never been demonstrated that, without the aid of such transmission, the evolution of the organic world becomes unintelligible.

The inheritance of acquired characters has never been proved, either by means of direct observation or by experiment[42]. It must be admitted that there are in existence numerous descriptions of cases which tend to prove that such mutilations as the loss of fingers, the scars of wounds, etc., are inherited by the offspring, but in these descriptions the previous history is invariably obscure, and hence the evidence loses all scientific value.

As a typical example of the scientific value of such cases I may mention the frequently quoted instance of the cow, which lost its left horn from suppuration, induced by some 'unknown cause,' and which afterwards produced two calves with a rudimentary left horn in each case. But as Hensen[43] has rightly remarked, the loss of the cow's horn may have arisen from a congenital malformation, which would certainly be transmitted, but which was not an acquired character.

The only cases worthy of scientific discussion are the well-known experiments upon guinea-pigs, conducted by the French physiologist Brown-Séquard. But the explanation of his results is, in my opinion, open to discussion. In these cases we have to do with the apparent transmission of artificially produced malformations. The division of important nerves, or of the spinal cord, or the removal of parts of the brain, produced certain symptoms which reappeared in the descendants of the mutilated animals. Epilepsy was produced by dividing the great sciatic nerve; the ear became deformed when the sympathetic nerve was severed in the throat; and prolapsus of the eye-ball followed the removal of a certain part of the brain—the corpora restiformia. All these effects were said to be transmitted to the descendants as far as the fifth or sixth generation.

But we must inquire whether these cases are really due to heredity and not to simple infection. In the case of epilepsy, at any rate, it is easy to imagine that the passage of some specific organism through the reproductive cells may take place, as in the case of syphilis. We are, however, entirely ignorant of the nature of the former disease. This suggested explanation may not perhaps apply to the other cases: but we must remember that animals which have been subjected to such severe operations upon the nervous system have sustained a great shock, and if they are capable of breeding, it is only probable that they will produce weak descendants, and such as are easily affected by disease. Such a result does not however explain why the offspring should suffer from the same disease as that which was artificially induced in the parents. But this does not appear to have been by any means invariably the case. Brown-Séquard himself says, 'The changes in the eye of the offspring were of a very variable nature, and were only occasionally exactly similar to those observed in the parents.'

There is no doubt, however, that these experiments demand careful consideration, but before they can claim scientific recognition, they must be subjected to rigid criticism as to the precautions taken, the number and nature of the control experiments, etc.

Up to the present time such necessary conditions have not been sufficiently observed. The recent experiments themselves are only described in short preliminary notices, which, as regards their accuracy, the possibility of mistake, the precautions taken, and the exact succession of individuals affected, afford no data upon which a scientific opinion can be founded. Until the publication of a complete series of experiments, we must say with Du Bois Reymond[44], 'The hereditary transmission of acquired characters remains an unintelligible hypothesis, which is only deduced from the facts which it attempts to explain.'

We therefore naturally ask whether the hypothesis is really necessary for the explanation of known facts.

At the first sight it certainly seems to be necessary, and it appears rash to attempt to dispense with its aid. Many phenomena only appear to be intelligible if we assume the hereditary transmission of such acquired characters as the changes which we ascribe to the use or disuse of particular organs, or to the direct influence of climate. Furthermore, how can we explain instinct as hereditary habit unless it has gradually arisen by the accumulation, through heredity, of habits which were practised in succeeding generations?

I will now attempt to prove that even these cases, so far as they depend upon clear and indubitable facts, do not force us to accept the supposition of the transmission of acquired characters.

It seems difficult and well nigh impossible to deny the transmission of acquired characters when we remember the influence which use and disuse have exercised upon certain special organs. It is well known that Lamarck attempted to explain the structure of the organism as almost entirely due to this principle alone. According to his theory the long neck of the giraffe arose by constant stretching after the leaves of trees, and the web between the toes of a water-bird's foot by the extension of the toes, in an attempt to oppose as large a surface of water as possible in swimming. There can be no doubt that those muscles which are frequently used increase in size and strength, and that glands which often enter into activity become larger and not smaller, and that their functional powers increase. Indeed, the whole effect which exercise produces upon the single parts of the body is dependent upon the fact that frequently used organs increase in strength. This conclusion also refers to the nervous system, for a pianist who performs with lightning rapidity certain pre-arranged, highly complex, and combined movements of the muscles of his hands and fingers has, as Du Bois Reymond pointed out, not only exercised the muscles, but also those ganglionic centres of the brain which determine the combination of muscular movement. Other functions of the brain, such as memory, can be similarly increased and strengthened by exercise, and the question to be settled is whether characters acquired in this way by exercise and practice can be transmitted to the following generations. Lamarck's theory assumes that such transmission takes place, for without it no accumulation or increase of the characters in question would be possible, as a result of their exercise during any number of successive generations.

Against this we may urge that whenever, in the course of nature, an organ becomes stronger by exercise, it must possess a certain degree of importance for the life of the individual, and when this is the case it becomes subject to improvement by natural selection, for only those individuals which possess the organ in its most perfect form will be able to reproduce them. The perfection of form of an organ does not however depend upon the amount of exercise undergone by it during the life of the organism, but primarily and principally upon the fact that the germ from which the individual arose was predisposed to produce a perfect organ. The increase to which any organ can attain by exercise during a single life is bounded by certain limits, which are themselves fixed by the primary tendencies of the organ in question. We cannot by excessive feeding make a giant out of the germ destined to form a dwarf; we cannot, by means of exercise, transform the muscles of an individual destined to be feeble into those of a Hercules, or the brain of a predestined fool into that of a Leibnitz or a Kant, by means of much thinking. With the same amount of exercise the organ which is destined to be strong, will attain a higher degree of functional activity than one that is destined to be weak. Hence natural selection, in destroying the least fitted individuals, destroys those which from the germ were feebly disposed. Thus the result of exercise during the individual life does not acquire so much importance, for, as compared with differences in predisposition, the amount of exercise undergone by all the individuals of a species becomes relatively uniform. The increase of an organ in the course of generations does not depend upon the summation of the exercise taken during single lives, but upon the summation of more favourable predispositions in the germs.

In criticizing these arguments, it may be questioned whether the single individuals of a species which is undergoing modification do, as a matter of fact, exercise themselves in the same manner and to the same extent. But the consideration of a definite example clearly shows that this must be the case. When the wild duck became domesticated, and lived in a farm-yard, all the individuals were compelled to walk and stand more than they had done previously, and the muscles of the legs were used to a correspondingly greater degree. The same thing happens in the wild state, when any change in the conditions of life compels an organ to be more largely used. No individual will be able to entirely avoid this extra use, and each will endeavour to accommodate itself to the new conditions according to its power. The amount of this power depends upon the predisposition of the germ; and natural selection, while it apparently decides between individuals of various degrees of strength, is in truth operating upon the stronger and weaker germs.

But the very conclusions which have been drawn from the increase of activity which has arisen from exercise, must also be drawn from the instances of atrophy or degeneration following from the disuse of organs.

Darwin long ago called attention to the fact that the degeneration of an organ may, under certain circumstances, be beneficial to the species. For example, he first proved in the instance of Madeira, that the loss of wings may be of advantage to many beetles inhabiting oceanic islands. The individuals with imperfectly developed or atrophied wings have an advantage, because they are not carried out to sea by the frequent winds. The small eyes, buried in fur, possessed by moles and other subterranean mammals, can be similarly explained by means of natural selection. So also, the complete disappearance of the limbs of snakes is evidently a real advantage to animals which creep through narrow holes and clefts; and the degeneration of the wings in the ostrich and penguin is, in part, explicable as a favourable modification of the organ of flight into an organ for striking air or water respectively.

But when the degeneration of disused organs confers no benefits upon the individual, the explanation becomes less simple. Thus we find that the eyes of animals which inhabit dark caves (such as insects, crabs, fish, Amphibia, etc.) have undergone degeneration; yet this can hardly be of direct advantage to the animals, for they could live quite as well in the dark with well-developed eyes. But we are here brought into contact with a very important aspect of natural selection, viz. the power of conservation exerted by it. Not only does the survival of the fittest select the best, but it also maintains it[45]. The struggle for existence does not cease with the foundation of a new specific type, or with some perfect adaptation to the external or internal conditions of life, but it becomes, on the contrary, even more severe, so that the most minute differences of structure determine the issue between life and death.

The sharpest sight possessed by birds is found in birds of prey, but if one of them entered the world with eyes rather below the average in this respect, it could not, in the long run, escape death from hunger, because it would always be at a disadvantage as compared with others.

Hence the sharp sight of these birds is maintained by means of the continued operation of natural selection, by which the individuals with the weakest sight are being continually exterminated. But all this would be changed at once, if a bird of prey of a certain species were compelled to live in absolute darkness. The quality of the eyes would then be immaterial, for it could make no difference to the existence of the individual, or the maintenance of the species. The sharp sight might, perhaps, be transmitted through numerous generations; but when weaker eyes arose from time to time, these would also be transmitted, for even very short-sighted or imperfect eyes would bring no disadvantage to their owner. Hence, by continual crossing between individuals with the most varied degrees of perfection in this respect, the average of perfection would gradually decline from the point attained before the species lived in the dark.

We do not at present know of any bird living in perfect darkness, and it is improbable that such a bird will ever be found; but we are acquainted with blind fish and Amphibia, and among these the eyes are present it is true, but they are small and hidden under the skin. I think it is difficult to reconcile the facts of the case with the ordinary theory that the eyes of these animals have simply degenerated through disuse. If disuse were able to bring about the complete atrophy of an organ, it follows that every trace of it would be effaced. We know that, as a matter of fact, the olfactory organ of the frog completely degenerates when the olfactory nerve is divided; and that great degeneration of the eye may be brought about by the artificial destruction of the optic centre in the brain. Since, therefore, the effects of disuse are so striking in a single life, we should certainly expect, if such effects can be transmitted, that all traces of an eye would soon disappear from a species which lives in the dark.

The caverns in Carniola and Carinthia, in which the blind *Proteus* and so many other blind animals live, belong geologically to the Jurassic formation; and although we do not exactly know when for example the *Proteus* first entered them, the low organization of this amphibian certainly indicates that it has been sheltered there for a very long period of time, and that thousands of generations of this species have succeeded one another in the caves.

Hence there is no reason to wonder at the extent to which the degeneration of the eye has been already carried in the *Proteus*; even if we assume that it is merely due to the cessation of the conserving influence of natural selection.

But it is unnecessary to depend upon this assumption alone, for when a useless organ degenerates, there are also other factors which demand consideration, namely, the higher development of other organs which compensate for the loss of the degenerating structure, or the increase in size of adjacent parts. If these newer developments are of advantage to the species, they finally come to take the place of the organ which natural selection has failed to preserve at its point of highest perfection.

In the first place, a certain form of correlation, which Roux[46] calls 'the struggle of the parts in the organism,' plays a most important part. Cases of atrophy, following disuse, appear to be always attended by a corresponding increase of other organs: blind animals always possess very strongly developed organs of touch, hearing, and smell, and the degeneration of the wing-muscles of the ostrich is accompanied by a great increase in the strength of the muscles of the leg. If the average amount of food which an animal can assimilate every day remains constant for a considerable time, it follows that a strong influx towards one organ must be accompanied by a drain upon others, and this tendency will increase, from generation to generation, in proportion to the development of the growing organ, which is favoured by natural selection in its increased blood-supply, etc.; while the operation of natural selection has also determined the organ which can bear a corresponding loss without detriment to the organism as a whole.

Without the operation of natural selection upon different individuals, the struggle between the organs of a single individual would be unable to encourage a predisposition in the germ towards the degeneration or non-development of a useless organ, and it could only limit and degrade the development of an organ in the lifetime of the individual. If, therefore, acquired characters are not transmitted, the disposition to develope such an organ would be present in the same degree in each successive generation, although the realization would be less perfect. The complete disappearance of a rudimentary organ can only take place by the operation of natural selection; this principle will lead to its elimination, inasmuch as the disappearing structure takes the place and the nutriment of other useful and important organs. Hence the process of natural selection tends to entirely remove the former. The predisposition towards a weaker development of the organ is thus advantageous, and there is every reason for the belief that the advantages would continue to be gained, and that therefore the processes of natural selection would remain in operation, until the germ had entirely lost all tendency towards the development of the organ in question. The extreme slowness with which this process takes place, and the extraordinary persistence of rudimentary organs, at any rate in the embryo, together with their gradual but finally complete disappearance, can be clearly seen in the limbs of certain vertebrates and arthropods. The blind-worms have no limbs, but a rudimentary shoulder-girdle is present close under the skin, and the interesting fact has been quite recently established[47] that the fore-limbs are present in the embryo in the form of short stumps, which entirely disappear at a later stage. In most snakes all traces of limbs have been lost in the adult, but we do not yet know for certain whether they are also wanting in the embryo. I might further mention the very different stages of degeneration witnessed in the limbs of various salamanders; and the anterior limbs of *Hesperornis*—the remarkable toothed bird from the cretaceous rocks—which, according to Marsh[48], consists only of a very thin and relatively small humerus, which was probably concealed beneath the skin. The water-fleas (*Daphnidae*) possess in the embryonic state three complete and almost equal pairs of jaws, but two of these entirely disappear, and do not develope into jaws in any species. In the same way, the embryo of the maggot-like legless larva of bees and wasps possesses three pairs of ancestral limbs.

There are, however, cases in which, apparently, acquired variations of characters are transmitted without natural selection playing any active part in the change. Such a case is afforded by the short-sightedness so common in civilized nations.

This affection is certainly hereditary in some cases, and it may well have been explained as an example of the transmission of acquired changes. It has been argued that acquired short-sightedness can be in a slight degree transmitted, and that each successive generation has developed a further degree of the disease by habitually holding books etc. close to the eyes, so that the inborn predisposition to short-sightedness is continually accumulating.

But we must remember that variations in the refraction of the human eye have been for a long time independent of the preserving control of natural selection. In the struggle for existence, a blind man would certainly disappear before those endowed with sight, but myopia does not prevent any one from gaining a living.

A short-sighted lynx, hawk, or gazelle, or even a short-sighted Indian, would be eliminated by natural selection, but a short-sighted European of the higher class finds no difficulty in earning his bread.

Those fluctuations on either side of the average which we call myopia and hypermetropia, occur in the same manner, and are due to the same causes, as those which operate in producing degeneration in the eyes of cave-dwelling animals. If, therefore, we not infrequently meet with families in which myopia is hereditary, such results may be attributed to the transmission of an accidental disposition on the part of the germ, instead of to the transmission of acquired short-sightedness. A very large proportion of short-sighted people do not owe their affliction to inheritance at all, but have acquired it for themselves; for there is no doubt that a normal eye may be rendered myopic in the course of a life-time by continually looking at objects from a very short distance, even when no hereditary predisposition towards the disease can be shown to exist. Such a change would of course appear more readily if there was also a corresponding predisposition on the part of the eye. But I should not explain this widely spread predisposition towards myopia as due to the transmission of acquired short-sightedness, but to the greater variability of the eye, which necessarily results from the cessation of the controlling influence of natural selection.

This suspension of the preserving influence of natural selection may be termed *Panmixia*, for all individuals can reproduce themselves and thus stamp their characters upon the species, and not only those which are in all respects, or in respect to some single organ, the fittest. In my opinion, the greater number of those variations which are usually attributed to the direct influence of external conditions of life, are to be ascribed to panmixia. For example, the great variability of most domesticated animals essentially depends upon this principle.

A goose or a duck must possess strong powers of flight in the natural state, but such powers are no longer necessary for obtaining food when it is brought into the poultry-yard, so that a rigid selection of individuals with well-developed wings, at once ceases among its descendants. Hence in the course of generations, a deterioration of the organs of flight must necessarily ensue, and the other members and organs of the bird will be similarly affected.

This example very clearly indicates that the degeneration of an organ does not depend upon its disuse; for although our domestic poultry very rarely make use of their wings, the muscles of flight have not disappeared, and, at any rate in the goose, do not seem to have undergone any marked degeneration.

The numerous and exact observations conducted by Darwin upon the weight and measurement of the bones in domestic fowls, seem to me to possess a significance beyond that which he attributed to them.

If the weight of the wing-bones of the domestic duck bears a smaller proportion to the weight of the leg-bones than in the wild duck, and if, as Darwin rightly assumes, this depends not only upon the diminution of the wings, but also upon the increase of the legs, it by no means follows that this latter increase in organs which are now more frequently used, is dependent upon hereditary influences alone.

It is quite possible that it depends, on the one hand, upon the suspension of natural selection, or panmixia (and these effects would be transmitted), and on the other hand upon the direct influence of increased use during the course of a single life. We do not yet know with any accuracy, the amount of change which may be produced by increased use in the course of a single life. If it is desired to prove that use and disuse produce hereditary effects without the assistance of natural selection, it will be necessary to domesticate wild animals (for example the wild duck) and preserve all their descendants, thus excluding the operation of natural selection. If then all individuals of the second, third, fourth and later generations of these tame ducks possess identical variations, which increase from generation to generation, and if the nature of these changes proves that they must have been due to the effect of use or disuse, then perhaps the transmission of such effects may be admitted; but it must always be remembered that domestication itself influences the organism,—not only directly, but also indirectly, by the increase of variability as a result of the suspension of natural selection. Such experiments have not yet been carried out in sufficient detail[49].

It is usually considered that the origin and variation of instincts are also dependent upon the exercise of certain groups of muscles and nerves during a single life-time; and that the gradual improvement which is thus caused by practice, is accumulated by hereditary transmission. I believe that this is an entirely erroneous view, and I hold that all instinct is entirely due to the operation of natural selection, and has its foundation, not upon inherited experiences, but upon the variations of the germ.

Why, for instance, should not the instinct to fly from enemies have arisen by the survival of those individuals which are naturally timid and easily startled, together with the extermination of those which are unwary? It may be urged in opposition to this explanation that the birds of uninhabited islands which are not at first shy of man, acquire in a few generations an instinctive dread of him, an instinct which cannot have arisen in so short a time by means of natural selection. But, in this case are we really dealing with the origin of a new instinct, or only with the addition of one new perception ('Wahrnehmung,' Schneider)[50], of the same kind as those which incite to the instinct of flight—an instinct which had been previously developed in past ages but had never been called forth by man? Again, has any one ascertained whether the young birds of the second or third generation are frightened by man? May it not be that the experience of a single life-time plays a great part in the origin of the habit? For my part, I am inclined to believe that the habit of flying from man is developed in the first generation which encounters him as a foe[51]. We see how wary and cautious a flock of birds become as soon as a few shots have been fired at them, and yet shortly before this occurrence they were perhaps playing carelessly close to the sportsmen. Intelligence plays a considerable part in the life of birds, and it by no means follows that the transmission of individual habits explains the above-mentioned phenomena. The long-continued operation of natural selection may very well have been necessary before the perception of man could awake the instinct to flee in young, inexperienced birds. Unfortunately the observations upon these points are far too indefinite to enable us to draw conclusions.

There is again the frequently-quoted instance of the young pointer, 'which, untrained, and without any example which might have been imitated, pointed at a lizard in a subtropical jungle, just as many of its forefathers had pointed at partridges on the plain of St. Denis,' and which, without knowing the effect of a shot, sprang forward barking, at the first discharge, to bring in the game. This conduct must not be attributed to the inheritance of any mental picture, such as the effect of a shot, but to the inheritance of a certain reflex mechanism. The young pointer does not spring forward at the shot because he has inherited from his forefathers a certain association of ideas,—shot and game,—but because he has inherited a reflex mechanism, which impels him to start forward on hearing a report. We cannot yet determine without more experiments how such an impulse due to perception ('Wahrnehmungstrieb,' Schneider) has arisen; but, in my opinion, it is almost inconceivable that artificial breeding has had nothing to do with it; and that we are here concerned—not with the inheritance of the effects of training—but with some predisposition on the part of the germ, which has been increased by artificial selection.

The necessity for extreme caution in appealing to the supposed hereditary effects of use, is well shown in the case of those numerous instincts, which only come into play once in a lifetime, and which do not therefore admit of improvement by practice. The queen-bee takes her nuptial flight only once, and yet how many and complex are the instincts and the reflex mechanisms which come into play on that occasion. Again, in many insects the deposition of eggs occurs but once in a life-time, and yet such insects always fulfil the necessary conditions with unfailing accuracy, either simply dropping the eggs into water, or carefully fixing them on the surface of the earth beneath some stone, or laying them on a particular part of a certain species of plant; and in all these cases the most complicated actions are performed. It is indeed astonishing to watch one of the *Cynipidae* (*Rhodites rosae*) depositing her eggs in the tissue of a young bud. She first carefully examines the bud on all sides, and feels it with her legs and antennae. Then she slowly inserts her long ovipositor between the closely-rolled leaves of the bud, but if it does not reach exactly the right spot, she will withdraw and re-insert it many times, until at length, when the proper place has been found, she will slowly bore deep into the very centre of the bud, so that the egg will reach the exact spot, where the necessary conditions for its development alone exist.

But each *Cynips* lays eggs many times, and it may be argued that practice may have led to improvement in this case; we cannot however, as a matter of fact, expect much improvement in a process which is repeated, perhaps a dozen times, at short intervals of time, and which is of such an excessively complex nature.

It is the same with the deposition of eggs in most insects. How can practice have had any influence upon the origin of the instinct which leads one of our butterflies—(*Vanessa levana*)—to lay its green eggs in single file, as columns, which project freely from the stem or leaf, so that protection is gained by their close resemblance to the flower-buds of the stinging-nettle, which forms the food-plant of their caterpillars?

Of course the butterfly is not aware of the advantage which follows from such a proceeding; intelligence has no part in the process. The entire operation depends upon certain inherent anatomical and physiological arrangements:—on the structure of the ovary and oviducts, on the simultaneous ripening of a certain number of eggs, and on certain very complex reflex mechanisms which compel the butterfly to lay its eggs on certain parts of certain plants. Schneider is certainly right when he maintains that this mechanism is released by a sensation, arising from the perception (whether by sight or smell, or both together) of the particular plant or part of the plant upon which the eggs are to be laid[52]. At any rate, we cannot, in such cases, appeal to the effects of constant use and the transmission of acquired characters, as an explanation; and the origin of the impulse can only be understood as a result of the process of natural selection.

The protective cocoons by which the pupae of many insects are surrounded also belong to the same category, and improvement by practice is entirely out of the question, for they are only constructed once in the course of a life-time. And yet these cocoons are often remarkably complex: think, for instance, of the cocoon spun by the caterpillar of the emperor moth (*Saturnia carpini*), which is so tough that it can hardly be torn, and which the moth would be unable to leave, if an opening were not provided for the purpose; while, on the other hand, the pupa would not be defended against enemies if the opening were not furnished with a circle of pointed bristles, converging outwards, on the principle of the lobster pot, so that the moth can easily emerge, although no enemy can enter. The impulse which leads to the production of such a structure can only have arisen by the operation of natural selection—not, of course, during the history of a single species, but during the development of numerous, consecutive species—by gradual and unceasing improvements in the initial stages of cocoon-building. A number of species exists at the present day, of which the cocoons can be arranged in a complete series, becoming gradually less and less complex, from that described above, down to a loosely-constructed, spherical case in which the pupa is contained.

The cocoon spun by the larva of *Saturnia carpini* differs but little in complexity from the web of the spider, and if the former is constructed without assistance from the experience of the single individual—and this must certainly be admitted—it follows that the latter may be also built without the aid of experience, while there is neither reason nor necessity for appealing to the entirely unproved transmission of acquired skill in order to explain this and a thousand other operations.

It may be objected that, in man, in addition to the instincts inherent in every individual, special individual predispositions are also found, of such a nature that it is impossible that they can have arisen by individual variations of the germ. On the other hand, these predispositions—which we call talents—cannot have arisen through natural selection, because life is in no way dependent upon their presence, and there seems to be no way of explaining their origin except by an assumption of the summation of the skill attained by exercise in the course of each single life. In this case, therefore, we seem at first sight to be compelled to accept the transmission of acquired characters.

Now it cannot be denied that all predispositions may be improved by practice during the course of a life-time,—and, in truth, very remarkably improved. If we could explain the existence of great talent, such as, for example, a gift for music, painting, sculpture, or mathematics, as due to the presence or absence of a special organ in the brain, it follows that we could only understand its origin and increase (natural selection being excluded) by accumulation, due to the transmission of the results of practice through a series of generations. But talents are not dependent upon the possession of special organs in the brain. They are not simple mental dispositions, but combinations of many dispositions, and often of a most complex nature: they depend upon a certain degree of irritability, and a power of readily transmitting impulses along the nerve-tracts of the brain, as well as upon the especial development of single parts of the brain. In my opinion, there is absolutely no trustworthy proof that talents have been improved by their exercise through the course of a long series of generations. The Bach family shows that musical talent, and the Bernoulli family that mathematical power, can be transmitted from generation to generation, but this teaches us nothing as to the origin of such talents. In both families the high-water mark of talent lies, not at the end of the series of generations, as it should do if the results of practice are transmitted, but in the middle. Again, talents frequently appear in some single member of a family which has not been previously distinguished.

Gauss was not the son of a mathematician; Handel's father was a surgeon, of whose musical powers nothing is known; Titian was the son and also the nephew of a lawyer, while he and his brother, Francesco Vecellio, were the first painters in a family which produced a succession of seven other artists with diminishing talents. These facts do not, however, prove that the condition of the nerve-tracts and centres of the brain, which determine the specific talent, appeared for the first time in these men: the appropriate condition surely existed previously in their parents, although it did not achieve expression. They prove, as it seems to me, that a high degree of endowment in a special direction, which we call talent, cannot have arisen from the experience of previous generations, that is, by the exercise of the brain in the same specific direction.

It appears to me that talent consists in a happy combination of exceptionally high gifts, developed in one special direction. At present, it is of course impossible to understand the physiological conditions which render the origin of such combinations possible, but it is very probable that the crossing of the mental dispositions of the parents plays a great part in it. This has been admirably and concisely expressed by Goethe in describing his own characteristics—

Vom Vater hab' ich die Statur

Des Lebens ernstes Führen,

Vom Mütterchen die Frohnatur

Die Lust zum Fabuliren, etc.

The combination of talents frequently found in one individual, and the appearance of different remarkable talents in the various branches of one and the same family, indicate that talents are only special combinations of certain highly-developed mental dispositions which are found in every brain. Many painters have been admirable musicians, and we very frequently find both these talents developed to a slighter extent in a single individual. In the Feuerbach family we find a distinguished jurist, a remarkable philosopher, and a highly-talented artist; and among the Mendelssohns a philosopher as well as a musician.

The sudden and yet widespread appearance of a particular talent in correspondence with the general intellectual excitement of a certain epoch points in the same direction. How many poets arose in Germany during the period of sentiment which marked the close of the last century, and how completely all poetic gifts seem to have disappeared during the Thirty Years' War. How numerous were the philosophers that appeared in the epoch which succeeded Kant; while all philosophic talent seemed to have deserted the German nation during the sway of the antagonistic 'exact science,' with its contempt for speculation.

Wherever academies are founded, there the Schwanthalers, Defreggers, and Lenbachs emerge from the masses which had shown no sign of artistic endowment through long periods of time[53]. At the present day there are many men of science who, had they lived at the time of Bürger, Uhland, or Schelling, would probably have been poets or philosophers. And the man of science also cannot dispense with that mental disposition directed in a certain course, which we call talent, although the specific part of it may not be so obvious: we may, indeed, go further, and maintain that the Physicist and the Chemist are characterized by a combination of mental dispositions which differ from those of the Botanist and the Zoologist. Nevertheless, a man is not born a physicist or a botanist, and in most cases chance alone determines whether his endowments are developed in either direction.

Lessing has asked whether Raphael would have been a less distinguished artist had he been born without hands: we might also enquire whether he might not have been as great a musician as he was painter if, instead of living during the historical high-water mark of painting, he had lived, under favourable personal influences, at the time of highly-developed and widespread musical genius. A great artist is always a great man, and if he finds the outlet for his talent closed on one side, he forces his way through on the other.

From all these examples I wish to show that, in my opinion, talents do not appear to depend upon the improvement of any special mental quality by continued practice, but they are the expression, and to a certain extent the bye-product, of the human mind, which is so highly developed in all directions.

But if any one asks whether this high mental development, acquired in the course of innumerable generations of men, is not dependent upon the hereditary effects of use, I would remind him that human intelligence in general is the chief means and the chief weapon which has served and still serves the human species in the struggle for existence[54]. Even in the present state of civilization—distorted as it is by numerous artificial encroachments and unnatural conditions—the degree of intelligence possessed by the individual chiefly decides between destruction and life; and in a natural state, or still better in a state of low civilization, this result is even more striking.

Here again, therefore, we encounter the effects of natural selection, and to this power we must attribute, at any rate, a great part of the phenomena we have been discussing, and it cannot be shown that—in addition to its operation—the transmission of characters acquired by practice plays any part in nature.

I only know of one class of changes in the organism which is with difficulty explained by the supposition of changes in the germ; these are the modifications which appear as the direct consequence of some alteration in the surroundings. But our knowledge on this subject is still very defective, and we do not know the facts with sufficient precision to enable us to pronounce a final verdict as to the cause of such changes: and for this reason, I do not propose to consider the subject in detail.

These changes—such, for example, as are produced by a strange climate—have been always looked at under the supposition that they are transmitted and intensified from generation to generation, and for this reason the observations are not always sufficiently precise. It is difficult to say whether the changed climate may not have first changed the germ, and if this were the case the accumulation of effects through the action of heredity would present no difficulty. For instance, it is well known that increased nourishment not only causes a plant to grow more luxuriantly, but it alters the plant in some distinct way, and it would be wonderful indeed if the seeds were not also larger and better furnished with nutritive material. If the increased nourishment be repeated in the next generation, a still further increase in the size of the seed, in the luxuriance of the plant, and in all other changes which ensue, is at any rate conceivable if it is not a necessity. But this would not be an instance of the transmission of acquired characters, but only the consequence of a direct influence upon the germ-cells, and of better nourishment during growth.

A similar interpretation explains the converse change. When horses of normal size are introduced into the Falkland Islands, the next generation is smaller in consequence of poor nourishment and the damp climate, and after a few generations they have deteriorated to a marked extent. In such a case we have only to assume that the climate which is unfavourable and the nutriment which is insufficient for horses, affect not only the animal as a whole, but also its germ-cells. This would result in the diminution in size of the germ-cells, the effects upon the offspring being still further intensified by the insufficient nourishment supplied during growth. But such results would not depend upon the transmission by the germ-cells of certain peculiarities due to the unfavourable climate, which only appear in the full-grown horse.

It must be admitted that there are cases, such as the climatic varieties of certain butterflies, which raise some difficulties against this explanation. I myself, some years ago, experimentally investigated one such case[55], and even now I cannot explain the facts otherwise than by supposing the passive acquisition of characters produced by the direct influence of climate.

It must be remembered, however, that my experiments, which have been repeated upon several American species by H. W. Edwards, with results confirmatory of my own in all essential respects, were not undertaken with the object of investigating the question from this point of view alone. New experiments, under varying conditions, will be necessary to afford the true explanation of this aspect of the question; and I have already begun to undertake them.

Leaving on one side, for the moment, these doubtful, and insufficiently investigated cases, we may still maintain that the assumption that changes induced by external conditions in the organism as a whole, are communicated to the germ-cells after the manner indicated in Darwin's hypothesis of pangenesis,—is wholly unnecessary for the explanation of these phenomena. Still we cannot exclude the possibility of such a transmission occasionally occurring, for, even if the greater part of the effects must be attributed to natural selection, there might be a smaller part in certain cases which depends on this exceptional factor.

A complete and satisfactory refutation of such an opinion cannot be brought forward at present: we can only point out that such an assumption introduces new and entirely obscure forces, and that innumerable cases exist in which we can certainly exclude all assistance from the transmission of acquired characters. In most cases of variation in colour we have no explanation but the survival of the fittest[56], and the same holds good for all changes of form which cannot be influenced by the will of the animal. Very numerous adaptations, such, for instance, as occur in the eggs of animals,—the markings, and appendages which conceal them from enemies, the complex coverings which prevent them from drying up or protect them from the injurious influence of cold,—must have all arisen entirely independently of any expression of will, or of any conscious or unconscious action on the part of the animals. I will not mention here the case of plants, which as every one knows are unconscious, for they are beyond my province. In this matter, there can be no suggestion of adaptation depending upon a struggle between the various parts of the organism (Roux)[57]. Natural selection cannot operate upon the different epithelial cells which secrete the egg-shell of *Apus*, since it is of no consequence to the animal which secretes the egg-shell whether a good or a bad shell is produced. Natural selection first operates among the offspring, and the egg with a shell incapable of resisting cold or drought is destroyed. The different cells of the same individual are not selected, but the different individuals themselves.

In all such cases we have no explanation except the operation of natural selection, and if we cannot accept this, we may as well abandon any attempt at a natural explanation. But, in my opinion, there is no reason why natural selection should be considered inadequate to the task. It is true that the objection has been lately urged, that it is inconceivable that all the wonderful adaptations of the organism to its surroundings can have arisen through the selection of individuals; and that for this purpose an infinite number of individuals and infinite time would be required; and stress is laid upon the fact that the wished-for useful changes can only arise singly and very rarely among a great number of individuals.

This last objection to the modern conception of natural selection has apparently some weight, for, as a matter of fact, useful variations of a conspicuous kind seldom appear, and are often entirely absent for many generations. If we expect to find that qualitative changes take place by sudden leaps, we can never escape this difficulty. But, I think, we must not look for conspicuous variations—such as occur among domesticated animals and plants—in the process of the evolution of species as it goes on in nature. Natural selection does not deal with qualitative but quantitative changes in the individual, and the latter are always present.

A simple example will make this clearer. Let us suppose that it was advantageous to some species—for instance the ancestors of the giraffe—to lengthen some part of the body, such as the neck: this result could be obtained in a relatively short time, for the members of the species already possessed necks of varying length, and the variations which form the material for natural selection were already in existence. Now all the organs of every species vary in size, and any one of them will undergo constant and progressive increase, as soon as it acquires exceptional usefulness. But not only will the organ fluctuate as a whole, but also the parts composing it will become larger or smaller under given conditions, will increase or diminish by the operation of natural selection. I believe that qualitative variations always depend upon differences in the size and number of the component parts of the whole. A skin appears to be naked, when it is really covered with a number of small fine hairs: if these grow larger and increase in number, a thick covering is formed, and we say that the skin is woolly or furry. In the same way the skin of many worms and Crustacea is apparently colourless, but the microscope reveals the presence of a number of beautiful pigment spots; and not until these have increased enormously does the skin appear coloured to the naked eye. The presence or absence of colour and its quality when present are here dependent upon the quantity of the most minute particles, and on the distance at which the object in question is observed. Again, the first appearance of colour, or the change from a green to a yellow or red colour depends upon slight variations in the position or in the number of the oxygen atoms which enter into the chemical combination in question. Fluctuations in the chemical composition of the molecules of a unicellular organism (for example) must continually arise, just as fluctuations are always occurring in the number of pigment granules in a certain cell, or in the number of pigment cells in a certain region of the body, or even in the size of the various parts of the body.

All these quantitative relations are exposed to individual fluctuations in every species; and natural selection can strengthen the fluctuations of any part, and thus cause it to develope further in any given direction.

From this point of view, it becomes less astonishing and less inconceivable that organisms adapt themselves—as we see that they obviously do—in all their parts to any condition of existence, and that they behave like a plastic mass which can be moulded into almost any imaginable form in the course of time.

If we ask in what lies the cause of this variability, the answer must undoubtedly be that it lies in the germ-cells. From the moment when the phenomena which precede segmentation commence in the egg, the exact kind of organism which will be developed is already determined—whether it will be larger or smaller, more like its father or its mother, which of its parts will resemble the one and which the other, even to the minutest detail. In spite of this, there still remains a certain scope for the influence of external conditions upon the organism. But this scope is limited, and forms but a small area round the fixed central point which is determined by heredity. Abundant nourishment can make the body large and strong, but can never make a giant out of the germ-cell destined to become a dwarf. Unhealthy sedentary habits or insufficient nourishment makes the factory-hand pale and stunted; life on board ship, with plenty of exercise and sea air, gives the sailor bodily strength and a tanned skin; but when once the resemblance to father or mother, or to both, is established in the germ-cell it can never be effaced, let the habit of life be what it will.

But if the essential nature of the germ-cell dominates over the organism which will grow from it, so also the quantitative individual differences, to which I referred just now, are, by the same principle, established in the germ, and—whatever be the cause which determines their presence—they must be looked upon as inherent in it. It therefore follows that, although natural selection appears to operate upon the qualities of the developed organism alone, it in truth works upon peculiarities which lie hidden in the germ-cells. Just as the final development of any predisposition in the germ, and just as any character in the mature organism vibrates with a certain amplitude around a fixed central point, so the predisposition of the germ itself fluctuates, and it is on this that the possibility of an increase of the predisposition in question, and its average result, depends.

If we trace all the permanent hereditary variations from generation to generation back to the quantitative variations of the germ, as I have sought to do, the question naturally occurs as to the source from which these variations arose in the germ itself. I will not enter into this subject at any length on the present occasion, for I have already expressed my opinion upon it[58].

I believe however that they can be referred to the various external influences to which the germ is exposed before the commencement of embryonic development. Hence we may fairly attribute to the adult organism influences which determine the phyletic development of its descendants. For the germ-cells are contained in the organism, and the external influences which affect them are intimately connected with the state of the organism in which they lie hid. If it be well nourished, the germ-cells will have abundant nutriment; and, conversely, if it be weak and sickly, the germ-cells will be arrested in their growth. It is even possible that the effects of these influences may be more specialized; that is to say, they may act only upon certain parts of the germ-cells. But this is indeed very different from believing that the changes of the organism which result from external stimuli can be transmitted to the germ-cells and will re-develop in the next generation at the same time as that at which they arose in the parent, and in the same part of the organism.

We have an obvious means by which the inheritance of all transmitted peculiarities takes place, in *the continuity of the substance of the germ-cells, or germ-plasm*. If, as I believe, the substance of the germ-cells, the germ-plasm, has remained in perpetual continuity from the first origin of life, and if the germ-plasm and the substance of the body, the somatoplasm, have always occupied different spheres, and if changes in the latter only arise when they have been preceded by corresponding changes in the former, then we can, up to a certain point, understand the principle of heredity; or, at any rate, we can conceive that the human mind may at some time be capable of understanding it. We may at least maintain that it has been rendered intelligible, for we can thus trace heredity back to growth; we can thus look upon reproduction as an overgrowth of the individual, and can thus distinguish between a succession of species and a succession of individuals, because in the latter succession the germ-plasm remains similar, while in the succession of the former it becomes different. Thus individuals, as they arise, are always assuming new and more complex forms, until the interval between the simple unicellular protozoon and the most complex of all organisms—man himself—is bridged over.

I have not been able to throw light upon all sides of the question which we are here discussing. There are still some essential points which I must leave for the present; and, furthermore, I am not yet in a position to explain satisfactorily all the details which arise at every step of the argument. But it appeared to me to be necessary to state this weighty and fundamental question, and to formulate it concisely and definitely; for only in this way will it be possible to arrive at a true and lasting solution of the problem. We must however be clear on this point—that the understanding of the phenomena of heredity is only possible on the fundamental supposition of the continuity of the germ-plasm. The value of experiment in relation to this question is somewhat doubtful. A careful collection and arrangement of facts is far more likely to decide whether, and to what extent, the continuity of germ-plasm is reconcilable with the assumption of the transmission of acquired characters from the parent body to the germ, and from the germ to the body of the offspring. At present such transmission is neither proved as a fact, nor has its assumption been shown to be unquestionably necessary.

Footnotes for Essay II.

33. Pflüger, 'Ueber den Einfluss der Schwerkraft auf die Theilung der Zellen und auf die Entwicklung des Embryo,' Arch. f. Physiol. Bd. XXXII. p. 68, 1883.

34. Victor Hensen in his 'Physiologie der Zeugung,' Leipzig, 1881, p. 216.

35. That is for the preservation of its life.

36. Compare Weismann, 'Die Entstehung der Sexualzellen bei den Hydromedusen,' Jena, 1883.

37. It is doubtful whether *Magosphaera* should be looked upon as a mature form; but nothing hinders us from believing that species have lived, and are still living, in which the ciliated sphere has held together until the encystment, that is the reproduction, of the constituent single cells.

38. Or is an exception perhaps afforded by the nutritive cells of the egg, which occur in many animals?

39. Or more precisely, they must give up as many molecules as would correspond to the number of the kind of cell in question found in the mature organism.

40. See Darwin, 'The Variation of Animals and Plants under Domestication,' 1875, vol. ii. chapter xxvii. pp. 349-399.

41. To this class of phenomena of course belong those acts of will which call forth the functional activity of certain groups of cells. It is quite clear that such impulses do not originate in the constitution of the tissue in question, but are due to the operation of external causes. The activity does not arise directly from any natural disposition of the germ, but is the result of accidental external impressions. A domesticated duck uses its legs in a different manner from, and more frequently than a wild duck, but such functional changes are the consequence of changed external conditions, and are not due to the constitution of the germ.

42. Upon this subject Pflüger states—'I have made myself accurately acquainted with all facts which are supposed to prove the inheritance of acquired characters,—that is of characters which are not due to the peculiar organization of the ovum and spermatozoon from which the individual is formed, but which follow from the incidence of accidental external influences upon the organism at any time in its life. Not one of these facts can be accepted as a proof of the transmission of acquired characters.' *l. c.* p. 68.

43. 'Physiologie der Zeugung.'

44. See 'Ueber die Uebung,' Berlin, 1881.

45. This principle was, I believe, first pointed out by Seidlitz. Compare Seidlitz, 'Die Darwin'sche Theorie,' Leipzig, 1875, p. 198.

46. W. Roux, 'Der Kampf der Theile im Organismus,' Leipzig, 1881.

47. Compare Born in 'Zoolog. Anzeiger,' 1883, No. 150, p. 537.

48. O. C. Marsh, 'Odontornithes, a Monograph on the extinct toothed Birds of North America,' Washington, 1880.

49. C. Darwin, 'Variation of Animals and Plants under Domestication.' Vol. I.

50. Compare 'Der thierische Wille,' Leipzig, 1880.

51. Steller's interesting account of the Sea-cow (*Rhytina Stelleri*) proves that this suggestion is valid. This large mammal was living in great numbers in Behring Strait at the end of the last century, but has since been entirely exterminated by man. Steller, who was compelled by shipwreck to remain in the locality for a whole year, tells us that the animals were at first without any fear of man, so that they could be approached in boats and could thus be killed. After a few months however the survivors became wary, and did not allow Steller's men to approach them, so that they were difficult to catch.—A. W., 1888.

52. Compare Schneider, 'Der thierische Wille.'

53. [The author refers to the Academy of Arts at Munich. S. S.]

54. Compare Darwin's 'Descent of Man.'

55. 'Studien zur Descendenztheorie, I. Ueber den Saison-Dimorphismus der Schmetterlinge.' Leipzig, 1875. English edition translated and edited by Professor Meldola, 'Studies in the Theory of Descent,' Part I.

56. The colours which have been called forth by sexual selection must also be included here.

57. Wilhelm Roux, 'Der Kampf der Theile im Organismus.' Leipzig, 1881.

58. Consult 'Studien zur Descendenztheorie, IV. Über die mechanische Auffassung der Natur,' p. 303, etc. Translated and edited by Professor Meldola; see 'Studies in the Theory of Descent,' p. 677, &c.

III.

LIFE AND DEATH.

1883.

LIFE AND DEATH.

PREFACE.

The following paper was first printed as an academic lecture in the summer of the present year (1883), with the title 'Upon the Eternal Duration of Life' ('Über die Ewigkeit des Lebens'). In now bringing it before a larger public in an expanded and improved form, I have chosen a title which seemed to me to correspond better with the present contents of the paper.

The stimulus which led to this biological investigation was given in a memoir by Götte, in which this author opposes views which I had previously expressed. Although such an origin has naturally caused my paper to take the form of a reply, my intention was not merely to controvert the views of my opponent, but rather—using those opposed views as a starting-point—to throw new light upon certain questions which demand consideration; to give additional support to thoughts which I have previously expressed, and to penetrate, if possible, more deeply into the problem of life and death.

If, in making this attempt, the views of my opponent have been severely criticized, it will be acknowledged that the criticisms do not form the purpose of my paper, but only the means by which the way to a more correct understanding of the problems before us may be indicated.

A. W.

Freiburg i. Breisgau,

Oct. 18, 1883.

III.

LIFE AND DEATH.

In the previous essay, entitled 'The Duration of Life,' I have endeavoured to show that the limitation of life in single individuals by death is not, as has been hitherto assumed, an inevitable phenomenon, essential to the very nature of life itself; but that it is an adaptation which first appeared when, in consequence of a certain complexity of structure, an unending life became disadvantageous to the species. I pointed out that we could not speak of natural death among unicellular animals, for their growth has no termination which is comparable with death. The origin of new individuals is not connected with the death of the old; but increase by division takes place in such a way that the two parts into which an organism separates are exactly equivalent one to another, and neither of them is older or younger than the other. In this way countless numbers of individuals arise, each of which is as old as the species itself, while each possesses the capability of living on indefinitely, by means of division.

I suggested that the Metazoa have lost this power of unending life by being constructed of numerous cells, and by the consequent division of labour which became established between the various cells of the body. Here also reproduction takes place by means of cell-division, but every cell does not possess the power of reproducing the whole organism. The cells of the organism are differentiated into two essentially different groups, the reproductive cells—ova or spermatozoa, and the somatic cells, or cells of the body, in the narrower sense. The immortality of the unicellular organism has only passed over to the former; the others must die, and since the body of the individual is chiefly composed of them, it must die also.

I have endeavoured to explain this fact as an adaptation to the general conditions of life. In my opinion life became limited in its duration, not because it was contrary to its very nature to be unlimited, but because an unlimited persistence of the individual would be a luxury without a purpose. Among unicellular organisms natural death was impossible, because the reproductive cell and the individual were one and the same: among multicellular animals it was possible, and we see that it has arisen.

Natural death appeared to me to be explicable on the principle of utility, as an adaptation.

These opinions, to which I shall return in greater detail in a later part of this paper, have been opposed by Götte[59], who does not attribute death to utility, but considers it to be a necessity inherent in life itself. He considers that it occurs not only in the Metazoa or multicellular animals, but also in unicellular forms of life, where it is represented by the process of encystment, which is to be regarded as the death of the individual. This encystment is a process of rejuvenescence, which, after a longer or shorter interval, interrupts multiplication by means of fission. According to Götte, this process of rejuvenescence consists in the dissolution of the specific structure of the individual, or in the retrogression of the individual to a form of organic matter which is no longer living but which is comparable to the yolk of an egg. This matter is, by means of its internal energy, and in consequence of the law of growth which is inherent in its constitution, enabled to give rise to a new individual of the same species. Furthermore, the process of rejuvenescence among unicellular beings corresponds to the formation of germs in the higher organisms. The phenomena of death were transmitted by heredity from the unicellular forms to the Metazoa when they arose. Death does not therefore appear for the first time in the Metazoa, but it is an extremely ancient process which 'goes back to the first origin of organic beings' (l. c., p. 81).

It is obvious, from this short *résumé*, that Götte's view is totally opposed to mine. Inasmuch as only one of these views can be fundamentally right, it is worth while to compare the two; and although we cannot at present hope to explain the ultimate physiological processes which involve life and death, I think nevertheless that it is quite possible to arrive at definite conclusions as to the general causes of these phenomena. At any rate, existing facts have not been so completely thought out that it is useless to consider them once more.

The question—what do we understand by death? must be decided before we can speak of the origin of death. Götte says, 'we are not able to explain this general expression quite definitely and in all its details, because the moment of death, or perhaps more exactly the moment when death is complete, can in no case be precisely indicated. We can only say that in the death of the higher animals, all those phenomena which make up the life of the individual cease, and further that all the cells and elements of tissue which form the dead organism, die, and are resolved into their elements.'

This definition would suffice if it did not include that which is to be defined. For it assumes that under the expression 'dead organism' we must include those organisms which have brought to an end the whole of their vital functions, but of which the component cells and elements may still be living. This view is afterwards more accurately explained, and in fact there is no doubt that the cessation of the activity of life in the multicellular organism rarely implies any direct connection with the cessation of vital functions in all its constituents. The question however arises, whether it is right or useful to limit the conception of death to the cessation of the functions of the organism. Our conceptions of death have been derived from the higher organisms alone, and hence it is quite possible that the conception may be too limited. The limitation might perhaps be removed by accurate and scientific comparison with the somewhat corresponding phenomena among unicellular organisms, and we might then arrive at a more comprehensive definition. Science has without doubt the right to make use of popular terms and conceptions, and by a more profound insight to widen or restrict them. But the main idea must always be retained, so that nothing quite new or strange may appear in the widened conception. The conception of death, as it has been expressed with perfect uniformity in all languages, has arisen from observations on the higher animals alone; and it signifies not only the cessation of the vital functions of the whole organism, but at the same time the cessation of life in its single parts, as is shown by the impossibility of revival. The *post-mortem* death of the cells is also part of death, and was so, long before science established the fact that an organism is built up of numerous very minute living elements, of which the vital processes partially continue for some time after the cessation of those of the whole organism. It is precisely this incapacity on the part of the organism to reproduce the phenomena of life anew, which distinguishes genuine death from the arrest of life or trance; and the incapacity depends upon the fact that the death of the cells and tissues follows upon the cessation of the vital functions as a whole. I would, for this reason, define death as an arrest of life, from which no lengthened revival, either of the whole or any of its parts, can take place; or, to put it concisely, as a definite arrest of life. I believe that in this definition I have expressed the exact meaning of the conception which language has sought to convey in the word death. For our present purpose, the cause which gives rise to this phenomenon is of no importance,—whether it is simultaneous or successive in the various parts of the organism, whether it makes its appearance slowly or rapidly. For the conception itself it is also quite immaterial whether we are able to decide if death has really taken place in any particular case; however uncertain we might be, the state which we call death would be not less sharply and definitely limited. We might consider the caterpillar of *Euprepia flavia* to be dead when frozen in ice, but if it recovered after thawing and became an imago, we should say that it had only been apparently dead, that life stood still for a time, but had not ceased for ever. It is only the irretrievable loss of life in an organism which we call death, and we ought to hold fast to this conception, so that it will not slip from us, and become worthless, because we no longer know what we mean by it.

We cannot escape this danger if we look upon the *post-mortem* death of the cells of the body as a phenomenon which may accompany death, but which may sometimes be wanting. An experiment might be made in which some part of a dead animal, such as the comb of a cock, might be transplanted, before the death of the cells, to some other living animal: such a part might live in its new position, thus showing that single members may survive after the appearance of death, as I understand it. But the objection might be raised that in such a case the cock's comb has become a member of another organism, so that it would be lost labour to insert a clause in our definition of death which would include this phenomenon. The same objection might be raised if the transplantation took place a day or even a year before the death of the cock.

Götte is decidedly in error when he considers that the idea of death merely expresses an 'arrest of the sum of vital actions in the individual,' without at the same time including that definite arrest which involves the impossibility of any revival. Decomposition is not quite essential to our definition, inasmuch as death may be followed by drying-up[60], or by perpetual entombment in Siberian ice (as in the well-known case of the mammoth), or by digestion in the stomach of a beast of prey. But the notion of a dead body is indeed inseparably connected with that of death, and I believe that I was right in distinguishing between the division of an Infusorian into two daughter-cells, and the death of a Metazoon, which leaves offspring behind it, by calling attention to the absence of a dead body in the process of fission among Infusoria (See below.). The real proof of death is that the organized substance which previously gave rise to the phenomena of life, for ever ceases to originate such phenomena. This, and this alone, is what mankind has hitherto understood by death, and we must start from this definition if we wish to retain a firm basis for our considerations.

We must now consider whether this definition, derived from observation of higher animals, may be also applied without alteration to the lower, or whether the corresponding phenomena which arise in these latter, differ in detail from those of the higher animals, so that a narrower limitation of the above definition is rendered necessary.

Götte believes the process of encystment which takes place in so many unicellular animals (Monoplastides) to be the analogue of death. According to this authority, the individuals in question, not only undergo a kind of winter sleep—a period of latent life—but when surrounded by the cyst they lose their former specific organization; they become a 'homogeneous substance,' and are resolved into a germ, from which, by a process of development, a new individual of the same species once more arises. The division of the contents of the cyst, viz. its multiplication, is, according to this view, of secondary importance, and the essential feature in the process is the rejuvenescence of the individual. This rejuvenescence however is said to not only consist in the simple transformation of the old individual, but in its death, followed by the building up anew of another individual. 'The parent organism and its offspring are two successive living stages of the same substance—separated, and at the same time connected, by the condition of rejuvenescence which lies between them' (l. c., p. 79). An 'absolute continuity of life does not exist'; it is only the dead organic matter which establishes the connection, and the 'identity of this matter ensures heredity.'

It is certainly surprising that Götte should identify encystment with a cessation of life, and we may well inquire for the evidence which is believed to support such a view. The only evidence lies in a certain degree of degeneration in the structure of the individual, and in the cessation of the visible external phenomena of life, such as feeding and moving. Does Götte really believe that it is an incorrect interpretation of the facts to assume that a *vita minima* continues to exist in the protoplasm, after its complexity has diminished? Are we compelled to invoke a mystical explanation of the facts, by an appeal to such an indefinite principle as Götte's rejuvenescence? Would not the oxygen, dissolved in the water, affect the organic substance the life of which it formerly maintained, and would it not cause its decomposition, if it were in reality dead?

I, too, hold that the division of the encysted mass is of secondary importance, and that the encystment itself, without the resulting multiplication, is the original and essential part of the phenomenon. But it does not follow from this that the encystment should be considered as a process of rejuvenescence. What is there to be rejuvenated? Certainly not the substance of the animal, for nothing is added to it, and it can therefore acquire no new energy; and the forms of energy which it manifests cannot be changed, since the form of the matter is just the same after quitting the cyst as it was before. Rejuvenescence has also been mentioned in connection with the process of conjugation, but this is quite another thing. It is quite reasonable, at least in a certain sense, to maintain the connection of rejuvenescence with conjugation; for a fusion of the substance of two individuals takes place, to a greater or lesser extent, in conjugation, and the matter which composes each individual is therefore really altered. But in simple encystment, rejuvenescence can only be understood in the sense in which we speak of the fable of the Phœnix, which, when old, was believed to be consumed by fire, and to rise again from its own ashes as a young bird. I doubt whether this idea is in agreement with the physiology of to-day, or with the laws of the conservation of energy. It is easy to pull down an old house with rotten beams and crumbling walls, but it would be impossible to build it anew with the old material, even if we used new mortar, represented in Götte's hypothesis by water and oxygen. For these reasons I consider the idea of rejuvenescence of the encysted individual to be contrary to our present physiological knowledge.

It is much more simple and natural to regard encystment as adapted for the protection of certain individuals in a colony from destruction by being dried up or frozen, or for the protection of the individual during multiplication by division, when it is helpless, and would easily fall a prey to enemies, or to secure advantages in some other way[61]. The case of *Actinosphaerium*, mentioned by Götte, clearly demonstrates that rejuvenescence of the individual is not the only event which happens during encystment, for this would scarcely require six months. The long duration of latent life, from summer to the next spring, clearly proves that encystment is of the highest importance for the species, in order to maintain the life of the individual through the dangers of an unfavourable season[62].

When in this case, the specific organization degenerates to a certain extent, such changes depend in part upon the endeavour to diminish as far as possible the size of the organism—the pseudopodia being drawn in, while the vacuoles contract and completely disappear. The degeneration may also, perhaps, depend in part upon the secretion of the cyst itself, which implies a certain loss of substance[63]. But degeneration chiefly depends upon the fact that the encystment is accompanied by reproduction in the way of fission, which seems to begin with a simplification of the organization, that is, with a fusion of the numerous nuclei. It is well known that many unicellular animals contain several nuclei—in other words, that the nuclear substance is scattered in small parts throughout the whole cell. But when the animal prepares for division, these pieces of nuclear substance fuse into a single nucleus which itself undergoes division into two equal parts[64] during the division of the animal. It is evident that the equal division of the whole nuclear substance only becomes possible in this way.

There are, however, numerous cases which prove that the bodies of encysted animals may retain, during the whole process, exactly the same structure and differentiation, which were previously characteristic of them. Thus the large Infusorian *Tillina magna*, described by Gruber, can be seen through the thin-walled cyst to retain the characteristic structure of its ectoplasm, and the whole of its organization. Even the movements of the enclosed animal do not cease; it continues to rotate actively in the narrow cyst, as do the two or four parts into which it subsequently divides. Such observations prove that Götte's view that 'every characteristic of the previous organization is lost,' is quite out of the question[65] (l. c., p. 62).

For this reason I must strongly oppose Götte's view that an encysted individual is a germ, viz. an organic mass still unorganized which can only become an adult individual by means of a process of development. I believe that an encysted individual is one possessing a protective membrane, in structure more or less simplified as an adaptation to the narrow space within the cyst, and to a possible subsequent increase by division, in short one in which active life is reduced to a minimum, and sometimes even completely in abeyance, as happens when it is frozen.

It is evident from the above considerations that encystment in no way corresponds with that which every one, including myself, understands by death, because during encystment one and the same being is first apparently dead and then again alive; and we merely witness a condition of rest, from which active life will again emerge. This would remain true even if it were proved that life is, in reality, suspended for a time. But such proof is still wanting, and Götte was apparently only influenced by theoretical considerations, when he imagined that death intervened where unprejudiced observers have only recognised a condition of rest. He apparently entirely overlooked the fact that it is possible to test his views; for all unicellular beings are in reality capable of dying: we can kill them, for example, by boiling, and they are then really dead and cannot be revived. But this state of the organism differs chemically and physically from the encysted condition, although we do not know all the details of the difference. The encysted animal, when placed in fresh water, presently originates a living individual, but the one killed by boiling only results in decomposition of the dead organic matter. Hence we see that the same external conditions give rise to different results in two different states of the organism. It cannot be right to apply the same term to two totally different states. There is only one phenomenon which can be called death, although it may be produced by widely different causes. But if the encysted condition is not identical with the death which we can produce at will, then natural death, viz. that arising from internal causes, does not exist at all among unicellular organisms.

These facts refute Götte's peculiar view, which depends on the existence of natural death among the Monoplastid organisms; upon proof of the contradictory, his whole theory collapses. But there is nevertheless a certain interest in following it further, for we shall thus reach many ideas worthy of consideration.

First, the question arises as to how death could have been transmitted from the Monoplastides[66] to the Polyplastides, a process which must have taken place according to Götte. I will for the present omit the fact that I cannot accept the supposition that the process of encystment represents death. We may then inquire whether death has taken the place of encystment among the Polyplastides, or, if this is not the case, whether any process comparable to encystment exists among the Polyplastides.

Götte believes that death is always connected with reproduction, and is a consequence of the latter in both Protozoa and Metazoa. Reproduction has, in his opinion, a directly 'fatal effect,' and the reproducing individual must die. Thus the may-fly and the butterfly die directly after laying their eggs, and the male bee dies immediately after pairing; the Orthonectides expire after expelling their germ-cells, while *Magosphaera* resolves itself into germ-cells, and nothing persists except these elements. It is but a step from this latter organism to the unicellular animals which transform themselves as a whole into germ-cells; but in order to achieve this they must undergo the process of rejuvenescence, which Götte assumes to be the same as death.

These views contain many fallacies quite apart from the soundness or unsoundness of their foundation. The process of encystment, as Götte thinks, represents, in the Monoplastides, true reproduction to which multiplication by means of division has been secondarily added. This encystment cannot be dispensed with, for internal causes determine that it must occasionally interrupt the process of multiplication by simple division. But, on the other hand, Götte also considers the division of the contents of the cyst to be a secondary process. The essential characteristic of encystment is a simple process of rejuvenescence without multiplication. Hence we are forced to accept a primitive condition in which simple division as well as the division of the encysted individual were absent, and in which reproduction consisted only in an often-repeated process of rejuvenescence among existing individuals, without any increase in their number. Such a condition is inconceivable because it would involve a rapid disappearance of the species, and the whole consideration clearly shows us that division of un-encysted individuals must have existed from the first, and that this, and not a vague and mysterious rejuvenescence, has always been the real and primitive reproduction of the Monoplastides. The fact that encystment does not always lead to the division of the contents of the cyst proves, in my opinion, that not reproduction but preservation against injury from without, was the primitive meaning of encystment. It is possible that at the present time there are but few Monoplastides which are able to go through an infinite number of divisions without the interposition of the resting condition implied by encystment; although it has not yet been demonstrated for all species[67]. But it is not right to conclude from this that there is an internal necessity which leads to encystment, that is to say to identify this process with rejuvenescence. It is much more probable that encystment is merely an adaptation to continual changes in the external conditions of life, such as drought and frost, and perhaps also the want of food which arises from the over-population of small areas. The same phenomenon is known in certain low Crustacea—the *Daphnidae*—which possess an ephippium or protective case for their winter-eggs. This case is only developed after a certain definite number of generations has been run through, an event which may happen at any time in the year in species living in pools which are liable to be often dried-up; but only in the autumn in such as live in lakes which are never dry. No one ever doubted that the periodical formation of the ephippium in certain generations was an adaptation to changes in the external conditions of life.

Even if the process of rejuvenescence in the Monoplastides were really equivalent to the death of the higher animals, we could not conclude from this that it is necessarily associated with reproduction. Encystment alone is not reproduction, and it first becomes a form of reproduction when it is associated with the division of the encysted animal. Simple division was the true and original form of reproduction in Monoplastides, and even now it is the principal and fundamental form.

Hence we see that among the Monoplastides reproduction is not connected with death, even if we accept Götte's view and allow that encystment represents death. I shall return later on to the relation between death and reproduction in the Metazoa; but the question first arises whether encystment, if it is not death, has any analogue in the higher animals, and further whether death takes that place in their development which is occupied by encystment in the Monoplastides.

Among the higher Metazoa there can be no doubt as to what we mean by death, but the precise nature of that which dies is not equally evident, and the popular conception is not sufficient for us. It is necessary to distinguish between the mortal and the immortal part of the individual—the body in its narrower sense (*soma*) and the germ-cells. Death only affects the former; the germ-cells are potentially immortal, in so far as they are able, under favourable circumstances, to develope into a new individual, or, in other words, to surround themselves with a new body (*soma*)[68].

But how is it with the lowest Polyplastides in which there is no antithesis between the somatic and germ-cells, and among which each of the component cells of the multicellular body has retained all the animal functions of the Monoplastides, even including reproduction?

Götte believes that the natural death of these organisms (which he rightly calls Homoplastides) consists in 'the dissolution of the cell-colony.' As an example of such dissolution Götte takes Häckel's *Magosphaera planula*, a marine free-swimming organism in the form of a sphere composed of a single layer of ciliated cells, imbedded in a jelly. (For figure see below.) This organism cannot however be 'considered as a genuine perfect Polyplastid, for at a certain time the component cells part from one another and then continue to live independently in the condition of Monoplastides.' These free amoebiform organisms increase considerably in size, encyst, and finally undergo numerous divisions—a kind of segmentation within the cyst. The result of the division is a sphere of ciliated cells similar to that with which the cycle began. In fact, *Magosphaera* is not a perfect Polyplastid, but a transitional form between Polyplastides and Monoplastides, as the discoverer of the group of animals of which it is the only representative, indicated, when he named the group 'Catallacta.'

Development of Magosphaera Planula (after Häckel).

1. Encysted amoeboid form. 2 and 3. Two stages in the division of the same. 4. Free ciliated sphere, the cells of which are connected by a gelatinous mass. 5. One of the ciliated cells which has become free by the breaking up of the sphere. 6. The same in the amoeboid form. 7. The same grown to a larger size.

According to Götte, the natural death of *Magosphaera* consists, as in the undoubted Protozoa, in a process of rejuvenescence by encystment. The dissolution of the ciliated sphere into single cells 'cannot be identical with natural death. For the regular and complete separation of the *Magosphaera*-cells proves that their individuality has not been completely subordinated to that of the whole colony, and it proves that the latter is not completely individualised[69].'

Nothing can be said against this, if we agree in identifying death with the encystment of the Monoplastides. Now we could, as Götte rightly remarks, derive the lower forms of Polyplastides from *Magosphaera* if 'the connection between the cells of the ciliated sphere were retained until encystment, viz. until the reproduction of the single cells had taken place[70].' After this had been accomplished, Götte considers that death would consist 'in the complete separation of the cells from one another, accompanied in all probability by their simultaneous change into germ-cells.' The fallacy in this is evident; if death is represented in one case by the encystment during which single cells change into germ-cells, then this must apply to the other case also, for nothing has changed except the duration of the cell-colony. The nature of encystment cannot be affected by the fact that the cells separate from one another a little earlier or a little later. If it is true that death is represented by encystment among the Monoplastides, then the same conclusion must also hold for the Polyplastides; or rather death must be represented in them by the process of rejuvenescence, which Götte considers to be the essential part of encystment. Götte ought not to identify death with the dissolution of the cell-colony of which the lowest and highest Polyplastides are alike composed; but he should seek it in the process of rejuvenescence which takes place within the germ-cells. If it is essential to the nature of reproduction that the cells set apart for that purpose should pass through a process of rejuvenescence, which is equivalent to death, then this must be true for the reproductive cells of all organisms. If these conclusions hold good, there is nothing to prevent us from assuming that such a process of rejuvenescence actually occurs in the higher animals. Götte evidently holds this view, as is plainly shown in the last pages of his essay. He there attempts to bring his views of the death and rejuvenescence of the germ into harmony with his previously developed idea of the derivation of death among the Polyplastides from the dissolution of the cell-colonies. Götte still clings to the view which he propounded in describing the development of *Bombinator*, according to which the egg-cell of the higher Metazoa must pass through a process of rejuvenescence representing death, before it can become a germ.

According to Götte's[71] idea 'the egg of a *Bombinator igneus* before fertilization cannot be considered to be a cell either wholly or in part; and this is equally true of it at its origin and after its complete development; it is only an essentially homogeneous organic mass enclosed by a membrane which has been deposited externally.' This mass is 'unorganised and not living[72],' and 'during the first phenomena of its development all vital powers must be excluded.' In this way the continuity of life between two successive individuals is always interrupted; or, as Götte says in his last essay:—'The continuity of life between individuals of which one is derived from the other by means of reproduction, exists neither in the rejuvenescence of the Monoplastides nor in the condition of the germ among the Polyplastides—a condition which is derived from the former[73].'

This is quite logical, although in my opinion it is both unproved and incorrect. But, on the other hand, it is certainly illogical for Götte to derive the death of the Metazoa in a totally different way, i. e. from the dissolution of their cell-colonies. It is quite plain that the death of the Metazoa does not especially concern the reproductive cells, but the individual which bears them; Götte must therefore seek for some other origin of death—an origin which will enable it to reach the body (*soma*)—as opposed to the germ-cells. If there still remained any doubt about the failure to establish a correspondence between death and the encystment of the Monoplastides, we have here, at any rate, a final demonstration of the failure!

But there is yet another great fallacy concealed in this derivation of the death of the Polyplastides.

Among the lowest Polyplastides, where all the cells still remain similar, and where each cell is also a reproductive cell, the dissolution of the cell-colony is, according to Götte, to be regarded as death, inasmuch as 'the integrity of the mother-individual absolutely comes to an end' (l. c., p. 78). The dissolution of a cell-colony into its component living elements can only be called death in the most figurative sense, and can have nothing to do with the real death of the individuals; it only consists in a change from a higher to a lower stage of individuality. Could we not kill a *Magosphaera* by boiling or by some other artificial means, and would not the state which followed be death? Even if we define death as an arrest of life, the dissolution of *Magosphaera* into many single cells which still live, is not death, for life does not cease in the organic matter of which the sphere was composed, but expresses itself in another form. It is mere sophistry to say that life ceases because the cells are no longer combined into a colony. Life does not in truth cease for a moment. Nothing concrete dies in the dissolution of *Magosphaera*; there is no death of a cell-colony, but only of a conception. The Homoplastides, that is cell-colonies built up of equal cells, have not yet gained any natural death, because each of their cells is, at the same time, a somatic as well as a reproductive cell: and they cannot be subject to natural death, or the species would become extinct.

It is more to the purpose that Götte has sought for an illustration of death among those remarkable parasites, the Orthonectides, because in them we do at any rate meet with real death. They are indeed very low organisms; but nevertheless they stand far above *Magosphaera*, even if the latter were hypothetically perfected up to the level of a true Homoplastid, for the cells which compose the body of the Orthonectides are not all similar, but are so far differentiated that they are even arranged in the primitive germ-layers, and a form results which has rightly been compared with that of the Gastrula. It is true they are not quite so simple as Götte[74] figures them, for they not only consist of ectoderm and germ-cells, but, according to Julin[75], the endoderm is arranged in two layers—the germ-cells and a layer which forms during development a strong muscular coat; and in the second female form the egg-cells are surrounded by a tolerably thick granular tissue.

Orthonectides (after Julin).

8. First female form: the cap-like anterior part has become detached and the egg-cells (*eiz*) are escaping.
9. Second female form: *eiz* = egg-cells; outside these are the muscular layer (*m*) and the ectoderm (*ekt*).
10 and 11. Two fragments of such a female broken to pieces by spontaneous division: the egg-cells are embedded in a granular mass, and undergo embryonic development in it at a later period; the whole is surrounded by ciliated cells.
12. Male discharging the spermatozoa by the breaking up of the ectoderm (*ekt*); *sp* spermatozoa; *m* muscle.

There is nevertheless no doubt that in the first female form, when sexually mature, the greater part, not only of the endoderm but of the whole body, is made up of ova, so that the animal resembles a thin-walled sac full of eggs. The ova escape by the bursting of the thin ectoderm, and when they have all escaped, the thin disintegrated membrane, composed of ciliated cells, is no longer in a condition to live, and dies at once. This is the course of events as described by Götte, and he is probably correct in his interpretation. This is the real death of the Orthonectides, and if we regard them as low primitive forms (Mesozoa), here for the first time in the ascending series we meet with natural death. But the causes of this are scarcely so clear as Götte seems to think when he ascribes it to the effect of reproduction—an effect which is 'not only empirically necessary, but absolutely unavoidable.' Such a necessity is explained by the fact that the endoderm consists entirely of germ-cells. Now the life of the organism, being dependent upon the mutual action of both layers, must cease as soon as the whole endoderm is extruded during reproduction.

Arguments such as these pass over the presence of a mesoderm; but apart from this omission, it does not appear to me so self-evident from a purely physiological standpoint, that the ectodermal sheath with its muscle layer must die after the extrusion of the germ-cells.

In those females to which Götte refers in this passage, the whole sheath remains at first quite uninjured, with the exception of a small cap at the anterior end, which is pushed off to give exit to the ova; and inasmuch as the sheath continues to swim about in the nutritive fluids after this has taken place, the proof is at any rate wanting that it cannot support itself quite as well as before, although it has lost the germ-cells.

Then why does it die? My answer to this is simple:—because it has lived its time; because its length of life is limited to a period which corresponds with the time necessary for complete reproduction. The physical constitution of the body is so regulated that it remains capable of living until the extrusion of the reproductive cells, and then dies, however favourable external conditions may be for its further support.

The correctness of this explanation is shown by a consideration of the males and the second form of females; for in these cases the body falls to pieces, not as a consequence of reproduction, but as a preparation for it!

Götte only mentions the second female form in a note, in which he says, it appears 'that in the second female form of these animals the whole body breaks into many pieces, and the superficial layer gradually atrophies, so that it dies before the eggs are extruded.' In Julin's account[76], upon which Götte bases his statements, there are, however, some not unimportant differences. For instance, the eggs are not extruded at all, but embryonic development takes place within the body of the mother, which has previously undergone spontaneous division into several pieces. In this case, the eggs differ from those of the other female form, inasmuch as they do not constitute the whole of the endoderm, but are embedded (as was stated above) in a fairly voluminous granular mass at the expense of which, or at least by means of which, they are nourished; for they increase considerably in size during their development. But not only this granular mass, but all the layers of the body of the mother, even the ectoderm, persist during the embryonic development of the offspring. Indeed, the ectoderm must continue to grow during the division of the mother animal, for it gradually covers the products of division on all sides, and, by means of its cilia, causes the animal to swim about in the fluids of its host. After some time the cilia are lost, and the separate parts into which the mother animal has divided, fix themselves upon some part of the body-cavity of the host; the young become free, and the remains of the body of the mother probably disappear by dissolution and resorption[77]. In this case the remains of the mother animal seem to be, to some extent, consumed by the embryos,—a process which sometimes, although very rarely, happens elsewhere. We can scarcely consider this as a primitive arrangement, or look upon it as a proof that 'reproduction' has a necessarily fatal effect upon the Polyplastid organism.

In the male, the mass of spermatozoa does not swell out the body to such an extent that its walls must give way and thus permit an exit, but the large ectoderm cells atrophy spontaneously at the time of maturity, and as they fall off, exit is given to the spermatozoa here and there. In this instance also the dissolution of the body is not a consequence of reproduction, but reproduction can only take place when the dissolution of the body has preceded it!

In this remarkable arrangement we cannot discern anything except an evident adaptation of the life of the body-cells to reproductive purposes, and this adaptation was rendered possible because, after the evacuation of the sexual cells, the body ceased to be of any value for the maintenance of the species.

But even if we assume, that the death of the Orthonectides is, in Götte's sense, a consequence of reproduction, inasmuch as, in the two forms of females as well as in the male, the extrusion of a mass of developed germ-cells or embryos deprives the organism of the physiological possibility of living longer, how can we explain the necessity of death as a direct consequence of reproduction in all Polyplastides? Is the body—the *soma*—of the Metazoa so imperfectly developed, as compared with the reproductive cells, that the extrusion of the latter involves the death of the former? As a matter of fact in the majority of cases the relations are reversed; the number of body-cells usually exceeds the germ-cells a hundred- or a thousand-fold, and the body is, as regards nutrition, so completely independent of the reproductive cells, that it need not be in the least disadvantageously affected by their extrusion. And if the Orthonectid-like ancestors of the Metazoa were compelled to give up their insignificant somatic part after the extrusion of their germ-cells, because it could now no longer support itself, does it therefore follow that the somatic cells had for ever lost the power of surviving, even when their phyletic descendants were surrounded by more favourable conditions? Had they to inherit 'the necessity of death' for all time? Whence came this great change in the nature of organisms which, before the differentiation of Homoplastids into Heteroplastids, were endowed with the immortality of unicellular beings?

And it must be remembered that it is only an assumption which places the Orthonectides among the lowest Metazoa (Heteroplastids). I do not intend to greatly emphasize this point, but the formation of the Gastrula by embole, and the absence of a mouth and alimentary canal, shows that these parasites are extremely degenerate, and the same may be said of almost all endoparasites. The Gastrula, as an independent organism, was without doubt primitively provided with both mouth and stomach, and the mass of ova filling the female Orthonectid is an adaptation to a parasitic life, which on the one side renders the possession of a stomach a superfluity, and on the other demands the production of a great number of germ-cells[78]. It is certain that the Orthonectides, as at present constituted, cannot have lived in the free condition, and also that their adaptation to parasitism cannot have arisen at the beginning of the phyletic development of Metazoa, because they inhabit star-fishes and Nemertines—both relatively highly developed Metazoa. Hence it is, at any rate, doubtful whether the Orthonectides can claim to pass as typical forms of the lowest Heteroplastids, and whether their reproduction can be looked upon 'as typical for the unknown ancestors of all Polyplastids' (l. c., p. 45). If, however, we accept some organism resembling these Orthonectides as the most ancient Heteroplastid, being a free-living organism, it must have had a stomach, and the cells surrounding it must—as a whole or in part—have possessed the power of digesting; at any rate, they cannot all have been germ-cells, and therefore it is improbable that death would be the direct result of the extrusion of the germ-cells.

Let us now consider the manner in which Götte has endeavoured to explain the transmission of the cause of death—which first appeared in the Orthonectides—from these organisms to all later Metazoa, until the very highest forms are reached. Exact proofs of this supposition are unfortunately wanting, and the evidence is confined to the collection of a number of cases in which death and reproduction take place nearly or quite simultaneously. These would prove nothing, even if *post hoc* were always *propter hoc*; and there are, opposed to them, a number of cases in which reproduction and death take place at different times. In obtaining evidence for 'the fatal influence of reproduction,' is it possible to point to every case of sudden death after the act of oviposition or fertilization? These cases occur among many of the higher animals, especially in Insects, and were collected by me in an earlier work[79]. It is obvious that such cases are exceptional, but in a restricted sense it is quite true, as far as these individual instances are concerned, that death appears as a consequence of reproduction. The male bee, which invariably dies while pairing, is undoubtedly killed in consequence of a very powerful nervous shock; and the female Psychid, which has laid all her eggs at once, dies of 'exhaustion'—however we may attempt to explain the term on physiological principles.

Can we conclude from these cases that the effects of reproduction are, in Götte's sense, universally fatal; that reproduction is the positive and 'exclusive explanation of natural death'? (l. c., p. 32.) I need not linger over these isolated examples, but I turn at once to the foundation of the whole conclusion—a foundation which is obviously unable to support the superstructure erected on it. Götte formally derives the idea that death is a necessary condition of reproduction, from a very heterogeneous collection of facts. When we examine this collection we find that the process which is taken to be death is not the same thing in all these instances, while the same is true of the influence of reproduction by which death is supposed to be caused. The whole conception arises out of the process of encystment, which is regarded as the building-up of reproductive material—that is, as true reproduction; and since, according to Götte's view, the formation of germs is always intimately connected with an arrest of life, and since, by his own definition, this stand-still of life is equivalent to death, it follows that, with such a theory, reproduction, in its essential nature, must be inseparably connected with death. It is necessary at this juncture to remember what Götte means by the process of rejuvenescence, and to point out that he is dealing with something quite different from 'the fatal influence of reproduction,' which was just now mentioned with regard to insects. 'Rejuvenescence,' bound up as it is with encystment and reproduction, is, according to Götte, 'a re-coining of the specific protoplasm, by means of which the identity of its substance is fixed by heredity,' a 'marvellous process in which phenomena the most important in the whole life of the animal, and in fact of all organisms—reproduction and death—have their roots' (l. c., p. 81). Whether such re-coining really takes place or not, at any rate I claim to have shown above that it does not correspond with death in the Metazoa, and—if it is represented at all in these latter—that it ought to be looked for in the reproductive cells; and indeed, in another passage, Götte himself has placed the process in these cells.

While, among the Monoplastids, according to Götte, the causes of the supposed death lie hidden in this mysterious change of the organism into reproductive material, Götte asserts that among the Polyplastids (such as *Magosphaera*, hypothetically perfected so as to form a genuine Polyplastid), the causes of death operate so that the organism breaks up into its component cells, all these being still reproductive cells—a process which, unlike 'rejuvenescence,' has nothing mysterious about it, and which is certainly not genuine death. In the Orthonectid-like animals death does not occur as a consequence of the dispersal of the reproductive cells, but rather because the part of the animal which remains is so small and effete that, being unable to support itself, it necessarily dies. From this point at least the object of death and the conception of it remain the same, but now the idea of reproduction undergoes a change. When the Rhabdite females of *Ascaris* are eaten up by their offspring, is this mode of death connected with the 'rejuvenescence of protoplasm'? (l. c., p. 34.) Is there any deep underlying relationship between such an end and the essential nature of reproduction? The same question may be asked with regard to the 'Redia or the Sporocyst of Trematodes, which are converted into slowly dying sacs during the growth of the Cercariae within them.' We cannot speak of the 'fatal influence of reproduction' among tape-worms just because 'in the ripe segments the whole organization degenerates under the influence of the excessive growth of the uterus.' It certainly degenerates, but only so far as the developing mass of eggs demands. In fact, at a sufficiently high temperature, death does not occur, and such mature segments of tape-worms creep about of their own accord. We cannot fail to recognize that in this as well as in the above-mentioned cases we have to do with adaptation to certain very special conditions of existence—an adaptation leading to an immense development of reproductive cells in a mother organism which can no longer take in nourishment, or which has become entirely superfluous because its duty to its species is already fulfilled. If this is an example of death inherent in the essential nature of reproduction, then so is the death of a mature segment of a tapeworm in the gastric juices of the pig that eats it.

With Götte, the conception of reproduction, like the conception of death, is a protean form, which he welcomes in any shape, if only he can use it as evidence. If death is a necessary consequence of reproduction, its cause must be always essentially the same, and might be expressed in one of the following suggestions:—(1) in the necessity for a 're-coining' of the protoplasm of the germ-cells; but here death could only affect the germ-cells themselves: (2) perhaps in the withdrawal of nourishment by the mass of developing reproductive material, just as death occurs sometimes among men by the excessive drain on the system caused by morbid tumours: (3) or in consequence of the development of the offspring in the body of the mother; this however would only affect the females, and could therefore have no deep and general significance: (4) from the extrusion of the sexual cells,—ova or spermatozoa,—and in the impossibility of further nourishment which is consequent upon this extrusion—(Orthonectides?): or (5) finally in an excessively powerful nervous shock brought about by the ejection of the reproductive cells.

But no one of these alternatives is the universal and inevitable cause of death. This proves irrefutably that death does not proceed as an intrinsic necessity from reproduction, although it may be connected with the latter, sometimes in one way and sometimes in another. But we must not overlook the fact that in many cases death is not connected with reproduction at all; for many Metazoa survive for a longer or shorter period after the reproductive processes have ceased.

In fact, I believe I have definitely shown that no process exists among unicellular animals which is at all comparable with the natural death of the higher organisms. Natural death first appeared among multicellular beings, and among these first in the Heteroplastids. Furthermore, it was not introduced from any absolute intrinsic necessity inherent in the nature of living matter, but on grounds of utility, that is from necessities which sprang up, not from the general conditions of life, but from those special conditions which dominate the life of multicellular organisms. If this were not so, unicellular beings must also have been endowed with natural death. I have already expressed these ideas elsewhere[80], and have briefly indicated how far, in my opinion, natural death is expedient for multicellular organisms. I found the essential reason for confining the life of the Metazoa to a fixed and limited period, in the wear and tear to which an individual is exposed in the course of a life-time. For this reason, 'the longer the individual lived, the more defective and crippled it would become, and the less perfectly would it fulfil the purpose of its species' (l. c., p. 24). Death seemed to me to be expedient since 'worn-out individuals are not only valueless to the species, but they are even harmful, for they take the place of those which are sound' (l. c., p. 24).

I still adhere entirely to this explanation; not of course in the sense that an actual physical struggle has ever taken place between the mortal and immortal varieties of any species. If Götte understood me thus, he may be justified by the brief explanations given in the essay to which I have alluded; but when he also attributes to me the opinion that such hypothetically immortal Metazoa had but a very limited period for reproduction, I fail to see what part of the essay in question can be brought forward in support of his statement. Only under some such supposition can I be reproached with having assumed the existence of a process of natural selection which could never be effective, because any advantage which accrued to the species from the shortening of the duration of life could not make itself felt in a more rapid propagation of the short-lived individuals. The statement 'that in this and in every other case it is a sufficient explanation of the processes of natural selection to render it probable that any kind of advantage is gained'[81] is indeed erroneous. The explanation ought rather to be 'that the forms in question would for ever transmit their characters to a greater number of descendants than the other forms.' I have not however as yet attempted to think out in detail such processes of natural selection as would limit the somatic part of the Metazoan body to a short term of existence, and I only wished to emphasize the general principle lying at the basis of the whole process, without stating the precise manner in which it operates.

If I now attempt to take this course, and to reconstruct theoretically the gradual appearance of natural death in the Metazoa, I must begin by again alluding to Götte's criticisms in reference to the operation of natural selection.

I consider death as an adaptation, and believe that it has arisen by the operation of natural selection. Götte[82], however, concludes from this that 'the first origin of hereditary and consequently (for the organization in question) necessary death, is not explained but already assumed.' 'The operation and significance of the principle of utility consists in selecting the fittest from among the structures and processes which are at hand, and not in directly creating new ones. Every new structure arises at first, quite independently of any utility, from certain material causes present in a number of individuals, and when it has proved useful and is transmitted, it extends, according to the laws of the survival of the fittest, in the group of animals in which it appeared. This extension will undergo further increase with every advance in utility which results from further structural changes, until it extends over the whole group. So that usefulness effects the preservation and the distribution of new structures, but has nothing whatever to do with the causes of their primary origin and their consequent transmission to all other individuals. Indeed, on these hereditary causes the necessity of the structures in question depends, so that their usefulness in no way explains their necessity.'

'These conclusions, when applied to the origin of natural death called forth by internal causes, would show that it became inevitable and hereditary in a number of the originally immortal Metazoa, before there could be any question as to the benefits derived from its influence. Such influence must have consisted in the fact that more descendants survived the struggle for existence and were able to enter upon reproduction among the individuals which had inherited the predisposition to die than among the potentially immortal beings which would be damaged in the struggle for existence, and would therefore be exposed to still further injuries. The existing necessity for natural death in all Metazoa might therefore be derived in an unbroken line of descent from the first mortal Metazoan, of which the death became inevitable from internal causes, before the principle of utility could operate in favour of its dissemination.'

In reply to this I would urge: that it has been very often maintained that natural selection can produce nothing new, but can only bring to the front something which existed previously to the exercise of choice; but this argument is only true in a very limited sense. The complex world of plants and animals which we see around us contains much that we should call new in comparison with the primitive beings from which, as we believe, everything has developed by means of natural selection. No leaves or flowers, no digestive system, no lungs, legs, wings, bones or muscles were present in the primitive forms, and all these must have arisen from them according to the principle of natural selection. These primitive forms were in a certain sense predestined to develope them, but only as possibilities, and not of necessity; nor were they preformed in them. The course of development, as it actually took place, first became a necessity by the action of natural selection, that is by the choice of various possibilities, according to their usefulness in fitting the organism for its external conditions of life. If we once accept the principle of natural selection, then we must admit that it really can create new structures, instincts, etc., not suddenly or discontinuously, but working by the smallest stages upon the variations that appear. These changes or variations must be looked upon as very insignificant, and are, as I have of late attempted to show[83], quantitative in their nature; and it is only by their accumulation that changes arise which are sufficiently striking to attract our attention, so that we call them 'new' organs, instincts, etc.

These processes may be compared to a man on a journey who proceeds from a certain point on foot by short stages, at any given time, and in any direction. He has then the choice of an infinite number of routes over the whole earth. If such a man begins his wanderings in obedience to the impulse of his own will, his own pleasure or interest,—proceeding forwards, to the right or left, or even backwards, with longer or shorter pauses, and starting at any particular time,—it is obvious that the route taken lies in the man himself and is determined by his own peculiar temperament. His judgment, experience, and inclination will influence his course at each turn of his journey, as new circumstances arise. He will turn aside from a mountain which he considers too lofty to be climbed; he will incline to the right, if this direction appears to afford a better passage over a swollen stream; he will rest when he reaches a pleasant halting-place, and will hurry on when he knows that enemies beset him. And in spite of the perfectly free choice open to him, the course he takes is in fact decided by both the place and time of his starting and by circumstances which—always occurring at every part of the journey—impel him one way or the other; and if all the factors could be ascertained in the minutest detail, his course could be predicted from the beginning.

Such a traveller represents a species, and his route corresponds with the changes which are induced in it by natural selection. The changes are determined by the physical nature of the species, and by the conditions of life by which it is surrounded at any given time. A number of different changes may occur at every point, but only that one will actually develope which is the most useful, under existing external conditions. The species will remain unaltered as long as it is in perfect equilibrium with its surroundings, and as soon as this equilibrium is disturbed it will commence to change. It may also happen that, in spite of all the pressure of competing species, no further change occurs because no one of the innumerable very slight changes, which are alone possible at any one time, can help in the struggle; just as the traveller who is followed by an overpowering enemy, is compelled to succumb when he has been driven down to the sea. A boat alone could save him, without it he must perish; and so it sometimes happens that a species can only be saved from destruction by changes of a conspicuous kind, and these it is unable to produce.

And just as the traveller, in the course of his life, can wander an unlimited distance from his starting-point, and may take the most tortuous and winding route, so the structure of the original organism has undergone manifold changes during its terrestrial life. And just as the traveller at first doubts whether he will ever get beyond the immediate neighbourhood of his starting-point, and yet after some years finds himself very far removed from it—so the insignificant changes which distinguish the first set of generations of an organism lead on through innumerable other sets, to forms which seem totally different from the first, but which have descended from them by the most gradual transition. All this is so obvious that there is hardly any need of a metaphor to explain it, and yet it is frequently misunderstood, as shown by the assertion that natural selection can create nothing new: the fact being that it so adds up and combines the insignificant small deviations presented by natural variation, that it is continually producing something new.

If we consider the introduction of natural death in connection with the foregoing statements, we may imagine the process as taking place in such a way that,—with the differentiation of Heteroplastids from Homoplastids, and the appearance of division of labour among the homogeneous cell-colonies,—natural selection not only operated upon the physiological peculiarities of feeding, moving, feeling, or reproduction, but also upon the duration of the life of single cells. At this developmental stage there would, at any rate, be no further necessity for maintaining the power of limitless duration. The somatic cells might therefore assume a constitution which excluded the possibility of unending life, provided only that such a constitution was advantageous for them.

It may be objected that cells of which the ancestors possessed the power of living for ever, could not become potentially mortal (that is subject to death from internal causes) either suddenly or gradually, for such a change would contradict the supposition which attributes immortality to their ancestors and to the products of their division. This argument is valid, but it only applies so long as the descendants retain the original constitution. But as soon as the two products of the fission of a potentially immortal cell acquire different constitutions by unequal fission, another possibility arises. Now it is conceivable that one of the products of fission might preserve the physical constitution necessary for immortality, but not the other; just as it is conceivable that such a cell—adapted for unending life—might bud off a small part, which would live a long time without the full capabilities of life possessed by the parent cell; again, it is possible that such a cell might extrude a certain amount of organic matter (a true excretion) which is already dead at the moment it leaves the body. Thus it is possible that true unequal cell-division, in which only one half possesses the condition necessary for increasing, may take place; and in the same way it is conceivable that the constitution of a cell determines the fixed duration of its life, examples of which are before us in the great number of cells in the higher Metazoa, which are destroyed by their functions. The more specialized a cell becomes, or in other words, the more it is intrusted with only one distinct function, the more likely is this to be the case: who then can tell us, whether the limited duration of life was brought about in consequence of the restricted functions of the cell or whether it was determined by other advantages[84]? In either case we must maintain that the disadvantages arising from a limited duration of the cells are more than compensated for by the advantages which result from their highly effective specialized functions. Although no one of the functions of the body is necessarily attended by the limited duration of the cells which perform it, as is proved by the persistence of unicellular forms, yet any or all of them might lead to such a limitation of existence without in any way injuring the species, as is proved by the Metazoa. But the reproductive cells cannot be limited in this way, and they alone are free from it. They could not lose their immortality, if indeed the Metazoa are derived from the immortal Protozoa, for from the very nature of that immortality it cannot be lost. From this point of view the body, or *soma*, appears in a certain sense as a secondary appendage of the real bearer of life,—the reproductive cells.

Just as it was possible for the specific somatic cells to be differentiated from among the chemico-physical variations which presented themselves in the protoplasm, by means of natural selection, until finally each function of the body was performed by its own special kind of cell; so it might be possible for only those variations to persist the constitution of which involved a cessation of activity after a certain fixed time. If this became true of the whole mass of somatic cells, we should then meet with natural death for the first time. Whether we ought to regard this limitation of the life of the specific somatic cells as a mere consequence of their differentiation, or at the same time as a consequence of the powers of natural selection especially directed to such an end,—appears doubtful. But I am myself rather inclined to take the latter view, for if it was advantageous to the somatic cells to preserve the unending life of their ancestors—the unicellular organisms, this end might have been achieved, just as it was possible at a later period, in the higher Metazoa, to prolong both the duration of life and of reproduction a hundred- or a thousand-fold. At any rate, no reason can be given which would demonstrate the impossibility of such an achievement.

With our inadequate knowledge it is difficult to surmise the immediate causes of such a selective process. Who can point out with any feeling of confidence, the direct advantages in which somatic cells, capable of limited duration, excelled those capable of eternal duration? Perhaps it was in a better performance of their special physiological tasks, perhaps in additional material and energy available for the reproductive cells as a result of this renunciation of the somatic cells; or perhaps such additional power conferred upon the whole organism a greater power of resistance in the struggle for existence, than it would have had, if it had been necessary to regulate all the cells to a corresponding duration.

But we are not at present able to obtain a clear conception of the internal conditions of the organism, especially when we are dealing with the lowest Metazoa, which seem to be very rarely found at the present day, and of which the vital phenomena we only know as they are exhibited by two species, both of doubtful origin. Both species have furthermore lost much of their original nature, both in structure and function, as a result of their parasitic mode of life. Of the Orthonectides and Dicyemidae we know something, but of the reproduction in the single free non-parasitic form, discovered by F. E. Schulze and named by him *Trichoplax adhaerens*, we know nothing whatever, and of its vital phenomena too little to be of any value for the purpose of this essay.

At this point it is advisable to return once more to the derivation of death in the Metazoa from the Orthonectides, as Götte endeavoured to derive it, when he overlooked the fact that, according to his theory, natural death is inherited from the Monoplastids and cannot therefore have arisen anew in the Polyplastids. According to this theory, death must necessarily have appeared in the lowest Metazoa as a result of the extrusion of the germ-cells, and by continual repetition must have become hereditary. We must not however forget that, in this case, the cause of death is exclusively external, by which I mean that the somatic cells which remained after the extrusion of the reproductive cells, were unable to feed any longer or at any rate to an adequate extent; and that the cause of their death did not lie in their constitution, but in the unfavourable conditions which surrounded them. This is not so much a process of natural death as of artificial death, regularly appearing in each individual at a corresponding period, because, at a certain time of life, the organism becomes influenced by the same unfavourable conditions. It is just as if the conditions of life invariably led to death by starvation at a certain stage in the life of a certain species. But we know that death arises from purely internal causes among the higher Metazoa, and that it is anticipated by the whole organisation as the normal end of life. Hence nothing is gained by this explanation founded on the Orthonectides, and we should have to seek further and in a later stage of the development of the Metazoa, for the internal causes of true natural death.

Another theory might be based upon the supposition that natural death has been derived, in the course of time, from an artificial death which always appeared at the same stage of each individual life—as we have supposed to be the case in the Orthonectides. I cannot agree with this view, because it involves the transmission of acquired characters, which is at present unproved and must not be assumed to occur until it has been either directly or indirectly demonstrated[185]. I cannot imagine any way in which the somatic cells could communicate this assumed death by starvation to the reproductive cells in such a manner that the somatic cells of the resulting offspring would spontaneously die of hunger in the same manner and at a corresponding time as those of the parent. It would be as impossible to imagine a theoretical conception of such transmission as of the supposed instance of kittens being born without a tail after the parent's tail had been docked; although to make the cases parallel the kittens' tails ought to be lost at the same period of life as that at which the parent lost hers. And in my opinion we do not add to the intelligibility of such an idea by assuming the artificial removal of tails through hundreds of generations. Such changes, and indeed all changes, are, as I think, only conceivable and indeed possible when they arise from within, that is, when they arise from changes in the reproductive cells. But I find no difficulty in believing that variations in these cells took place during the transition from Homoplastids to Heteroplastids, variations which formed the material upon which the unceasing process of natural selection could operate, and thus led to the differentiation of the previously identical cells of the colony into dissimilar ones—some becoming perishable somatic cells, and others the immortal reproductive cells.

It is at any rate a delusion to believe that we have explained natural death, by deriving it from the starvation of the *soma* of the Orthonectides, by the aid of the unproved assumption of the transmission of acquired variations. We must first explain why these organisms produce only a limited number of reproductive cells which are all extruded at once, so that the *soma* is rendered helpless. Why should not the reproductive cells ripen in succession as they do indirectly among the Monoplastides, that is to say in a succession of generations, and as they do directly in great numbers among the Metazoa? There would then be no necessity for the *soma* to die, for a few reproductive cells would always be present, and render the persistence of the individual possible. In fact, the whole arrangement—the formation of reproductive cells at one time only, and their sudden extrusion,—presupposes the mortality of the somatic cells, and is an adaptation to it, just as this mortality itself must be regarded as an adaptation to the simultaneous ripening and sudden extrusion of the generative cells. In short, there is no alternative to the supposition stated above, viz. that the mortality of the somatic cells arose with the differentiation of the originally homogeneous cells of the Polyplastids into the dissimilar cells of the Heteroplastids. And this is the first beginning of natural death.

Probably at first the somatic cells were not more numerous than the reproductive cells, and while this was the case the phenomenon of death was inconspicuous, for that which died was very small. But as the somatic cells relatively increased, the body became of more importance as compared with the reproductive cells, until death seems to affect the whole individual, as in the higher animals, from which our ideas upon the subject are derived. In reality, however, only one part succumbs to natural death, but it is a part which in size far surpasses that which remains and is immortal,—the reproductive cells.

Götte combats the statement that the idea of death necessarily implies the existence of a corpse. Hence he maintains that the cellular sac which is left after the extrusion of the reproductive cells among the Orthonectides, and which ultimately dies, is not a corpse; 'for it does not represent the whole organism, any more than the isolated ectoderm of any other Heteroplastid' (l. c., p. 48). But it is only a popular notion that a corpse must represent the entire organism. In cases of violent death this idea is correct, because then the reproductive cells are also killed. But as soon as we recognise that the reproductive cells on the one side, and the somatic cells on the other, form respectively the immortal and mortal parts of the Metazoan organism, then we must acknowledge that only the latter,—that is, the *soma* without the reproductive cells,—suffers natural death. The fact that all the reproductive cells have not left the body (as sometimes happens) before natural death takes place, does not affect this conception. Among insects, for instance, it may happen that natural death occurs before all the reproductive cells have matured, and these latter then die with the *soma*. But this does not make any difference to their potential immortality, any more than it modifies the scientific conception of a corpse. The idea of natural death involves that of a corpse, which consists of the *soma*, and when the latter happens to contain reproductive cells, these do not succumb to a natural death, which can never apply to them, but to an accidental death. They are killed by the death of the *soma* just as they might be killed by any other accidental cause of death.

The scientific conception of a corpse is not affected, whether the dead *soma* remains whole for some time, or falls to pieces at once. I cannot therefore agree with Götte when he denies that an Orthonectid possesses 'the possibility of becoming a corpse' (in his sense of the word) because 'its death consists in the dissolution of the structure of the organism.' When the young of the Rhabdites form of *Ascaris nigrovenosa* bore through the body-walls of their parent, cause it to disintegrate and finally devour it, the whole organism disappears, and it would be difficult to say whether a corpse exists in the popular sense of the word. But, scientifically speaking, there is certainly a corpse; the real *soma* of the animal dies, and this, however subdivided, must be considered as a corpse. The fact that natural death is so difficult to define without any accurate conception of what is meant by a corpse, proves the necessity for arriving at a scientific idea as to the meaning of the latter. There is no death without a corpse—whether the latter be small or large, whole or in pieces.

If we compare the bodies of the higher Metazoa with those of the lower, we see at once that not only has the structure of the body increased in size and complexity as far as the *soma* is concerned, but we also see that another factor has been introduced, which exercises a most important influence in lengthening the duration of life. This is the replacement of cells by multiplication. Somatic cells have acquired (at any rate in most tissues) the power of multiplying, after the body is completely developed from the reproductive cells. The cells which have undergone histological differentiation can increase by fission, and thus supply the place of those which are being continually destroyed in the course of metabolism. The difference between the higher and lower Metazoa in this respect lies in the fact that there is only one generation of somatic cells in the latter, and these are used up in the process of metabolism at almost the same time that the reproductive cells are extruded, while among the former there are successive generations of somatic cells. I have elsewhere endeavoured to render the duration of life in the animal kingdom intelligible by the application of this principle, and have attempted to show that its varying duration is determined in different species by the varying number of somatic cell-generations[86]. Of course, the varying duration of each cell-generation materially influences the total length of life, and experience teaches us that the duration of cell-generations varies, not only in the lowest Metazoa as compared with the highest, but even in the various kinds of cells in one and the same species of animal.

We must, for the present, leave unanswered the question—upon what changes in the physical constitution of protoplasm does the variation in the capacity for cell-duration depend; and what are the causes which determine the greater or smaller number of cell-generations. I mention this obvious difficulty because it is the custom to meet every attempt to search deeper into the common phenomena of life with the reproach that so much is still left unexplained. If we must wait for the explanation of these processes until we have ascertained the molecular structure of cells, together with the changes that occur in this structure and the consequences of the changes, we shall probably never understand either the one or the other. The complex processes of life can only be followed by degrees, and we can only hope to solve the great problem by attacking it from all sides.

Therefore it is, in my opinion, an advance if we may assume that length of life is dependent upon the number of generations of somatic cells which can succeed one another in the course of a single life; and, furthermore, that this number, as well as the duration of each single cell-generation, is predestined in the germ itself. This view seems to me to derive support from the obvious fact that the duration of each cell-generation, and also the number of generations, undergo considerable increase as we pass from the lowest to the highest Metazoa.

In an earlier work[87] I have attempted to show how exactly the duration of life is adapted to the conditions by which it is surrounded; how it is lengthened or shortened during the formation of species, according to the conditions of life in each of them; in short, how it is throughout an adaptation to these conditions. A few points however were not touched upon in the work referred to, and these require discussion; their consideration will also throw some light upon the origin of natural death and the forms of life affected by it.

I have above explained the limited duration of the life of somatic cells in the lower Metazoa—Orthonectides—as a phenomenon of adaptation, and have ascribed it to the operation of natural selection, at the same time pointing out that the existence of immortal Metazoan organisms is conceivable. If the Monoplastides are able to multiply by fission, through all time, then their descendants, in which division of labour has induced the antithesis of reproductive and somatic cells, might have done the same. If the Homoplastid cells reproduced their kind uninterruptedly, equal powers of duration must have been possible for the two kinds of Heteroplastid cells; they too might have been immortal so far as immortality only depends upon the capacity for unlimited reproduction.

But the capacity for existence possessed by any species is not only dependent upon the power within it; it is also influenced by the conditions of the external world, and this renders necessary the process which we call adaptation. Thus it is just as inconceivable that either a homogeneous or a heterogeneous cell-colony possessing the physiological value of a multicellular individual should continue to grow to an unlimited extent by continued cell-division, as it is inconceivable that a unicellular being should increase in size to an unlimited extent. In the latter case the process of cell-division imposes a limit upon the size attained by growth. In the former, the requirements of nutrition, respiration, and movement must prescribe a limit to the growth of the cell-colony which constitutes the individual of the higher species, just as in the case of the unicellular Monoplastides, and it does not affect the argument if we consider this limitation to be governed by the process of natural selection. It would only be possible to regulate the relations of the single cells of the colony to each other by fixing the number of cells within narrow limits. During the development of *Magosphaera*—one of the Homoplastides—the cells arrange themselves in the form of a hollow sphere, lying in a gelatinous envelope. But the fact that reproduction does not follow the simple unvarying rhythm of unicellular organisms is of more importance; for a rhythm of a higher order appears, in which each cell of the colony separates from its neighbours, when it has reached a certain size, and proceeds by very rapid successive divisions to give rise to a certain number of parts which arrange themselves as a new colony. The number of divisions is controlled by the number of cells to which the colony is limited, and at first this number may have been very small. With the introduction of this secondary higher rhythm during reproduction, the first germ of the Polyplastides became evident; for then each process of fission was not, as in unicellular organisms, equivalent to all the others; for in a colony of ten cells the first fission differs from the second, third, or tenth, both in the size of the products of division and also in remoteness from the end of the process. This secondary fission is what we know as segmentation.

It seems to me of little importance whether the first process of segmentation takes place in the water or within a cyst, although it is quite possible that the necessity for some protective structure appeared at a very early period, in order to shield the segmenting cell from danger.

It is impossible to accept Götte's conception of the germ (Keim), and at this point the question arises as to its true meaning. I should propose to include under this term every cell, cytode, or group of cells which, while not possessing the structure of the mature individual of the species, possesses the power of developing into it under certain circumstances. The emphasis is now laid upon the expression development, which is something opposed to simple growth, without change of form. A cell which becomes a complete individual by growth alone is not a germ but an individual, although a very small one. For example, the small encapsuled Heliozoon, which arises as the product of multiple fission, is not a germ in our sense of the word. It is an individual, provided with all the characteristic marks of its species, and it has only to protrude the retracted processes (pseudopodia) and to take in the expelled water (formation of vacuoles) in order to become capable of living in a free state. In this sense of the word, germs are confined to the Polyplastides, but are found in many Monoplastides. There is nevertheless, in my opinion, a profound and significant difference between the germs of these two groups. And this lies not so much in the morphological as in the developmental significance of these structures. As far as I have been able to compare the facts, I may state that the germs of the Monoplastides are entirely of secondary origin, and have never formed the phyletic origin of the species in which they are found. For instance, the spore-formation of the Gregarines resulted from a gradually increasing process of division, which was concentrated into the period of encystment; and it was induced by a necessity for rapid multiplication due to the parasitic life and unfavourable surroundings of these animals. If Gregarines were free-living animals, they would not need this method of reproduction. The encysted animal would probably divide into eight, four, or two parts, or perhaps, like many Infusoria[88], it would not divide at all, so that the whole reproduction would depend on simple fission alone during the free state.

The original mode of reproduction among the Monoplastides was undoubtedly simple fission. This became connected with encystment, which originally took place without multiplication; and only when the divisions in the cyst became excessively numerous did such minute plastids appear that a genuine process of development had to be undergone in order to produce complete individuals. Here we have the general conception of the germ as I defined it. Its limitations are naturally not very sharply defined, for it is impossible to draw an absolute distinction between simple growth and true development accompanied by changes in form and structure. For instance, Häckel's *Protomyxa aurantiaca* divides within its cyst into numerous plastids, which might be spoken of as germs. But the changes of form which they undergo before they become young *Protomyxae* are very small, and for the most part depend upon the expansion of the body, which existed in the capsule as a contracted pear-shaped mass. It is therefore more correct to speak only of the simple growth of the products of the fission of the parent organism, and to look upon these products as young *Protomyxae* rather than germs. On the other hand, the young animals which creep out of the germs (the 'spores') of *Gregarina gigantea*, described by E. van

Beneden, differ essentially from the adult, and pass through a series of developmental stages before they assume the characteristic form of a Gregarine.

This is true development[89]. But such a method of germ-formation and development are found most frequently, although not exclusively, among the parasitic Monoplastides, and this fact alone serves to indicate their secondary origin. It is a form of ontogenetic development differing from that of the Polyplastides in that it does not revert to a phyletically primitive condition of the species, but, on the contrary, exhibits stages which first appear in the phyletic development of the specific form. The Psorosperms were only formed after the Gregarines had become established as a group. The amoeboid organisms which creep out of them are in no way to be regarded as the primitive forms of the Gregarines, even if the latter may have resembled them, but they are coenogenetic forms produced by the necessity for a production of numerous and very minute germs. The necessity for a process of genuine development perhaps depends upon the small amount of material contained in one of these germs, and on other conditions, such as change of host, change of medium, etc. It therefore results that the fundamental law of biogenesis does not apply to the Monoplastides; for these forms are either entirely without a genuine ontogeny and only possess the possibility of growth, or else they are only endowed with a coenogenetic ontogeny[90].

Some authorities may be inclined to limit the above proposition, and to maintain that we must admit the possibility that we are likely to occasionally meet with an ontogeny of which the stages largely correspond with the most important stages in the phyletic development of the species, and that the ontogenetic repetition of the phylogeny, although not the rule, may still occur as a rare exception in the Protozoa.

A careful consideration of the subject indicates, however, that the occurrence of such an exception is very improbable. Such an ontogeny would, for instance, occur if one of the lowest Monoplastides, such as a Moneron, were to develope into a higher form, such as one of the Flagellata, with mouth, eye-spot, and cortical layer, under such external conditions that it would be advantageous for the existence of its species that it should no longer reproduce itself by simple fission, but that the periodical formation of a cyst (which was perhaps previously part of the life-history) should be associated with the occurrence of numerous divisions within the cyst itself, and with the formation of germs. We must suppose either that these germs were so minute that the young animals could not become Flagellata directly, or that it was advantageous for them to move and feed as Monera at an early period, and to assume the more complex structure of the parent by gradual stages. In other words, the phyletic development would proceed hand in hand with the ontogeny corresponding to it, although not from any internal cause, but as an adaptation to the existing conditions of life. But the supposed transformation of the species also depended upon these same conditions of life, which must therefore have been of such a nature as to bring about simultaneously, by an intercalation of germs and by a genuine development, the evolution of the form in question in the last stage of its ontogeny, and the maintenance of its original condition during the initial stage. Such a combination of circumstances can have scarcely ever happened. Against the occurrence of such a transformation as we have supposed, it might be argued, indeed, that the assumed production of very numerous germs does not occur among free-living Monoplastides. Those which have acquired parasitic habits must be younger phyletic forms, for their first host—whether a lowly or a highly organized Metazoon—must have appeared before they could gain access to it and adapt themselves to the conditions of a parasitic life, and by this time the Flagellate Infusoria were already established. It is by far less probable that the persistence or rather the intercalation of the ancestral form would occur in an ontogenetic cycle, consisting of a series of stages, and not of two only, as in our example. For as soon as reproduction can be effected by the simple fission of the adult, not only is there no reason why the earlier phyletic stages should be again and again repeated, but such recapitulation is simply impossible. We cannot, therefore, conclude that the anomalous early stages of a Monoplastid such as *Acineta* correspond with an early form of phyletic development.

Supposing, for instance, that the Acinetaria were derived from the Ciliata, then this transformation must have taken place in the course of the continued division of the ciliate ancestor—partially connected with encystment, but for the most part independently of it. Of the myriads of generations which such a process of development may have occupied, perhaps the first set moved with suctorial processes, while the second gradually adopted sedentary habits, and throughout the whole of the long series, each succeeding generation must have been almost exactly like its predecessor, and must always have consisted of individuals which possessed the characters of the species.

This does not exclude the possibility that in spite of an assumed sedentary mode of life, the need for locomotion and for obtaining food in fresh places may have arisen at some period of life. But whenever formation of swarm-spores takes place instead of simple fission, this does not depend upon the persistence of an ancestral form in the ontogenetic cycle, but is due to the intercalation of an entirely new ontogenetic stage, which happens to resemble an ancestral form, in the possession of cilia, etc.

I imagine that I have now sufficiently explained the above proposition, that the repetition of the phylogeny in the ontogeny does not and cannot occur among unicellular organisms.

With the Polyplastides the opposite is the case. There is no species, as far as we know, which does not—either in each individual, or after long cycles which comprise many individuals (alternation of generations)—invariably revert to the Monoplastid state. This applies from the lowest forms, such as *Magosphaera* and the Orthonectides, up to the very highest. In the latter a great number of intermediate phyletic stages always occur, although some have been omitted as the result of concentration in the ontogeny, while others have sometimes been intercalated.

Sexual reproduction is the obvious cause of this very important arrangement. Even if this is an hypothesis rather than a fact we must nevertheless accept it unconditionally, because it is a method of reproduction found everywhere. It is the rule in every group of the animal kingdom, and is only absent in a few species in which it is replaced by parthenogenesis. In these latter instances sexual reproduction may be local, and entirely absent in certain districts only (*Apus*), or it may be only apparently wanting; in some cases where it is undoubtedly absent, it is equally certain that it was present at an earlier period (*Limnadia Hermanni*). We cannot as yet determine whether its loss will not involve the degeneration and ultimate extinction of the species in question.

If the essential nature of sexual reproduction depends upon the conjugation of two equivalent but dissimilar morphological elements, then we can understand that a multicellular being can only attain sexual reproduction when a unicellular stage is present in its development; for the coalescence of entire multicellular organisms in such a manner that fusion would only take place between equivalent cells, would seem to be impracticable. In the necessity for sexual reproduction, there is therefore also implied the necessity for reverting to the original condition of the Polyplastides—that of a single cell—and upon this alone depends the fundamental law of biogenesis. This law is therefore confined to the Polyplastides, and does not apply to the Monoplastides; and Götte's suggestion that the latter fall back into the primitive condition of the organism during their encystment (rejuvenescence), finds no support in this aspect of the question.

I have on a previous occasion[91] referred the utility of death to the ultimate fact that the unending life of the Metazoan body would be a useless luxury, and to the fact that the individuals would necessarily become injured in the course of time, and would be therefore 'not only valueless to the species, but ... even harmful, for they take the place of those which are sound' (l. c., p. 24). I might also have said that such damaged individuals would sooner or later fall victims to some accidental death, so that there would be no possibility of real immortality. I now propose to examine this statement a little more closely, and to return to a question which has already been alluded to before.

It is obvious that the advantages above set forth did not form the motive which impelled natural selection to convert the immortal life of the Monoplastides into the life of limited duration possessed by the Heteroplastides, or more correctly, which led to the restriction of potential immortality to the reproductive cells of the latter. It is at any rate theoretically conceivable that a struggle might arise between the mortal and immortal individuals of a certain Metazoan species, and that natural selection might secure the success of the former, because the longer the immortal individuals lived, the more defective they became, and as a result gave rise to weaker offspring in diminished numbers. Probably no one would be bold enough to suggest such a crude example of natural selection. And yet I venture to think that the principle of natural selection is here also to be taken into account, and even plays, although in a negative rather than a positive way, a very essential part in determining the duration of life in the Metazoa.

When the somatic cells of the first Heteroplastides ceased to be immortal, such a loss would not in any way have precluded them from regaining this condition. Just as, with the differentiation of the first somatic cells of the lowest Heteroplastides, their duration was limited to that of a single cell-generation,—so it must have been possible for them, at a later period and if the necessity arose, to lengthen their duration over two, three, or more generations. And if my theory of the duration of life in the Metazoa is well founded, these cells have as a matter of fact increased their duration, to an extent about equal to that of the organism to which they belong. There is no ground whatever for the assumption that it is impossible to fix the number of cell-generations at infinity,—as actually happens in the case of the reproductive cells,—but on the other hand it has already been shown to be obvious that such an extension is opposed to the principle of utility. It could never be to the advantage of a species to produce crippled individuals, and therefore the infinite duration of individuals has never reappeared among the Metazoa. So far the limited duration of Metazoan life may be attributed to the worthlessness or even the injurious nature of individuals, which although immortal, were nevertheless liable to wear and tear. This fact explains why immortality has never reappeared, it explains the predominance of death, but it was not the single primary cause of this phenomenon. The perishable and vulnerable nature of the *soma* was the reason why nature made no effort to endow this part of the individual with a life of unlimited length.

Götte considers that death is inherent in reproduction, and in a certain sense this is true, but not in the general way supposed by him.

I have endeavoured to show above that it is most advantageous for the preservation of the species among the lowest Metazoa, that the body should consist of a relatively small number of cells, and that the reproductive cells should ripen simultaneously and all escape together. If this conclusion be accepted, the uselessness of a prolonged life to the somatic cells is obvious, and the occurrence of death at the time of the extrusion of the reproductive cells is explained. In this manner death (of the *soma*) and reproduction are here made to coincide.

This relation of reproduction to death still exists in a great number of the higher animals. But such an association, together with the simultaneous ripening of the reproductive cells, has not been maintained continuously in the past. As the *soma* becomes larger and more highly organized, it is able to withstand more injuries, and its average duration of life will extend: *pari passu* with these changes it will become increasingly advantageous not only for the number of reproductive cells to be multiplied, but also for the time during which they are produced to be prolonged. In this manner a lengthening of the reproductive period arises, at first continuously and then periodically. It is beyond my present purpose to consider in detail the conditions upon which this lengthening depends, but I would emphasize the fact that a lengthening of life is connected with the increase in the duration of reproduction, while on the other hand there is no reason to expect life to be prolonged beyond the reproductive period; so that the end of this period is usually more or less coincident with death.

A further prolongation of life could only take place when the parent begins to undertake the duty of rearing the young. The most primitive form of this is found among those animals, which do not expel their reproductive cells as soon as they are ripe but retain them within their bodies, so that the early stages of development take place under the shelter of the parent organism. Associated with such a process there is frequently a necessity for the germs to reach a certain spot, where alone their further development can take place. Thus a segment of a tapeworm lives until it has brought the embryos into a position which affords the possibility of their passive transference to the stomach of their special host. But the duration of life is first materially lengthened when the offspring begin to be really tended, and as a general rule the increase in length is exactly proportional to the time which is demanded by the care of the young. Accurately conducted observations are wanting upon this precise point, but the general tendency of the facts, as a whole, cannot be doubted. Those insects of which the care for their offspring terminates with the deposition of eggs at the appropriate time, place, etc., do not survive this act; and the duration of life in such imagos is shorter or longer according as the eggs are laid simultaneously or ripen gradually. On the other hand, insects—such as bees and ants—which tend their young, have a life which is prolonged for years.

But the lengthening of the reproductive period alone may result in a marked increase in the length of life, as is proved by the queen-bee. In all these cases it is easy to imagine the operation of natural selection in producing such alterations in the duration of life, and indeed we might accurately calculate the amount of increase which would be produced in any given case if the necessary data were available, viz. the physiological strength of the body, and its relations to the external world, such as, for instance, the power of obtaining food at various periods of life, the expenditure of energy necessary for this end, and the statistics of destruction, that is, the probabilities in favour of the accidental death of a single individual at any given time. These statistics must be known both for the imagos, larvae, and eggs; for the lower they are for the imagos, and the higher for the larvae and eggs, the more advantageous will it be, *ceteris paribus*, for the number of eggs produced by the imago to be increased, and the more probable it would therefore be that a long reproductive period, involving a lengthening of the life of the imago, would be introduced. But we are still far from being able to apply mathematics to the phenomena of life; the factors are too numerous, and no attempt has been made as yet to determine them with accuracy.

But we must at least admit the principle that both the lengthening and shortening of life are possible by means of natural selection, and that this process is alone able to render intelligible the exact adaptation of the length of life to the conditions of existence.

A shortening of the normal duration of life is also possible; this is shown in every case of sudden death, after the deposition of the whole of the eggs at a single time. This occurs among certain insects, while nearly allied forms of which the oviposition lasts over many days therefore possess a correspondingly long imago-life. The *Ephemeridae* and Lepidoptera afford many examples of this, and in an earlier work I have collected some of them[92]. The humming-bird hawk-moth flies about for weeks laying an egg here and there, and, like the allied poplar hawk-moth and lime hawk-moth, probably dies when it has deposited all the eggs which can be matured with the amount of nutriment at its disposal. Many other Lepidoptera, such as the majority of butterflies, fly about for weeks depositing their eggs, but others, such as the emperor-moths and lappet-moths, lay their eggs one after another and then die. The eggs of the parthenogenetic *Psychidae* are laid directly after the imago has left the cocoon, and death ensues immediately, so that the whole life of the imago only lasts for a few hours. No one could look upon this brief life as a primitive arrangement among Lepidoptera, any more than we do upon the absence of wings in the female *Psychidae*; shortening of life here is therefore clearly explicable.

In such cases have we any right to speak of the fatal effect of reproduction? We may certainly say that these insects die of exhaustion; their vital strength is used up in the last effort of laying eggs, and in the case of the males, in the act of copulation. Reproduction is here certainly the most apparent cause of death, but a more remote and deeper cause is to be found in the limitation of vital strength to the length and the necessary duties of the reproductive period. The fact that there are female Lepidoptera which, like the emperor-moths, do not feed in the imago-state, proves the truth of this statement. They still possess a mouth and a complete alimentary canal, but they have no spiral 'tongue,' and do not take food of any kind, not even a drop of water. They live in a torpid condition for days or weeks until fertilization is accomplished, and then they lay their eggs and die. The habit of extracting honey from flowers—common to most hawk-moths and butterflies—would not have thus fallen into disuse, if the store of nutriment, accumulated in the form of the fat-bodies, during the life of the caterpillar, had not been exactly sufficient to maintain life until the completion of oviposition. The fact that the habit of taking food has been thus abandoned is a proof that the duration of life beyond the reproductive period would not be to the advantage of the species.

The protraction of existence into old age among the higher Metazoa proves that death is not a necessary consequence of reproduction. It seems to me that Götte's statement 'that the appearances of senility must not be regarded as the general cause of death' is not in opposition to my opinions but rather to those which receive general acceptance. I have myself pointed out that 'death is not always preceded by senility or a period of old age[93].'

The materials are wanting for a comprehensive investigation of the causes which first introduced this period among the higher Metazoa; in fact the most fundamental data are absent, for we do not even know the part of the animal kingdom in which it first appeared: we cannot even state the amount by which the duration of life exceeds that of the period of reproduction, or what is the value to the species of this last stage in the life of the individual.

It is in these general directions that we must seek for the significance of old age. It is obviously of use to man, for it enables the old to care for their children, and is also advantageous in enabling the older individuals to participate in human affairs and to exercise an influence upon the advancement of intellectual powers, and thus to influence indirectly the maintenance of the race. But as soon as we descend a step lower, if only as far as the apes, accurate facts are wanting, for we are, and shall probably long be, ignorant of the total duration of their life, and the point at which the period of reproduction ceases.

I must here break off in the midst of these considerations, rather than conclude them, for much still remains to be said. I hope, nevertheless, that I have thrown new light upon some important points, and I now propose to conclude with the following short abstract of the results of my enquiry.

I. Natural death occurs only among multicellular beings; it is not found among unicellular organisms. The process of encystment in the latter is in no way comparable with death.

II. Natural death first appears among the lowest Heteroplastid Metazoa, in the limitation of all the cells collectively to one generation, and of the somatic or body-cells proper to a restricted period: the somatic cells afterwards in the higher Metazoa came to last several and even many generations, and life was lengthened to a corresponding degree.

III. This limitation went hand in hand with a differentiation of the cells of the organism into reproductive and somatic cells, in accordance with the principle of division of labour. This differentiation took place by the operation of natural selection.

IV. The fundamental biogenetic law applies only to multicellular beings; it does not apply to unicellular forms of life. This depends on the one hand upon the mode of reproduction by fission which obtains among the Monoplastides (unicellular organisms), and on the other upon the necessity, induced by sexual reproduction, for the maintenance of a unicellular stage in the development of the Polyplastides (multicellular organisms).

V. Death itself, and the longer or shorter duration of life, both depend entirely on adaptation. Death is not an essential attribute of living matter; it is neither necessarily associated with reproduction, nor a necessary consequence of it.

In conclusion, I should wish to call attention to an idea which is rather implied than expressed in this essay:—it is, that reproduction did not first make its appearance coincidently with death. Reproduction is in truth an essential attribute of living matter, just as is the growth which gives rise to it. It is as impossible to imagine life enduring without reproduction as it would be to conceive life lasting without the capacity for absorption of food and without the power of metabolism. Life is continuous and not periodically interrupted: ever since its first appearance upon the earth, in the lowest organisms, it has continued without break; the forms in which it is manifested have alone undergone change. Every individual alive to-day—even the very highest—is to be derived in an unbroken line from the first and lowest forms.

Footnotes for Chapter III.

59. 'Ueber den Ursprung des Todes,' Hamburg and Leipzig, 1883.

60. As in the case of the bodies of monks on the Great St. Bernard, or the dried-up bodies in the well-known Capuchine Monastery at Palermo.

61. Professor Gruber informs me that among the Infusoria of the harbour of Genoa, he has observed a species which encysts upon one of the free-swimming Copepoda. He has often found as many as ten cysts upon one of these Copepods, and has observed the escape of their contents whenever the water under the cover-glass began to putrefy. Here advantage is probably gained in the rapid transport of the cyst by the Crustacean.

62. The views of most biologists who have worked at this subject agree in all essentials with that expressed above. Bütschli says (Bronn's 'Klassen und Ordnungen des Thierreichs,' Protozoa, p. 148): 'The process of encystment does not appear to have originally borne any direct relation to reproduction: it appears on the contrary to have taken place originally,—as it frequently does at the present day,—either for the protection of the organism against injurious external influences, such as desiccation or the fatal effects of impure water, etc.; and also to enable the organism, after taking up an unusually abundant supply of food, to assimilate it in safety.' Balbiani ('Journ. de Micrographie,' Tom. V. 1881, p. 293) says in reference to the Infusoria, 'Un petit nombre d'espèces, au lieu de se multiplier à l'état de vie active, se reproduisent dans une sorte d'état de repos, dit état d'enkystement. Ces sortes de kystes peuvent être désignés sous le nom de kystes de reproduction, par opposition avec d'autres kystes, dans lesquels les Infusoires se renferment

pour se soustraire à des conditions devenues défavorables du milieu qu'ils habitent, le manque d'air, le dessèchement, etc.—ceux-ci sont des kystes de conservation....'

63. This is of importance in so far as single individuals might be thus compelled to encyst even when the existing external conditions of life do not require it. The substance which *Actinosphaerium*, for example, employs in the secretion of its thick siliceous cyst must have been gradually accumulated by means of a process peculiar to the species. We can scarcely be in error if we assume that the silica accumulated in the organism cannot increase to an unlimited extent without injury to the other vital processes and that the secretion of the cyst must take place as soon as the accumulation has exceeded a certain limit. Thus we can understand that encystment may occur without any external necessity. Similarly, certain Entomostraca (*e. g. Moina*) produce winter-eggs in a particular generation, and these are formed even when the animals are kept in a room protected from cold and desiccation.

64. Upon this point Professor Gruber intends to publish an elaborate memoir.

65. This view has not even been proved for *Actinosphaerium*, upon which Götte chiefly relies. The observations which we now possess merely indicate that the animal contracts to the smallest volume possible. Compare F. E. Schulze, 'Rhizopodenstudien,' I, Arch. f. mikr. Anat. Bd. 10, p. 328; and Karl Brandt, 'Ueber Actinosphaerium Eichhornii,' Inaug. Diss.; Halle, 1877.

66. The conception of Protozoa and Metazoa does not correspond exactly with that of unicellular and multicellular beings, for which Götte has proposed the names Mono- and Polyplastides.

67. Among the Rhizopoda encystment is only known in fresh-water forms, and not in a single one of the far more numerous marine forms which possess shells (see Bütschli, 'Protozoa,' p. 148); the marine Rhizopoda are not exposed to the effects of desiccation or frost, and thus the strongest motives for the process of encystment do not exist, at least among forms possessing a shell.

68. I trust that it will not be objected that the germ-cells cannot be immortal, because they frequently perish in large numbers, as a result of the natural death of the individual. There are certain definite conditions under which alone a germ-cell can render its potential immortality actual, and these conditions are for the most part fulfilled with difficulty (fertilization, etc.). It follows from this fact that the germ-cells must always be produced in numbers which reach some very high multiple of the necessary number of offspring, if these latter are to be ensured for the species. If in the natural death of the individual the germ-cells must also die, the *natural* death of the *soma* becomes a cause of *accidental* death to the germ-cells.

69. l. c., p. 78.

70. l. c., p. 47.

71. 'Entwicklungsgeschichte der Unke,' Leipzig, 1875, p. 65.

72. Id., p. 842.

73. 'Ursprung des Todes,' p. 79.

74. l. c., p. 42.

75. 'Contributions à l'histoire des Mesozoaires. Recherches sur l'organisation et le développement embryonnaire des Orthonectides,' Arch. de Biologie, vol. iii. 1882.

76. l. c., p. 37.

77. Julin does not enter into further details on this point, and it is not quite clear at what precise time the cells of the ectoderm atrophy; but this is irrelevant to the origin of death, since the granular mass surrounding the egg-cells at any rate belongs to the *soma* of the mother.

78. Leuckart finds such a great resemblance between the newly born young of *Distoma* and the Orthonectides, that he is inclined to believe that the latter are Trematodes, 'which in spite of sexual maturity have not developed further than the embryonic condition of the *Distoma*' ('Zur Entwicklungsgeschichte des Leberegels,' Zool. Anzeiger, 1881, No. 99). In reference to the Dicyemidae, which resemble the Orthonectides in their manner of living and in their structure, Gegenbaur has stated his opinion that they belong to a 'stage in the development of Platyhelminthes' (Grundriss d. vergleich. Anatomie). Giard includes both in the 'phylum Vermes,' and regards them as much degenerated by parasitism; and Whitman—the latest investigator of the Dicyemids—speaks of them in a similar manner in his excellent work 'Contributions to the Life-history and Classification of Dicyemids' (Leipzig, 1882).

79. 'Dauer des Lebens;' translated as the first essay in this volume.

80. See the first essay upon 'The Duration of Life,' p. 22 et seq.

81. 'Ursprung des Todes,' p. 29.

82. l. c., p. 5.

83. See the preceding essay 'On Heredity.'

84. The problem is very easily solved if we seek assistance from the principle of panmixia developed in the second essay 'On Heredity.' As soon as natural selection ceases to operate upon any character, structural or functional, it begins to disappear. As soon, therefore, as the immortality of somatic cells became useless they would begin to lose this attribute. The process would take place more quickly, as the histological differentiation of the somatic cells became more useful and complete, and thus became less compatible with their everlasting duration.—A. W. 1888.

85. See the preceding essay 'On Heredity.'

86. See the first essay on 'The Duration of Life.'

87. See the first essay on 'The Duration of Life.'

88. These assumptions can be authenticated among the Infusoria. The encysted *Colpoda cucullus*, Ehrbg. divides into two, four, eight, or sixteen parts; *Otostoma Carteri*, into two, four, or eight; *Tillina magna*, Gruber, into four or five; *Lagynus* sp. Gruber, into two; *Amphileptus meleagris*, Ehrbg. into two or four. The last two species and many others frequently do not divide at all during the encysted condition. But while any further increase in the number of divisions within the cyst does not occur in free-swimming Infusoria, the interesting case of *Ichthyophthirius multifiliis*, Fouquet, shows that parasitic habits call forth a remarkable increase in the number of divisions. This animal divides into at least a thousand daughter individuals.

89. True development also takes place in the above-mentioned *Ichthyophthirius*. While in other Infusoria the products of fission exactly resemble the parent, in *Ichthyophthirius* they have a different form; the sucking mouth is wanting while provisional clasping cilia are at first present. In this case therefore the word germ may be rightly applied, and *Ichthyophthirius* affords an interesting example of the phyletic origin of germs among the lower Flagellata and Gregarines. Cf. Fouquet, 'Arch. Zool. Expérimentale,' Tom. V. p. 159. 1876.

90. Bütschli, long ago, doubted the application of the fundamental law of biogenesis to the Protozoa (cf. 'Ueber die Entstehung der Schwärmsprösslings der Podophrya quadripartita,' Jen. Zeit. f. Med. u. Naturw. Bd. X. p. 19, Note). Gruber has more recently expressed similar views, and in fact denies the presence of development in the Protozoa, and only recognizes growth ('Dimorpha mutans, Z. f. W. Z.' Bd. XXXVII. p. 445). This proposition must however be restricted, inasmuch as a development certainly occurs, although one which is coenogenetic and not palingenetic.

91. See the first essay on 'The Duration of Life,' p. 23 *et seq.*

92. See Appendix to the first essay on 'The Duration of Life,' pp. 43-46.

93. See the first essay on 'The Duration of Life,' p. 21.

IV.

THE CONTINUITY OF THE GERM-PLASM

AS THE FOUNDATION OF A THEORY OF HEREDITY.

1885.

CONTINUITY OF THE GERM-PLASM, &c.

PREFACE.

The ideas developed in this essay were first expressed during the past winter in a lecture delivered to the students of this University (Freiburg), and they were shortly afterwards—in February and the beginning of March—written in their present form. I mention this, because I might otherwise be reproached for a somewhat partial use of the most recent publications on related subjects. Thus I did not receive Oscar Hertwig's paper—'Contributions to the Theory of Heredity' (Zur Theorie der Vererbung), until after I had finished writing my essay, and I could not therefore make as much use of it as I should otherwise have done. Furthermore, the paper by Kölliker on 'The Significance of the Nucleus in the Phenomena of Heredity' (Die Bedeutung der Zellkerne für die Vorgänge der Vererbung), did not appear until after the completion of my manuscript. The essential treatment of the subject would not, however, have been altered if I had received the papers at an earlier date, for as far as the most important point—the significance of the nucleus—is concerned, my views are in accordance with those of both the above-named investigators; while the points upon which our views do not coincide had already received attention in the manuscript.

A. W.

Freiburg I. Breisgau,

June 16, 1885.

CONTINUITY OF THE GERM-PLASM, &c.

CONTENTS.

IV.

THE CONTINUITY OF THE GERM-PLASM AS THE FOUNDATION OF A THEORY OF HEREDITY.

Introduction.

When we see that, in the higher organisms, the smallest structural details, and the most minute peculiarities of bodily and mental disposition, are transmitted from one generation to another; when we find in all species of plants and animals a thousand characteristic peculiarities of structure continued unchanged through long series of generations; when we even see them in many cases unchanged throughout whole geological periods; we very naturally ask for the causes of such a striking phenomenon: and enquire how it is that such facts become possible, how it is that the individual is able to transmit its structural features to its offspring with such precision. And the immediate answer to such a question must be given in the following terms:—'A single cell out of the millions of diversely differentiated cells which compose the body, becomes specialized as a sexual cell; it is thrown off from the organism and is capable of reproducing all the peculiarities of the parent body, in the new individual which springs from it by cell-division and the complex process of differentiation.' Then the more precise question follows: 'How is it that such a single cell can reproduce the *tout ensemble* of the parent with all the faithfulness of a portrait?'

The answer is extremely difficult; and no one of the many attempts to solve the problem can be looked upon as satisfactory; no one of them can be regarded as even the beginning of a solution or as a secure foundation from which a complete solution may be expected in the future. Neither Häckel's[94], 'Perigenesis of the Plastidule,' nor Darwin's[95] 'Pangenesis,' can be regarded as such a beginning. The former hypothesis does not really treat of that part of the problem which is here placed in the foreground, viz. the explanation of the fact that the tendencies of heredity are present in single cells, but it is rather concerned with the question as to the manner in which it is possible to conceive the transmission of a certain tendency of development into the sexual cell, and ultimately into the organism arising from it. The same may be said of the hypothesis of His[96], who, like Häckel, regards heredity as the transmission of certain kinds of motion. On the other hand, it must be conceded that Darwin's hypothesis goes to the very root of the question, but he is content to give, as it were, a provisional or purely formal solution, which, as he himself says, does not claim to afford insight into the real phenomena, but only to give us the opportunity of looking at all the facts of heredity from a common standpoint. It has achieved this end, and I believe it has unconsciously done more, in that the thoroughly logical application of its principles has shown that the real causes of heredity cannot lie in the formation of gemmules or in any allied phenomena. The improbabilities to which any such theory would lead are so great that we can affirm with certainty that its details cannot accord with existing facts. Furthermore, Brooks'[97] well-considered and brilliant attempt to modify the theory of Pangenesis, cannot escape the reproach that it is based upon possibilities, which one might certainly describe as improbabilities. But although I am of opinion that the whole foundation of the theory of Pangenesis, however it may be modified, must be abandoned, I think, nevertheless, its author deserves great credit, and that its production has been one of those indirect roads along which science has been compelled to travel in order to arrive at the truth. Pangenesis is a modern revival of the oldest theory of heredity, that of Democritus, according to which the sperm is secreted from all parts of the body of both sexes during copulation, and is animated by a bodily force; according to this theory also, the sperm from each part of the body reproduces the same part[98].

If, according to the received physiological and morphological ideas of the day, it is impossible to imagine that gemmules produced by each cell of the organism are at all times to be found in all parts of the body, and furthermore that these gemmules are collected in the sexual cells, which are then able to again reproduce in a certain order each separate cell of the organism, so that each sexual cell is capable of developing into the likeness of the parent body; if all this is inconceivable, we must enquire for some other way in which we can arrive at a foundation for the true understanding of heredity. My present task is not to deal with the whole question of heredity, but only with the single although fundamental question—'How is it that a single cell of the body can contain within itself all the hereditary tendencies of the whole organism?' I am here leaving out of account the further question as to the forces and the mechanism by which these tendencies are developed in the building-up of the organism. On this account I abstain from considering at present the views of Nägeli, for as will be shown later on, they only slightly touch this fundamental question, although they may certainly claim to be of the highest importance with respect to the further question alluded to above.

Now if it is impossible for the germ-cell to be, as it were, an extract of the whole body, and for all the cells of the organism to despatch small particles to the germ-cells, from which the latter derive their power of heredity; then there remain, as it seems to me, only two other possible, physiologically conceivable, theories as to the origin of germ-cells, manifesting such powers as we know they possess. Either the substance of the parent germ-cell is capable of undergoing a series of changes which, after the building-up of a new individual, leads back again to identical germ-cells; or the germ-cells are not derived at all, as far as their essential and characteristic substance is concerned, from the body of the individual, but they are derived directly from the parent germ-cell.

I believe that the latter view is the true one: I have expounded it for a number of years, and have attempted to defend it, and to work out its further details in various publications. I propose to call it the theory of 'The Continuity of the Germ-plasm,' for it is founded upon the idea that heredity is brought about by the transference from one generation to another, of a substance with a definite chemical, and above all, molecular constitution. I have called this substance 'germ-plasm,' and have assumed that it possesses a highly complex structure, conferring upon it the power of developing into a complex organism. I have attempted to explain heredity by supposing that in each ontogeny, a part of the specific germ-plasm contained in the parent egg-cell is not used up in the construction of the body of the offspring, but is reserved unchanged for the formation of the germ-cells of the following generation.

It is clear that this view of the origin of germ-cells explains the phenomena of heredity very simply, inasmuch as heredity becomes thus a question of growth and of assimilation,—the most fundamental of all vital phenomena. If the germ-cells of successive generations are directly continuous, and thus only form, as it were, different parts of the same substance, it follows that these cells must, or at any rate may, possess the same molecular constitution, and that they would therefore pass through exactly the same stages under certain conditions of development, and would form the same final product. The hypothesis of the continuity of the germ-plasm gives an identical starting-point to each successive generation, and thus explains how it is that an identical product arises from all of them. In other words, the hypothesis explains heredity as part of the underlying problems of assimilation and of the causes which act directly during ontogeny: it therefore builds a foundation from which the explanation of these phenomena can be attempted.

It is true that this theory also meets with difficulties, for it seems to be unable to do justice to a certain class of phenomena, viz. the transmission of so-called acquired characters. I therefore gave immediate and special attention to this point in my first publication on heredity[99], and I believe that I have shown that the hypothesis of the transmission of acquired characters—up to that time generally accepted—is, to say the least, very far from being proved, and that entire classes of facts which have been interpreted under this hypothesis may be quite as well interpreted otherwise, while in many cases they must be explained differently. I have shown that there is no ascertained fact, which, at least up to the present time, remains in irrevocable conflict with the hypothesis of the continuity of the germ-plasm; and I do not know any reason why I should modify this opinion to-day, for I have not heard of any objection which appears to be feasible. E. Roth[100] has objected that in pathology we everywhere meet with the fact that acquired local disease may be transmitted to the offspring as a predisposition; but all such cases are exposed to the serious criticism that the very point that first needs to be placed on a secure footing is incapable of proof, viz. the hypothesis that the causes which in each particular case led to the predisposition, were really acquired. It is not my intention, on the present occasion, to enter fully into the question of acquired characters; I hope to be able to consider the subject in greater detail at a future date. But in the meantime I should wish to point out that we ought, above all, to be clear as to what we really mean by the expression 'acquired character.' An organism cannot acquire anything unless it already possesses the predisposition to acquire it: acquired characters are therefore no more than local or sometimes general variations which arise under the stimulus provided by certain external influences. If by the long-continued handling of a rifle, the so-called 'Exercierknochen' (a bony growth caused by the pressure of the weapon in drilling) is developed, such a result depends upon the fact that the bone in question, like every other bone, contains within itself a predisposition to react upon certain mechanical stimuli, by growth in a certain direction and to a certain extent. The predisposition towards an 'Exercierknochen' is therefore already present, or else the growth could not be formed; and the same reasoning applies to all other 'acquired characters.'

Nothing can arise in an organism unless the predisposition to it is pre-existent, for every acquired character is simply the reaction of the organism upon a certain stimulus. Hence I should never have thought of asserting that predispositions cannot be transmitted, as E. Roth appears to believe. For instance, I freely admit that the predisposition to an 'Exercierknochen' varies, and that a strongly marked predisposition may be transmitted from father to son, in the form of bony tissue with a more susceptible constitution. But I should deny that the son could develope an 'Exercierknochen' without having drilled, or that, after having drilled, he could develope it more easily than his father, on account of the drilling through which the latter first acquired it. I believe that this is as impossible as that the leaf of an oak should produce a gall, without having been pierced by a gall-producing insect, as a result of the thousands of antecedent generations of oaks which have been pierced by such insects, and have thus 'acquired' the power of producing galls. I am also far from asserting that the germ-plasm—which, as I hold, is transmitted as the basis of heredity from one generation to another—is absolutely unchangeable or totally uninfluenced by forces residing in the organism within which it is transformed into germ-cells. I am also compelled to admit that it is conceivable that organisms may exert a modifying influence upon their germ-cells, and even that such a process is to a certain extent inevitable. The nutrition and growth of the individual must exercise some influence upon its germ-cells; but in the first place this influence must be extremely slight, and in the second place it cannot act in the manner in which it is usually assumed that it takes place. A change of growth at the periphery of an organism, as in the case of an 'Exercierknochen,' can never cause such a change in the molecular structure of the germ-plasm as would augment the predisposition to an 'Exercierknochen,' so that the son would inherit an increased susceptibility of the bony tissue or even of the particular bone in question. But any change produced will result from the reaction of the germ-cell upon changes of nutrition caused by alteration in growth at the periphery, leading to some change in the size, number, or arrangement of its molecular units. In the present state of our knowledge there is reason for doubting whether such reaction can occur at all; but, if it can take place, at all events the quality of the change in the germ-plasm can have nothing to do with the quality of the acquired character, but only with the way in which the general nutrition is influenced by the latter. In the case of the 'Exercierknochen' there would be practically no change in the general nutrition, but if such a bony growth could reach the size of a carcinoma, it is conceivable that a disturbance of the general nutrition of the body might ensue. Certain experiments on plants—in which Nägeli showed that they can be submitted to strongly varied conditions of nutrition for several generations, without the production of any visible hereditary change—show that the influence of nutrition upon the germ-cells must be very slight, and that it may possibly leave the molecular structure of the germ-plasm altogether untouched. This conclusion is also supported by comparing the uncertainty of these results with the remarkable precision with which heredity acts in the case of those characters which are known to be transmitted. In fact, up to the present time, it has never been proved that any changes in general nutrition can modify the molecular structure of the germ-plasm, and far less has it been rendered by any means probable that the germ-cells can be affected by acquired changes which have no influence on general nutrition. If we consider that each so-called predisposition (that is, a power of reacting upon a certain stimulus in a certain way, possessed by any organism or by one of its parts) must be innate, and further that each acquired character is only the predisposed reaction of some part of an organism upon some external influence; then we must admit that only one of the causes which produce any acquired character can be transmitted, the one which was present before the character itself appeared, viz. the predisposition; and we must further admit that the latter arises from the germ, and that it is quite immaterial to the following generation whether such predisposition comes into operation or not. The continuity of the germ-plasm is amply sufficient to account for such a phenomenon, and I do not believe that any objection to my hypothesis, founded upon the actually observed phenomena of heredity, will be found to hold. If it be accepted, many facts will appear in a light different from that which has been cast upon them by the hypothesis which has been hitherto received,—a hypothesis which assumes that the organism produces germ-cells afresh, again and again, and that it produces them entirely from its own substance. Under the former theory the germ-cells are no longer looked upon as the product of the parent's body, at least as far as their essential part—the specific germ-plasm—is concerned: they are rather considered as something which is to be placed in contrast with the *tout ensemble* of the cells which make up the parent's body, and the germ-cells of succeeding generations stand in a similar relation to one another as a series of generations of unicellular organisms, arising by a continued process of cell-division. It is true that in most cases the generations of germ-cells do not arise immediately from one another as complete cells, but only as minute particles of germ-plasm. This latter substance, however, forms the foundation of the germ-cells of the next generation, and stamps them with their specific character. Previous to the publication of my theory, G. Jäger[101], and later M. Nussbaum[102], have expressed ideas upon heredity which come very near to my own[103]. Both of these writers started with the hypothesis that there must be a direct connexion between the germ-cells of succeeding generations, and they tried to establish such a continuity by supposing that the germ-cells of the offspring are separated from the parent germ-cell before the beginning of embryonic development, or at least before any histological differentiation has taken place. In this form their suggestion cannot be maintained, for it is in conflict with numerous facts. A continuity of the germ-*cells* does not now take place, except in very rare instances; but this fact does not prevent us from adopting a theory of the continuity of the germ-*plasm*, in favour of which much weighty evidence can be brought forward. In the following pages I shall attempt to develope further the theory of which I have just given a short account, to defend it against any objections which have been brought forward, and to draw from it new conclusions which may perhaps enable us more thoroughly to appreciate facts which are known, but imperfectly understood. It seems to me that this theory of the continuity of the germ-plasm deserves at least to be examined in all its details, for it is the simplest theory upon the subject, and the one which is most obviously suggested by the facts of the case, and we shall not be justified in forsaking it for a more complex theory until proof that it can be no longer maintained is forthcoming. It does not presuppose anything except facts which can be observed at any moment, although they may not be understood,—such as assimilation, or the development of like organisms from like germs; while every other theory of heredity is founded on hypotheses which cannot be proved. It is nevertheless possible that continuity of the germ-plasm does not exist in the manner in which I imagine that it takes place, for no one can at present decide whether all the ascertained facts agree with and can be explained by it. Moreover the ceaseless activity of research brings to light new facts every day, and I am far from maintaining that my theory may not be disproved by some of these. But even if it should have to be abandoned at a later period, it seems to me that, at the present time, it is a necessary stage in the advancement of our knowledge, and one which must be brought forward and passed through, whether it prove right or wrong, in the future. In this spirit I offer the following considerations, and it is in this spirit that I should wish them to be received.

I. The Germ-plasm.

I must first define precisely the exact meaning of the term germ-plasm.

In my previous writings in which the subject has been alluded to, I have simply spoken of germ-plasm without indicating more precisely the part of the cell in which we may expect to find this substance—the bearer of the characteristic nature of the species and of the individual. In the first place such a course was sufficient for my immediate purpose, and in the second place the number of ascertained facts appeared to be insufficient to justify a more exact definition. I imagined that the germ-plasm was that part of a germ-cell of which the chemical and physical properties—including the molecular structure—enable the cell to become, under appropriate conditions, a new individual of the same species. I therefore believed it to be some such substance as Nägeli[104], shortly afterwards, called idioplasm, and of which he attempted, in an admirable manner, to give us a clear understanding. Even at that time one might have ventured to suggest that the organized substance of the nucleus is in all probability the bearer of the phenomena of heredity, but it was impossible to speak upon this point with any degree of certainty. O. Hertwig[105] and Fol[106] had shown that the process of fertilization is attended by a conjugation of nuclei, and Hertwig had even then distinctly said that fertilization generally depends upon the fusion of two nuclei; but the possibility of the co-operation of the substance of the two germ-cells could not be excluded, for in all the observed cases the sperm-cell was very small and had the form of a spermatozoon, so that the amount of its cell-body, if there is any, coalescing with the female cell, could not be distinctly seen, nor was it possible to determine the manner in which this coalescence took place. Furthermore, it was for some time very doubtful whether the spermatozoon really contained true nuclear substance, and even in 1879 Fol was forced to the conclusion that these bodies consist of cell-substance alone. In the following year my account of the sperm-cells of *Daphnidae* followed, and this should have removed every doubt as to the cellular nature of the sperm-cells and as to their possession of an entirely normal nucleus, if only the authorities upon the subject had paid more attention to these statements[107]. In the same year (1880) Balfour summed up the facts in the following manner—'The act of impregnation may be described as the fusion of the ovum and spermatozoon, and the most important feature in this act appears to be the fusion of a male and female nucleus[108].' It is true that Calberla had already observed in *Petromyzon*, that the tail of the spermatozoon does not penetrate into the egg, but remains in the micropyle; but on the other hand the head and part of the 'middle-piece' which effect fertilization, certainly contain a small fraction of the cell-body in addition to the nuclear substance, and although the amount of the former which thus enters the egg must be very small, it might nevertheless be amply sufficient to transmit the tendencies of heredity. Nägeli and Pflüger rightly asserted, at a later date, that the amount of the substance which forms the basis of heredity is necessarily very small, for the fact that hereditary tendencies are as strong on the paternal as on the maternal side, forces us to assume that the amount of this substance is nearly equal in both male and female germ-cells. Although I had not published anything upon the point, I was myself inclined to ascribe considerable importance to the cell-substance in the process of fertilization; and I had been especially led to adopt this view because my investigations upon *Daphnidae* had shown that an animal produces large sperm-cells with an immense cell-body whenever the economy of its organism permits. All *Daphnidae* in which internal fertilization takes place (in which the sperm-cells are directly discharged upon the unfertilized egg), produce a small number of such large sperm-cells (*Sida, Polyphemus, Bythotrephes*); while all species with external fertilization (*Daphnidae, Lynceinae*) produce very small sperm-cells in enormous numbers, thus making up for the immense chances against any single cell being able to reach an egg. Hence the smaller the chances of any single sperm-cell being successful, the larger is the number of such cells produced, and a direct result of this increase in number is a diminution in size. But why should the sperm-cells remain or become so large in the species in which fertilization is internal? The idea suggests itself that the species in this way gains some advantage, which must be given up in the other cases; although such advantage might consist in assisting the development of the fertilized ovum and not in any increase of the true fertilizing substance. At the present time we are indeed disposed to recognize this advantage in still more unimportant matters, but at that time the ascertained facts did not justify us in the assertion that fertilization is a mere fusion of nuclei, and M. Nussbaum[109] quite correctly expressed the state of our knowledge when he said that the act of fertilization consisted in 'the union of identical parts of two homologous cells.'

Pflüger's discovery of the 'isotropism' of the ovum was the first fact which distinctly pointed to the conclusion that the bodies of the germ-cells have no share in the transmission of hereditary tendencies. He showed that segmentation can be started in different parts of the body of the egg, if the latter be permanently removed from its natural position. This discovery constituted an important proof that the body of the egg consists of a uniform substance, and that certain parts or organs of the embryo cannot be potentially contained in certain parts of the egg, so that they can only arise from these respective parts and from no others. Pflüger was mistaken in the further interpretation, from which he concluded that the fertilized ovum has no essential relation to the organization of the animal subsequently formed by it, and that it is only the recurrence of the same external conditions which causes the germ-cell to develope always in the same manner. The force of gravity was the first factor, which, as Pflüger thought, determined the building up of the embryo: but he overlooked the fact that isotropism can only be referred to the body of the egg, and that besides this cell-body there is also a nucleus present, from which it was at least possible that regulative influences might emanate. Upon this point Born[110] first showed that the position of the nucleus is changed in eggs which are thus placed in unnatural conditions, and he proved that the nucleus must contain a principle which in the first place directs the formation of the embryo. Roux[111] further showed that, even when the effect of gravity is compensated, the development is continued unchanged, and he therefore concluded that the fertilized egg contains within itself all the forces necessary for normal development. Finally, O. Hertwig[112] proved from observations on the eggs of sea-urchins, that at any rate in these animals, gravity has no directive influence upon segmentation, but that the position of the first nuclear spindle decides the direction which will be taken by the first divisional plane of segmentation. These observations were however still insufficient to prove that fertilization is nothing more than the fusion of nuclei[113].

A further and more important step was taken when E. van Beneden[114] observed the process of fertilization in *Ascaris megalocephala*. Like the investigations of Nussbaum[115] upon the same subject, published at a rather earlier date, van Beneden's observations did not altogether exclude the possibility of the participation of the body of the sperm-cell in the real process of fertilization; still the fact that the nuclei of the egg-cell and the sperm-cell do not coalesce irregularly, but that their loops are placed regularly opposite one another in pairs and thus form one new nucleus (the first segmentation nucleus), distinctly pointed to the conclusion that the nuclear substance is the sole bearer of hereditary tendencies—that in fact fertilization depends upon the coalescence of nuclei. Van Beneden himself did not indeed arrive at these conclusions: he was prepossessed with the idea that fertilization depends upon the union of two sexually differentiated nuclei, or rather half-nuclei—the male and female pronuclei. He considered that only in this way could a single complete nucleus be formed, a nucleus which must of course be hermaphrodite, and he believed that the essential cause of further development lies in the fact that, at each successive division of nuclei and cells, this hermaphrodite nature of the nucleus is maintained by the longitudinal division of the loops of each mother-nucleus, causing a uniform distribution of the male and female loops in both daughter-nuclei.

But van Beneden undoubtedly deserves great credit for having constructed the foundation upon which a scientific theory of heredity could be built. It was only necessary to replace the terms male and female pronuclei, by the terms nuclear substance of the male and female parents, in order to gain a starting-point from which further advance became possible. This step was taken by Strasburger, who at the same time brought forward an instance in which the nucleus only of the male germ-cell (to the exclusion of its cell-body) reaches the egg-cell. He succeeded in explaining the process of fertilization in Phanerogams, which had been for a long time involved in obscurity, for he proved that the nucleus of the sperm-cell (the pollen-tube) enters the embryo-sac and fuses with the nucleus of the egg-cell: at the same time he came to the conclusion that the body of the sperm-cell does not pass into the embryo-sac, so that in this case fertilization can only depend upon the fusion of nuclei[116].

Thus the nuclear substance must be the sole bearer of hereditary tendencies, and the facts ascertained by van Beneden in the case of *Ascaris* plainly show that the nuclear substance must not only contain the tendencies of growth of the parents, but also those of a very large number of ancestors. Each of the two nuclei which unite in fertilization must contain the germ-nucleoplasm of both parents, and this latter nucleoplasm once contained and still contains the germ-nucleoplasm of the grandparents as well as that of all previous generations. It is obvious that the nucleoplasm of each antecedent generation must be represented in any germ-nucleus in an amount which becomes less as the number of intervening generations becomes greater; and the proportion can be calculated after the manner in which breeders, when crossing races, determine the proportion of pure blood which is contained in any of the descendants. Thus while the germ-plasm of the father or mother constitutes half the nucleus of any fertilized ovum, that of a grandparent only forms a quarter, and that of the tenth generation backwards only 1/1024, and so on. The latter can, nevertheless, exercise influence over the development of the offspring, for the phenomena of atavism show that the germ-plasm of very remote ancestors can occasionally make itself felt, in the sudden reappearance of long-lost characters. Although we are unable to give a detailed account of the way in which atavism happens, and of the circumstances under which it takes place, we are at least able to understand how it becomes possible; for even a very minute trace of a specific germ-plasm possesses the definite tendency to build up a certain organism, and will develope this tendency as soon as its nutrition is, for some reason, favoured above that of the other kinds of germ-plasm present in the nucleus. Under these circumstances it will increase more rapidly than the other kinds, and it is readily conceivable that a preponderance in the quantity of one kind of nucleoplasm may determine its influence upon the cell-body.

Strasburger—supported by van Beneden's observations, but in opposition to the opinions of the latter—had already explained, in a manner similar to that described above, the process by which the hereditary transmission of certain characters takes place, and to this extent our opinions coincide. The nature of heredity is based upon the transmission of nuclear substance with a specific molecular constitution. This substance is the specific nucleoplasm of the germ-cell, to which I have given the name of germ-plasm.

O. Hertwig[117] has also come to the same conclusion: at an earlier date he had looked upon the coalescence of nuclei as the most essential feature in the process of fertilization. He now believes that this former opinion has been confirmed by the recent discoveries which have been shortly described above.

Although I entirely agree with Hertwig, as far as the main question is concerned, I cannot share his opinions when he identifies Nägeli's idioplasm with the nucleoplasm of the germ-cell. Nägeli's idioplasm certainly includes the germ-plasm, if I may retain this expression for the sake of brevity. Nägeli in forming his hypothesis did indeed start with the germ-cells, but his idioplasm not only represents the nucleoplasm of the germ-cells, but also that of all the other cells of the organism; all these nucleoplasms taken together constitute Nägeli's idioplasm. According to Nägeli, the idioplasm forms a network which extends through the whole body, and represents the specific molecular basis which determines its nature. Although this latter suggestion—the general part of his theory—is certainly valid, and although it is of great importance to have originated the idea of idioplasm in this general sense, in contrast to the somato-plasm ('Nährplasma'), it is nevertheless true that we are not justified in retaining the details of his theory.

In the first place the idioplasm does not form a directly continuous network throughout the entire body; and, secondly, the whole organism is not penetrated by a single substance of homogeneous constitution, but each special kind of cell must contain the specific idioplasm or nucleoplasm which determines its nature. There are therefore in each organism a multitude of different kinds of idioplasm. Thus we should be quite justified in generally speaking of Nägeli's idioplasm as nucleoplasm, and *vice versa*.

It is perfectly certain that the idioplasm cannot form a continuous network through the whole organism, if it is seated in the nucleus and not in the cell-body. Even if the bodies of cells are everywhere connected by fine processes (as has been proved in animals by Leydig and Heitzmann, and in plants by various botanists), they do not form a network of idioplasm but of somato-plasm; a substance which, according to Nägeli, stands in marked contrast to idioplasm. Strasburger has indeed already spoken of a 'cyto-idioplasm,' and it is certainly obvious that the cell-body often possesses a specific character, but we must in all cases assume that such a character is impressed upon it by the influence of the nucleus, or, in other words, that the direction in which the cell-substance is differentiated in the course of development is determined by the quality of its nuclear substance. So far, therefore, the determining nuclear substance corresponds to the idioplasm alone, while the substance of the cell-body must be identified with the somato-plasm ('Nährplasma') of Nägeli. At all events, in practice, it will be well to restrict the term idioplasm to the regulative nuclear substance alone, if we desire to retain the well-chosen terms of Nägeli's theory.

But the second part of Nägeli's theory of the idioplasm is also untenable. It is impossible that this substance can have the same constitution everywhere in the organism and during every stage of its ontogeny. If this were so, how could the idioplasm effect the great differences which obtain in the formation of the various parts of the organism? In some passages of his work Nägeli seems to express the same opinion; e. g. on page 31 he says, 'It would be practicable to regard—although only in a metaphorical sense—the idioplasms of the different cells of an individual as themselves different, inasmuch as they possess specific powers of production: we should thus include among these idioplasms all the conditions of the organism which bring about the display of specific activity on the part of cells.' It can be clearly seen from the passages immediately preceding and succeeding the above-quoted sentence, that Nägeli, in speaking of these changes in the idioplasm, does not refer to material, but only to dynamical changes. On page 53 he lays special stress upon the statement that 'the idioplasm during its growth retains its specific constitution everywhere throughout the organism,' and it is only 'within these fixed structural limits that it changes its conditions of tension and movement, and thus alters the forms of growth and activity which are possible at each time and place.' Against such an interpretation weighty objections can be raised. At present I will only mention that the meaning of the phrase 'conditions of tension and movement' ought to be made clear, and that we ought to be informed how it is that mere differences in tension can produce as many different effects as could have been produced by differences of constitution. If any one were to assert that in *Daphnidae*, or in any other forms which produce two kinds of eggs, the power of developing only after a period of rest, possessed by the winter-eggs, is based upon the fact that their idioplasm is identical with that of the summer-eggs, but is in another condition of tension, I should think such a hypothesis would be well worth consideration, for the animals which arise from the winter-eggs are identical with those produced in summer: the idioplasm which caused their formation must therefore be identical in its constitution; and can only differ in the two cases, as water differs from ice. But the case is quite otherwise in the stages of ontogeny. How many different conditions of tension ought to be possessed by one and the same idioplasm in order to correspond to the thousand different structures and differentiations of cells in one of the higher organisms? In fact it would be hardly possible to form even an approximate conception of an explanation based upon mere 'conditions of tensions and movement.' But, furthermore, difference in effect should correspond, at any rate to some extent, with difference in cause: thus the idioplasm of a muscle-cell ought to differ more from that of a nerve-cell and of a digestive-cell in the same individual, than the idioplasm of the germ-cell of one individual differs from that of other individuals of the same species; and yet, according to Nägeli, the latter small difference in the effect is supposed to be due to difference of quality in the cause—the idioplasm, while the former fundamental difference in the histological differentiation of cells is supposed to follow from mere difference 'of tension and movement.'

Nägeli's hypothesis appears to be self-contradictory; for, although its author recognizes the truth of the fundamental law of development, and explains the stages of ontogeny as an abbreviated recapitulation of phyletic stages, he nevertheless explains the latter by a different principle from that which he employs to explain the former. According to Nägeli, the stages of phylogeny are based upon true qualitative differences in the idioplasm: the germ-plasm of a worm is qualitatively different from that of *Amphioxus*, a frog, or a mammal. But if such phyletic stages occur crowded together in the ontogeny of a single species, they are said to be based upon different 'conditions of tension and movement' of one and the same idioplasm! It seems to me to be necessary to conclude that if the idioplasm, in the course of phyletic development, undergoes any alteration in specific constitution, such alterations must also take place in ontogeny; so far at least as the phyletic stages are repeated. Either the whole phyletic development is based upon different 'conditions of tension and movement,' or if this—as I believe—is impossible, the stages of ontogeny must be based upon qualitative alterations in the idioplasm.

Involuntarily the question arises—how is it that such an acute thinker fails to perceive this contradiction? But the answer is not far to seek, and Nägeli himself indicates it when he adds these words to the sentence quoted above: 'It follows therefore that if a cell is detached as a germ-cell in any stage of ontogenetic development, and from any part of the organism, such a cell will contain all the hereditary tendencies of the parent individual.' In other words, if we are restricted to different 'conditions of tension and movement' as an explanation, it seems to follow as a matter of course that the idioplasm can re-assume its original condition, and therefore that the idioplasm of any cell in the body can again become the idioplasm of the germ-cell; for this to take place it is only necessary that the greater tension should become the less, or *vice versa*. But if we admit a real change in constitution, then the backward development of the idioplasm of the cells of the body into germ-cells appears to be very far from a matter of course, and he who assumes it must bring forward weighty reasons. Nägeli does not produce such reasons, but considers the metamorphosis of the idioplasm in ontogeny as mere differences in the 'conditions of tension and movement.' This phrase covers the weak part of his theory; and I look upon it as a valuable proof that Nägeli has also felt that the phenomena of heredity can only find their explanation in the hypothesis of the continuity of the germ-plasm; for his phrase is only capable of obscuring the question as to how the idioplasm of the cells of the body can be re-transformed into the idioplasm of germ-cells.

I am of the opinion that the idioplasm cannot be re-transformed, and I have defended this opinion for some years past[118], although I have hitherto laid especial stress on the positive aspect of the question, viz. on the continuity of the germ-plasm. I have attempted to prove that the germ-cells of an organism derive their essential nature from the fact that the germ-plasm of each generation is carried over into that which succeeds it; and I have tried to show that during the development of an egg into an animal, a part of the germ-substance—although only a minute part—passes over unchanged into the organism which is undergoing development, and that this part represents the basis from which future germ-cells arise. In this way it is to a certain extent possible to conceive how it is that the complex molecular structure of the germ-plasm can be retained unchanged, even in its most minute details, through a long series of generations.

But how would this be possible if the germ-plasm were formed anew in each individual by the transformation of somatic idioplasm? And yet if we reject the 'continuity of the germ-plasm' we are compelled to adopt this latter hypothesis concerning its origin. It is the hypothesis adopted by Strasburger, and we have therefore to consider how the subject presents itself from his point of view.

I entirely agree with Strasburger when he says, 'The specific qualities of organisms are based upon nuclei'; and I further agree with him in many of his ideas as to the relation between the nucleus and cell-body: 'Molecular stimuli proceed from the nucleus into the surrounding cytoplasm; stimuli which, on the one hand, control the phenomena of assimilation in the cell, and, on the other hand, give to the growth of the cytoplasm, which depends upon nutrition, a certain character peculiar to the species.' 'The nutritive cytoplasm assimilates, while the nucleus controls the assimilation, and hence the substances assimilated possess a certain constitution and nourish in a certain manner the cyto-idioplasm and the nuclear idioplasm. In this way the cytoplasm takes part in the phenomena of construction, upon which the specific form of the organism depends. This constructive activity of the cyto-idioplasm depends upon the regulative influence of the nuclei.' The nuclei therefore 'determine the specific direction in which an organism developes.'

The opinion—derived from the recent study of the phenomena of fertilization—that the nucleus impresses its specific character upon the cell, has received conclusive and important confirmation in the experiments upon the regeneration of Infusoria, conducted simultaneously by M. Nussbaum[119] at Bonn, and by A. Gruber[120] at Freiburg. Nussbaum's statement that an artificially separated portion of a *Paramaecium*, which does not contain any nuclear substance, immediately dies, must not be accepted as of general application, for Gruber has kept similar fragments of other Infusoria alive for several days. Moreover, Gruber had previously shown that individual Protozoa occur, which live in a normal manner, and are yet without a nucleus, although this structure is present in other individuals of the same species. But the meaning of the nucleus is made clear by the fact, published by Gruber, that such artificially separated fragments of Infusoria are incapable of regeneration, while on the other hand those fragments which contain nuclei always regenerate. It is therefore only under the influence of the nucleus that the cell substance re-developes into the full type of the species. In adopting the view that the nucleus is the factor which determines the specific nature of the cell, we stand on a firm foundation upon which we can build with security.

If therefore the first segmentation nucleus contains, in its molecular structure, the whole of the inherited tendencies of development, it must follow that during segmentation and subsequent cell-division, the nucleoplasm will enter upon definite and varied changes which must cause the differences appearing in the cells which are produced; for identical cell-bodies depend, *ceteris paribus*, upon identical nucleoplasm, and conversely different cells depend upon differences in the nucleoplasm. The fact that the embryo grows more strongly in one direction than in another, that its cell-layers are of different nature and are ultimately differentiated into various organs and tissues,—forces us to accept the conclusion that the nuclear substance has also been changed in nature, and that such changes take place during ontogenetic development in a regular and definite manner. This view is also held by Strasburger, and it must be the opinion of all who seek to derive the development of inherited tendencies from the molecular structure of the germ-plasm, instead of from preformed gemmules.

We are thus led to the important question as to the forces by which the determining substance or nucleoplasm is changed, and as to the manner in which it changes during the course of ontogeny, and on the answer to this question our further conclusions must depend. The simplest hypothesis would be to suppose that, at each division of the nucleus, its specific substance divides into two halves of unequal quality, so that the cell-bodies would also be transformed; for we have seen that the character of a cell is determined by that of its nucleus. Thus in any Metazoon the first two segmentation spheres would be transformed in such a manner that one only contained the hereditary tendencies of the endoderm and the other those of the ectoderm, and therefore, at a later stage, the cells of the endoderm would arise from the one and those of the ectoderm from the other; and this is actually known to occur. In the course of further division the nucleoplasm of the first ectoderm cell would again divide unequally, e.g. into the nucleoplasm containing the hereditary tendencies of the nervous system, and into that containing the tendencies of the external skin. But even then, the end of the unequal division of nuclei would not have been nearly reached; for, in the formation of the nervous system, the nuclear substance which contains the hereditary tendencies of the sense-organs, would, in the course of further cell-division, be separated from that which contains the tendencies of the central organs, and the same process would continue in the formation of all single organs, and in the final development of the most minute histological elements. This process would take place in a definitely ordered course, exactly as it has taken place throughout a very long series of ancestors; and the determining and directing factor is simply and solely the nuclear substance, the nucleoplasm, which possesses such a molecular structure in the germ-cell that all such succeeding stages of its molecular structure in future nuclei must necessarily arise from it, as soon as the requisite external conditions are present. This is almost the same conception of ontogenetic development as that which has been held by embryologists who have not accepted the doctrine of evolution: for we have only to transfer the primary cause of development, from an unknown source within the organism, into the nuclear substance, in order to make the views identical.

It appears at first sight that the knowledge which has been gained by studying the indirect division of nuclei is opposed to such a view, for we know that each mother-loop of the so-called nuclear plate divides longitudinally into two exactly equal halves, which can be stained and thus rendered visible.

In this way each resulting daughter-nucleus receives an equal supply of halves, and it therefore appears that the two nuclei must be completely identical. This at least is Strasburger's conclusion, and he regards such identity as a fundamental fact, which cannot be shaken, and with which all attempts at further explanation must be brought into accord.

How then can the gradual transformation of the nuclear substance be brought about? For such a transformation must necessarily take place if the nuclear substance is really the determining factor in development. Strasburger attempts to support his hypothesis by assuming that the inequality of the daughter-nuclei arises from unequal nutrition; and he therefore considers that the inequality is brought about after the division of the nucleus and of the cell. Strasburger has shown, in a manner which is above all criticism, that the nucleus derives its nutrition from the cell-body, but then the cell-bodies of the two *ex hypothesi* identical daughter-nuclei must be different from the first, if they are to influence their nuclei in different ways. But if the nucleus determines the nature of the cell, it follows that two identical daughter-nuclei which have arisen by division within one mother-cell cannot come to possess unequal cell-bodies. As a matter of fact, however, the cell-bodies of two daughter-cells often differ in size, in appearance, and in their subsequent history, and these facts are sufficient to prove that in such cases the division of the nucleus must have been unequal. It appears to me to be a necessary conclusion that, in such an instance, the mother-nucleus must have been capable of splitting into nuclear substances of differing quality. I think that, in his argument, Strasburger has over-estimated the support afforded by exact observations upon indirect nuclear division. Certainly the fact, discovered by Flemming, and more exactly studied by Balbiani and Pfitzner, that, in nuclear division, the loops split longitudinally, is of great and even of fundamental importance. Furthermore, the observations, conducted last year by van Beneden, on the process of fertilization in *Ascaris*, have given to Flemming's discovery a clearer and more definite meaning than could have been at first ascribed to it. The discovery proves, in the first place, that the nucleus always divides into two parts of equal quantity, and further that in every nuclear division, each daughter-nucleus receives the same amount of nuclear substance from the father as from the mother; but, as it seems to me, it is very far from proving that the quality of the parent nucleoplasms must always be equal in the daughter-nuclei. It is true that the fact seems to prove this; and if we remember the description of the most favourable instance which has been hitherto discovered, viz. the process of fertilization in the egg of *Ascaris*, as represented by van Beneden, the two longitudinal halves of each loop certainly impress the reader as being absolutely identical (compare, for instance, loc. cit. Plate

XIX, figs. 1, 4, 5). But we must not forget that we do not see the molecular structure of the nucleoplasm, but something which we can only look upon (when we remember how complex this molecular structure must be) as a very rough expression of its quantity. Our most powerful and best lenses just enable us to make out the form of separate stainable granules present in a loop which is about to divide: they appear as spheres and immediately after division as hemispheres. But according to Strasburger, these granules, the so-called microsomata, only serve for the nutrition of the nuclear substance proper, which lies between them unstainable, and therefore not distinctly visible. But even if these granules represent the true idioplasm, their division into two exactly equal parts would give us no proof of equality or inequality in their constitution: it would only give us an idea of their quantitative relations. We can only obtain proofs as to the quality of the molecular structure of the two halves by their effect on the bodies of the daughter-cells, and we know that these latter are frequently different in size and quality.

This point is so important that I must illustrate it by a few more examples. The so-called polar bodies (to be treated more in detail below) which are expelled during maturation from the eggs of so many animals, are true cells, as was first proved by Bütschli in Nematodes: their formation is due to a process of undoubted cell-division usually accompanied by a typical form of indirect nuclear division[121]. If any one is still in doubt upon this point, after the observations of Fol and Hertwig, he might easily be convinced of its truth by a glance at the figures (unfortunately too little known) which Trinchese[122] has published, illustrating this process in the eggs of certain gastropods. The eggs of *Amphorina coerulea* are in every way suitable for observation, being entirely translucent, and having large distinct nuclei which differ from the green cytoplasm in colour. In these eggs two polar bodies are formed one after the other: and each of them immediately re-divides: hence it follows that four polar bodies are placed at the pole of the egg. But how is it that these four cells perish, while the nucleus, remaining in the yolk and conjugating with the sperm-nucleus, makes use of the whole body of the egg and developes into the embryo? Obviously because the nature of the polar body is different from that of the egg-cell. But since the nature of the cell is determined by the quality of the nucleus, this quality must differ from the very moment of nuclear division. This is proved by the fact that the supernumerary spermatozoa which sometimes enter the egg do not conjugate with the polar bodies. According to Strasburger's theory, the objection might be urged that the different quality of the nuclei is here caused by the very different quantity of cytoplasm by which they are surrounded and nourished; but on the one hand the smallness of the cell-bodies which surround most polar globules must have some explanation, and this can only be found in the nature of the nucleus; and on the other hand the quantity of the cell-body which surrounds the polar globules of *Amphorina* is, as a matter of fact, somewhat larger than the sphere of green cytoplasm which surrounds the nucleus of the egg! The difference between the polar bodies and the egg-cell can thus only be explained on the supposition that, in the division of the nuclear spindle, two qualitatively different daughter-nuclei are produced.

There does not seem to be any objection to the view that the microsomata of the nuclear loops—assuming that these bodies represent the idioplasm—are capable of dividing into halves, equal in form and appearance, but unequal in quality. We know that this very process takes place in many egg-cells; thus in the egg of the earth-worm the first two segmentation spheres are equal in size and appearance, and yet the one forms the endoderm and the other the ectoderm of the embryo.

I therefore believe that we must accept the hypothesis that, in indirect nuclear division, the formation of unequal halves may take place quite as readily as the formation of equal halves, and that the equality or inequality of the subsequently produced daughter-cells must depend upon that of the nuclei. Thus during ontogeny a gradual transformation of the nuclear substance takes place, necessarily imposed upon it, according to certain laws, by its own nature, and such transformation is accompanied by a gradual change in the character of the cell-bodies.

It is true that we cannot gain any detailed knowledge of the nature of these changes in the nuclear substance, but we can very well arrive at certain general conclusions about them. If we may suppose, with Nägeli, that the molecular structure of the germ-idioplasm, or according to our terminology the germ-plasm, becomes more complicated according to the greater complexity of the organism developed from it, then the following conclusions will also be accepted,—that the molecular structure of the nuclear substance is simpler as the differences between the structures arising from it become less; that therefore the nuclear substance of the segmentation-cell of the earth-worm, which potentially contains the whole of the ectoderm, possesses a more complicated molecular structure than that of a single epidermic cell or nerve-cell. These conclusions will be admitted when it is remembered that every detail in the whole organism must be represented in the germ-plasm by its own special and peculiar arrangement of the groups of molecules (the micellae of Nägeli), and that the germ-plasm not only contains the whole of the quantitative and qualitative characters of the species, but also all individual variations as far as these are hereditary: for example the small depression in the centre of the chin noticed in some families. The physical causes of all apparently unimportant hereditary habits or structures, of hereditary talents, and other mental peculiarities, must all be contained in the minute quantity of germ-plasm which is possessed by the nucleus of a germ-cell;—not indeed as the preformed germs of structure (the gemmules of pangenesis), but as variations in its molecular constitution; if this be impossible, such characters could not be inherited. Nägeli has shown in his work, which is so rich in suggestive ideas, that even in so minute a space as the thousandth of a cubic millimetre, such an enormous number (400,000,000) of micellae may be present, that the most diverse and complicated arrangements become possible. It therefore follows that the molecular structure of the germ-plasm in the germ-cells of an individual must be distinguished from that of another individual by certain differences, although these may be but small; and it also follows that the germ-plasm of any species must differ from that of all other species.

These considerations lead us to conclude that the molecular structure of the germ-plasm in all higher animals must be excessively complex, and, at the same time, that this complexity must gradually diminish during ontogeny as the structures still to be formed from any cell, and therefore represented in the molecular constitution of its nucleoplasm, become less in number. I do not mean to imply that the nucleoplasm contains preformed structures which are gradually reduced in number as they are given off in various directions during the building-up of organs: I mean that the complexity of the molecular structure decreases as the potentiality for further development also decreases, such potentiality being represented in the molecular structure of the nucleus. The nucleoplasm, which in the grouping of its particles contains potentially a hundred different modifications of this substance, must possess far more numerous kinds and far more complex arrangements of such particles than the nucleoplasm which only contains a single modification, capable of determining the character of a single kind of cell. The development of the nucleoplasm during ontogeny may be to some extent compared to an army composed of corps, which are made up of divisions, and these of brigades, and so on. The whole army may be taken to represent the nucleoplasm of the germ-cell: the earliest cell-division (as into the first cells of the ectoderm and endoderm) may be represented by the separation of the two corps, similarly formed but with different duties: and the following cell-divisions by the successive detachment of divisions, brigades, regiments, battalions, companies, etc.; and as the groups become simpler so does their sphere of action become limited. It must be admitted that this metaphor is imperfect in two respects, first, because the quantity of the nucleoplasm is not diminished, but only its complexity, and secondly, because the strength of an army chiefly depends upon its numbers, not on the complexity of its constitution. And we must also guard against the supposition that unequal nuclear division simply means a separation of part of the molecular structure, like the detachment of a regiment from a brigade. On the contrary, the molecular constitution of the mother-nucleus is certainly changed during division in such a way that one or both halves receive a new structure which did not exist before their formation.

My opinion as to the behaviour of the idioplasm during ontogeny, not only differs from that of Nägeli, in that the latter maintains that the idioplasm only undergoes changes in its 'conditions of tension and movement,' but also because he imagines this substance to be composed of the preformed germs of structures ('Anlagen'). Nägeli's views are obviously bound up with his theory of a continuous network of idioplasm throughout the whole body; perhaps he would have adopted other conclusions had he been aware of the fact that the idioplasm must only be sought for in the nuclei. Nägeli's views as to ontogeny can be best seen in the following passages: 'As soon as ontogenetic development begins, the groups of micellae in the idioplasm which effect the first stage of development, enter upon active growth: such activity causes a passive growth of the other groups, and an increase in the whole idioplasm, perhaps to many times its former bulk. But the intensities of growth in the two series of groups are unequal, and consequently an increasing tension is produced which sooner or later, according to the number, arrangement, and energy of the active groups, necessarily renders the continuation of the process impossible. In consequence of such disturbance to the equilibrium, active growth now takes place in the next group, leading to fresh irritation, and this group then reacts more strongly than all the others upon the tension which first stimulated its activity. These changes are repeated until all the groups are gone through, and the ontogenetic development finally reaches the stage at which propagation takes place, and thus the original stage of the germ is reached.'

Hence, according to Nägeli, the different stages of ontogeny arise out of the activities of different parts of the idioplasm: certain groups of micellae in the idioplasm represent the germs ('Anlagen') of certain structures in the organism: when any such germ reacts under stimulation it produces the corresponding structure. It seems to me that this hypothesis bears some resemblance to Darwin's theory of pangenesis. I think that Nägeli's preformed germs of structures ('Anlagen') and his groups of such germs are highly elaborated equivalents of the gemmules of pangenesis, which, according to Darwin, manifest activity when their turn comes, or, according to Nägeli, when they react under stimulation. When a group of such germs, by their active growth or by their 'irritation,' have caused a similar active growth or a similar irritation in the next group, the former may come to rest, or may remain in a state of activity together with its successor, for a longer or shorter period. Its activity may even last for an unlimited time, as is the case in the formation of leafy shoots in many plants.

Here, again, we recognize the fact that Nägeli's whole hypothesis is intimately connected with the supposition that the entire mass of idioplasm is continuous throughout the organism. Sometimes one part of the idioplasm and sometimes another part is irritated, and then produces the corresponding organ. But if, on the other hand, the idioplasm does not represent a directly continuous mass, but is composed of thousands of single nucleoplasms which only act together through the medium of their cell-bodies, then we must substitute the conception of 'ontogenetic stages of development of the idioplasm' for the conception of germs of structure ('Anlagen'). The different varieties of nucleoplasm which arise during ontogeny represent, as it were, the germs of Nägeli ('Anlagen'), because, by means of their molecular structure, they create a specific constitution in the cell-bodies over which they have control, and also because they determine the succession of future nuclei and cells.

It is in this sense, and no other, that I can speak of the presence of preformed germs ('Anlagen') in the idioplasm. We may suppose that the idioplasm of the first segmentation nucleus is but slightly different from that of the second ontogenetic stage, viz. that of the two following segmentation nuclei. Perhaps only a few groups of micellae have been displaced or somewhat differently arranged. But nevertheless such groups of micellae were not the germs ('Anlagen') of a second stage which pre-existed in the first stage, for the two are distinguished by the possession of a different molecular structure. This structure in the second stage, under normal conditions of development, again brings about the change by which the different molecular structure of the third stage is produced, and so on.

It may be argued that von Baer's well-known and fundamental law of development is opposed to the hypothesis that the idioplasm of successive ontogenetic stages must gradually assume a simpler molecular structure. The organization of the species has, on the whole, increased immensely in complexity during the course of phylogeny: and if the phyletic stages are repeated in the ontogeny, it seems to follow that the structure of the idioplasm must become more complex in the course of ontogeny instead of becoming simpler. But the complexity of the whole organism is not represented in the molecular structure of the idioplasm of any single nucleus, but by that of all the nuclei present at any one time. It is true that the germ-cell, or rather the idioplasm of the germ-nucleus, must gain greater complexity as the organism which arises from it becomes more complex; but the individual nucleoplasms of each ontogenetic stage may become simpler, while the whole mass of idioplasms in the organism (which, taken together, represent the stage in question) does not by any means lose in complexity.

If we must therefore assume that the molecular structure of the nucleoplasm becomes simpler in the course of ontogeny, as the number of structures which it potentially contains become smaller, it follows that the nucleoplasm in the cells of fully differentiated tissues—such as muscle, nerve, sense-organs, or glands—must possess relatively the most simple molecular structure; for it cannot originate any fresh modification of nucleoplasm, but can only continue to produce cells of the same structure, although it does not always retain this power.

We are thus brought back to the fundamental question as to how the germ-cells arise in the organism. Is it possible that the nucleoplasm of the germ-cell, with its immensely complex molecular structure, potentially containing all the specific peculiarities of an individual, can arise from the nucleoplasm of any of the body-cells,—a substance which, as we have just seen, has lost the power of originating any new kind of cell, because of the continual simplification of its structure during development? It seems to me that it would be impossible for the simple nucleoplasm of the somatic cells to thus suddenly acquire the power of originating the most complex nucleoplasm from which alone the entire organism can be built up: I cannot see any evidence for the existence of a force which could effect such a transformation.

This difficulty has already been appreciated by other writers. Nussbaum's[123] theoretical views, which I have already mentioned, also depend upon the hypothesis that cells which have once become differentiated for the performance of special functions cannot be re-transformed into sexual cells: he also concludes that the latter are separated from all other cells at a very early period of embryonic development, before any histological differentiation has taken place. Valaoritis[124] has also recognised that the transformation of histologically differentiated cells into sexual cells is impossible. He was led to believe that the sexual cells of Vertebrata arise from the white blood corpuscles, for he looked upon these latter as differentiated to the smallest extent possible. Neither of these views can be maintained. The former, because the sexual cells of all plants and most animals are not, as a matter of fact, separated from the somatic cells at the beginning of ontogeny; the latter, because it is contradicted by the fact that the sexual cells of vertebrates do not arise from blood corpuscles, but from the germinal epithelium. But even if this fact had not been ascertained we should be compelled to reject Valaoritis' hypothesis on theoretical grounds, for it is an error to assume that white blood corpuscles are undifferentiated, and that their nucleoplasm is similar to the germ-plasm. There is no nucleoplasm like that of the germ-cell in any of the somatic cells, and no one of these latter can be said to be undifferentiated. All somatic cells possess a certain degree of differentiation, which may be rigidly limited to one single direction, or may take place in one of many directions. All these cells are widely different from the egg-cell from which they originated: they are all separated from it by many generations of cells, and this fact implies that their idioplasms possess a widely different structure from the idioplasm, or germ-plasm, of the egg-cell. Even the nuclei of the two first segmentation spheres cannot possess the same idioplasm as that of the first segmentation nucleus, and it is, of course, far less possible for such an idioplasm to be present in the nucleus of any of the later cells of the embryo. The structure of the idioplasm must necessarily become more and more different from that of the first segmentation nucleus, as the development of the embryo proceeds. The idioplasm of the first segmentation nucleus, and of this nucleus alone, is germ-plasm, and possesses a structure such that an entire organism can be produced from it. Many writers appear to consider it a matter of course that any embryonic cell can reproduce the entire organism, if placed under suitable conditions. But, when we carefully look into the subject, we see that such powers are not even possessed by those cells of the embryo which are nearest to the egg-cell—viz. the first two segmentation spheres. We have only to remember the numerous cases in which one of them forms the ectoderm of the animal while the other produces the endoderm, in order to admit the validity of this objection.

But if the first segmentation spheres are not able to develope into a complete organism, how can this be the case with one of the later embryonic cells, or one of the cells of the fully developed animal body? It is true that we speak of certain cells as being 'of embryonic character,' and only recently Kölliker[125] has given a list of such cells, among which he includes osteoblasts, cartilage cells, lymph corpuscles, and connective tissue corpuscles: but even if these cells really deserve such a designation, no explanation of the formation of germ-cells is afforded, for the idioplasm of the latter must be widely different from that of the former.

It is an error to suppose that we gain any further insight into the formation of germ-cells by referring to these cells of so-called 'embryonic character,' which are contained in the body of the mature organism. It is of course well known that many cells are characterized by very sharply defined histological differentiation, while others are but slightly differentiated; but it is as difficult to imagine that germ-cells can arise from the latter as from the former. Both classes of cells contain idioplasm with a structure different from that which is contained in the germ-cell, and we have no right to assume that any of them can form germ-cells until it is proved that somatic idioplasm is capable of undergoing re-transformation into germ-idioplasm.

The same argument applies to the cells of the embryo itself, and it therefore follows that those instances of early separation of sexual from somatic cells, upon which I have often insisted as indicating the continuity of the germ-plasm, do not now appear to be of such conclusive importance as at the time when we were not sure about the localization of the idioplasm in the nuclei. In the great majority of cases the germ-cells are not separated at the beginning of embryonic development, but only in some one of the later stages. A single exception is found in the pole-cells ('Polzellen') of Diptera, as was shown many years ago by Robin[126] and myself[127]. These are the first cells formed in the egg, and according to the later observations of Metschnikoff[128] and Balbiani[129], they become the sexual glands of the embryo. Here therefore the germ-plasm maintains a true unbroken continuity. The nucleus of the egg-cell directly gives rise to the nuclei of the pole-cells, and there is every reason to believe that the latter receive unchanged a portion of the idioplasm of the former, and with it the tendencies of heredity. But in all other cases the germ-cells arise by division from some of the later embryonic cells, and as these belong to a more advanced ontogenetic stage in the development of the idioplasm, we can only conclude that continuity is maintained, by assuming (as I do) that a small part of the germ-plasm persists unchanged during the division of the segmentation nucleus and remains mixed with the idioplasm of a certain series of cells, and that the formation of true germ-cells is brought about at a certain point in the series by the appearance of cells in which the germ-plasm becomes predominant. But if we accept this hypothesis it does not make any difference, theoretically, whether the germ-plasm becomes predominant in the third, tenth, hundredth, or millionth generation of cells. It therefore follows that cases of early separation of the germ-cells afford no proof of a direct persistence of the parent germ-cells in those of the offspring; for a cell the offspring of which become partly somatic and partly germ-cells cannot itself have the characters of a germ-cell; but it may nevertheless contain germ-idioplasm, and may thus transfer the substance which forms the basis of heredity from the germ of the parent to that of the offspring.

If we are unwilling to accept this hypothesis, nothing remains but to credit the idioplasm of each successive ontogenetic stage with a capability of re-transformation into the first stage. Strasburger accepts this view; and he believes that the idioplasm of the nuclei changes during the course of ontogeny, but returns to the condition of the first stage of the germ, at its close. But the rule of probability is against such a suggestion. Suppose, for instance, that the idioplasm of the germ-cell is characterized by ten different qualities, each of which may be arranged relatively to the others in two different ways, then the probability in favour of any given combination would be represented by the fraction $(1/2)^{10} = 1/1024$: that is to say, the re-transformation of somatic idioplasm into germ-plasm will occur once in 1024 times, and it is therefore impossible for such re-transformation to become the rule. It is also obvious that the complex structure of the germ-plasm which potentially contains, with the likeness of a faithful portrait, the whole individuality of the parent, cannot be represented by only ten characters, but that there must be an immensely greater number; it is also obvious that the possibilities of the arrangement of single characters must be assumed to be much larger than two; so that we get the formula $(1/p)$, where p represents the possibilities, and n the characters. Thus if n and p are but slightly larger than we assumed above, the probabilities become so slight as to altogether exclude the hypothesis of a re-transformation of somatic idioplasm into germ-plasm.

It may be objected that such re-transformation is much more probable in the case of those germ-cells which separate early from the somatic cells. Nothing can in fact be urged against the possibility that the idioplasm of (e. g.) the third generation of cells may pass back into the condition of the idioplasm of the germ-cell; although of course the mere possibility does not prove the fact. But there are not many cases in which the sexual cells are separated so early as the third generation: and it is very rare for them to separate at any time during the true segmentation of the egg. In *Daphnidae* (*Moina*) separation occurs in the fifth stage of segmentation[130], and although this is unusually early it does not happen until the idioplasm has changed its molecular structure six times. In *Sagitta*[131] the separation does not take place until the archenteron is being formed, and this is after several hundred embryonic cells have been produced, and thus after the germ-plasm has changed its molecular structure ten or more times. But in most cases, separation takes place at a much later stage; thus in Hydroids it does not happen until after hundreds or thousands of cell-generations have been passed through; and the same fact holds in the higher plants, where the production of germ-cells frequently occurs at the end of ontogeny. In such cases the probability of a re-transformation of somatic idioplasm into germ-plasm becomes infinitely small.

It is true that these considerations only refer to a rapid and sudden re-transformation of the idioplasm. If it could be proved that development is not merely in appearance but in reality a cyclical process, then nothing could be urged against the occurrence of re-transformation. It has been recently maintained by Minot[132] that all development is cyclical, but this is obviously incorrect, for Nägeli has already shown that direct non-cyclical courses of development exist, or at all events courses in which the earliest condition is not repeated at the close of development. The phyletic development of the whole organic world clearly illustrates a development of the latter kind; for although we may assume that organic development is not nearly concluded, it is nevertheless safe to predict that it will never revert to its original starting-point, by backward development over the same course as that which it has already traversed. No one can believe that existing Phanerogams will ever, in the future history of the world, retrace all the stages of phyletic development in precise inverse order, and thus return to the form of unicellular Algae or Monera; or that existing placental mammals will develope into Marsupialia, Monotremata, mammal-like reptiles, and the lower vertebrate forms, into worms and finally into Monera. But how can a course of development, which seems to be impossible in phylogeny, occur as the regular method of ontogeny? And quite apart from the question of possibility, we have to ask for proofs of the actual occurrence of cyclical development. Such a proof would be afforded if it could be shown that the nucleoplasm of those somatic cells which (e.g. in Hydroids) are transformed into germ-cells passes backwards through many stages of development into the nucleoplasm of the germ-cell. It is true that we can only recognise differences in the structure of the idioplasm by its effects upon the cell-body, but no effects are produced which indicate that such backward development takes place. Since the course of onward development is compelled to pass through the numerous stages which are implied in segmentation and the subsequent building-up of the embryo, etc., it is quite impossible to assume that backward development would take place suddenly. It would be at least necessary to suppose that the cells of embryonic character, which are said to be transformed into primitive germ-cells, must pass back through at any rate the main phases of their ontogeny. A sudden transformation of the nucleoplasm of a somatic cell into that of a germ-cell would be almost as incredible as the transformation of a mammal into an amoeba; and yet we are compelled to admit that the transformation must be sudden, for no trace of such retrogressive stages of development can be seen. If the appearance of the whole cell gives us any knowledge as to the structure of its nuclear idioplasm, we may be sure that the development of a primitive germ-cell proceeds without a break, from the moment of its first recognizable formation, to the ultimate production of distinct male or female sexual cells.

I am well aware that Strasburger has stated that, in the ultimate maturation of the sexual cells, the substance of the nuclei returns to a condition similar to that which existed at the beginning of ontogenetic development; still such a statement is no proof, but only an assumption made to support a theory. I am also aware that Nussbaum and others believe that, in the formation of spermatozoa in higher animals, a backward development sets in at a certain stage; but even if this interpretation be correct, such backward development would only lead as far as the primitive germ-cell, and would afford no explanation of the further transformation of the idioplasm of this cell into germ-plasm. But this latter transformation is just the point which most needs proof upon any theory except the one which assumes that the primitive germ-cell still contains unchanged germ-plasm. Every attempt to render probable such a re-transformation of somatic nucleoplasm into germ-plasm breaks down before the facts known of the Hydroids, in which only certain cells in the body, out of the numerous so-called embryonic cells, are capable of becoming primitive germ-cells, while the rest do not possess this power.

I must therefore consider as erroneous the hypothesis which assumes that the somatic nucleoplasm may be transformed into germ-plasm. Such a view may be called 'the hypothesis of the cyclical development of the germ-plasm.'

Nägeli has tried to support such an hypothesis on phyletic grounds. He believes that phyletic development follows from an extremely slow but steady change in the idioplasm, in the direction of greater complexity, and that such changes only become visible periodically. He believes that the passage from one phyletic stage to another is chiefly due to the fact that 'in any ontogeny, the very last structural change upon which the separation of germs depends, takes place in a higher stage, one or more cell-generations later' than it occurred in a lower stage. 'The last structural change itself remains the same, while the series of structural changes immediately preceding it is increased.' I believe that Nägeli, being a botanist, has been too greatly influenced by the phenomena of plant-life. It is certainly true that in plants, and especially in the higher forms, the germ-cells only make their appearance, as it were, at the end of ontogeny; but facts such as these do not hold in the animal kingdom: at any rate they are not true in the great majority of cases. In animals, as I have already mentioned several times, the germ-cells are separated from the somatic cells during embryonic development, sometimes even at its very commencement; and it is obvious that this latter is the original, phyletically oldest, mode of formation. The facts at our disposal indicate that the germ-cells only appear, for the first time, after embryological development, in those cases where the formation of asexually produced colonies takes place, either with or without alternation of generations; or in cases where alternation of generations occurs without the formation of such colonies. In a colony of polypes, the germ-cells are produced by the later generations, and not by the founder of the colony which was developed from an egg. This is also true of the colonies of Siphonophora, and the germ-cells appear to arise very late in certain instances of protracted metamorphosis (Echinodermata), but on the other hand, they arise during the embryonic development of other forms (Insecta) which also undergo metamorphosis. It is obvious that the phyletic development of colonies or stocks must have succeeded that of single individuals, and that the formation of germ-cells in the latter must therefore represent the original method. Thus the germ-cells originally arose at the beginning of ontogeny and not at its close, when the somatic cells are formed.

This statement is especially supported by the history of certain lower plants, or at any rate chlorophyll-containing organisms, and I think that these forms supply an admirable illustration of my theory as to the phyletic origin of germ-cells, as explained in my earlier papers upon the same subject.

The phyletic origin of germ-cells obviously coincides with the differentiation of the first multicellular organisms by division of labour[133]. If we desire to investigate the relation between germ-cells and somatic cells, we must not only consider the highly developed and strongly differentiated multicellular organisms, but we must also turn our attention to those simpler forms in which phyletic transitions are represented. In addition to solitary unicellular organisms, we know of others living in colonies of which the constituent units or cells (each of them equivalent to a unicellular organism) are morphologically and physiologically identical. Each unit feeds, moves, and under certain circumstances is capable of reproducing itself, and of thus forming a new colony by repeated division. The genus *Pandorina* (Fig. I), belonging to the natural order *Volvocineae*, represents such 'homoplastid' (Götte) organisms. It forms a spherical colony composed of ciliated cells, all of which are exactly alike: they are embedded in a colourless gelatinous mass. Each cell contains chlorophyll, and possesses a red eye-spot, and a pulsating vacuole. These colonies are propagated by the sexual and asexual (Fig. II) methods alternately, although in the former case the conjugating swarm-cells cannot be distinguished with certainty as male or female. In both kinds of reproduction, each cell in the colony acts as a reproductive cell; in fact, it behaves exactly like a unicellular organism.

I. *Pandorina morum* (after Pringsheim), a swarming colony.

II. A colony divided into sixteen daughter colonies: all the cells alike.

III. A young individual of *Volvox minor* (after Stein), still enclosed in the wall of the cell from which it has been parthenogenetically produced. The constituent cells are divided into somatic (*sz*), germ-cells (*kz*).

It is very interesting to find in another genus belonging to the same natural order, that the transition from the homoplastid to the heteroplastid condition, and the separation into somatic and reproductive cells, have taken place. In *Volvox* (Fig. III) the spherical colony consists of two kinds of cells, viz. of very numerous small ciliated cells, and of a much smaller number of large germ-cells without cilia. The latter alone possess the power of producing a new colony, and this takes place by the asexual and sexual methods alternately: in the latter a typical fertilization of large egg-cells by small spermatozoa occurs. The sexual differentiation of the germ-cells is not material to the question we are now considering; the important point is to ascertain whether here, at the very origin of heteroplastid organisms, the germ-cells, sexually differentiated or not, arise from the somatic cells *at the end of ontogeny*, or whether the substance of the parent germ-cell, during embryonic development, is *from the first* separated into somatic and germ-cells. The former interpretation would support Nägeli's view, the latter would support my own. But Kirchner[134] distinctly states that the germ-cells of *Volvox* are differentiated during embryonic development, that is, before the escape of the young heteroplastid organism from the egg-capsule. We cannot therefore imagine that the phyletic development of the first heteroplastid organism took place in a manner different from that which I have previously advocated on theoretical grounds, before this striking instance occurred to me. The germ-plasm (nucleoplasm) of some homoplastid organism (similar to *Pandorina*) must have become modified in molecular structure during the course of phylogeny, so that the colony of cells produced by its division was no longer made up of identical units, but of two different kinds. After this separation, the germ-cells alone retained the power of reproduction possessed by all the parent cells, while the rest only retained the power of producing similar cells by division. Thus *Volvox* seems to afford distinct evidence that in the phyletic origin of the heteroplastid groups, somatic cells were not, as Nägeli supposes, intercalated between the mother germ-cell and the daughter germ-cells in each ontogeny, but that the somatic cells arose directly from the former, with which they were previously identical, as they are even now in the case of *Pandorina*. Thus the continuity of the germ-plasm is established at least for the beginning of the phyletic series of development.

The fact, already often mentioned, that in most higher organisms the separation of germ-cells takes place later, and often very late, at the end of the whole ontogeny, proves that the time at which this separation of the two kinds of cells took place, must have been gradually changed. In this respect the well-established instances of early separation are of great value, because they serve to connect the extreme cases. It is quite impossible to maintain that the germ-cells of Hydroids or of the higher plants, exist from the time of embryonic development, as indifferent cells, which cannot be distinguished from others, and which are only differentiated at a later period. Such a view is contradicted by the simplest mathematical consideration; for it is obvious that none of the relatively few cells of the embryo can be excluded from the enormous increase by division, which must take place in order to produce the large number of daughter individuals which form a colony of polypes. It is therefore clear that all the cells of the embryo must for a long time act as somatic cells, and none of them can be reserved as germ-cells and nothing else: this conclusion is moreover confirmed by direct observation. The sexual bud of a *Coryne* arises at a part of the Polype which does not in any way differ from surrounding areas, the body wall being uniformly made up of two single layers of cells, the one forming the ectoderm and the other the endoderm. Rapid growth then takes place at a single spot, and some of the young cells thus produced are transformed into germ-cells, which did not previously exist as separate cells.

Strictly speaking I have therefore fallen into an inaccuracy in maintaining (in former works) that the germ-cells are themselves immortal; they only contain the undying part of the organism—the germ-plasm; and although this substance is, as far as we know, invariably surrounded by a cell-body, it does not always control the latter, and thus confer upon it the character of a germ-cell. But this admission does not materially change our view of the whole subject. We may still contrast the germ-cells, as the undying part of the Metazoan body, with the perishable somatic cells. If the nature and the character of a cell is determined by the substance of the nucleus and not by the cell-body, then the immortality of the germ-cells is preserved, although only the nuclear substance passes uninterruptedly from one generation to another.

G. Jäger[135] was the first to state that the body in the higher organisms is made up of two kinds of cells, viz., ontogenetic and phyletic cells, and that the latter, the reproductive cells, are not a product of the former (the body-cells), but that they arise directly from the parent germ-cell. He assumed that the formation of germ-cells takes place at the earliest stage of embryonic life, and he thus believed the connexion between the germ-plasm of the parent and of the offspring had received a satisfactory explanation. As I have previously mentioned in the introduction, Nussbaum also brought forward this hypothesis at a later period, and also based it upon a continuity of the germ-cells. He assumed that the fertilized egg is divided into the cells of the individual and into the cells which effect the preservation of the species, and he supported this view by referring to the few known cases of early separation of the sexual cells. He even maintained this hypothesis when I had proved in my investigations on Hydromedusae that the sexual cells are not always separated from the somatic cells during embryonic development, but often at a far later period. Not only is the hypothesis of a direct connexion between the germ-cells of the offspring and parent broken down by the facts known in the Hydroids, and in the Phanerogams[136] which resemble them in this respect, but even the instances of early separated germ-cells quoted by Jäger and Nussbaum do not as a matter of fact support their hypothesis. Among existing organisms it is extremely rare for the germ-cells to arise directly from the parent egg-cell (as in Diptera). If, however, the germ-cells are separated only a few cell-generations later, the postulated continuity breaks down; for an embryonic cell, of which the offspring are partly germ-cells and partly somatic cells, cannot itself possess the nature of a germ-cell, and its idioplasm cannot be identical with that of the parent germ-cell. In order to prove this, it is only necessary to refer to the arguments as to the ontogenetic stages of the idioplasm. In the above-mentioned instances, the continuity from the germ-substance of the parent to that of the offspring can only be explained by the supposition that the somatic nucleoplasm still contains some unchanged germ-plasm. I believe that the fundamental idea of Jäger and Nussbaum is quite correct: it is the same idea which has led me to the hypothesis of the continuity of the germ-plasm, viz., the conviction that heredity can only be understood by means of such an hypothesis. But both these writers have worked out the idea in the form of an hypothesis which does not correspond with the facts. That this is the case is also shown by the following words of Nussbaum—'the cell-material of the individual (somatic cells) can never produce a single sexual cell.' Such production undoubtedly takes place, not only in Hydroids and Phanerogams, but in many other instances. The germ-cells cannot indeed be produced by any indifferent cell of embryonic character, but by certain cells, and under circumstances which allow us to positively conclude that they have been predestined for this purpose from the beginning. In other words, the cells in question contain germ-plasm, and this alone enables them to become germ-cells.

As a result of my investigations on Hydroids[137], I concluded that the germ-plasm is present in a very finely divided and therefore invisible state in certain somatic cells, from the very beginning of embryonic development, and that it is then transmitted through innumerable cell-generations, to those remote individuals of the colony in which sexual products are formed. This conclusion is based upon the fact that germ-cells only occur in certain localized areas ('Keimstätten') in which neither germ-cells nor primitive germ-cells (the cells which are transformed into germ-cells at a later period) were previously present. The primitive germ-cells are also only formed in localized areas, arising from somatic cells of the ectoderm. The place at which germ-cells arise is the same in all individuals of the same species; but differs in different species. It can be shown that such differences correspond to different phyletic stages of a process of displacement, which tends to remove the localized area from its original position (the manubrium of the Medusa) in a centripetal direction. For the purposes of the present enquiry it is unnecessary to discuss the reasons for this change of position. The phyletic displacements of the localized areas are brought about during ontogeny by an actual migration of primitive germ-cells from the place where they arose to the position at which they undergo differentiation into germ-cells. But we cannot believe that primitive germ-cells would migrate if the germ-cells could be formed from any of the other young cells of indifferent character which are so numerous in Hydroids. Even when the localized area undergoes very slight displacement, e.g. when it is removed from the exterior to the interior of the mesogloea[138], the change is always effected by active migration of primitive germ-cells through the substance of the mesogloea. Although the localized area has been largely displaced in the course of phylogeny, the changes in position have always taken place by very gradual stages, and never suddenly, and all these stages are repeated in the ontogeny of all existing species, by the migration of the primitive germ-cells from the ancestral area to the place where the germ-cells now arise. Hartlaub[139] has recently added a further instance (that of *Obelia*) to the numerous minute descriptions of these phyletic displacements of the localized area, and ontogenetic migrations of the primitive germ-cells, which are given in my work already referred to. The instance of *Obelia* is of especial interest as the direction of displacement is here reversed, taking place centrifugally instead of in a centripetal direction.

But if displacements of the localized areas can only take place by the frequently roundabout method of the migration of primitive germ-cells, we are obliged to conclude that such is the only manner in which the change can be effected, and that other cells are unable to play the role of the primitive germ-cells. And if other cells are unable to take this part, it must be because nucleoplasm of a certain character has to be present in order to form germ-cells, or according to the terms of my theory, the presence of germ-plasm is indispensable for this purpose. I do not see how we can escape the conclusion that there is continuity of the germ-plasm; for if it were supposed that somatic idioplasm undergoes transformation into germ-plasm, such an assumption would not explain why the displacement occurs by small stages, and with extreme and constant care for the preservation of a connexion with cells of the ancestral area. This fact can only be explained by the hypothesis that cell-generations other than those which end in the production of the cells of the ancestral area, are totally incapable of transformation into germ-cells.

Strasburger has objected that the transmission of germ-plasm along certain lines, viz. through a certain succession of somatic cells, is impossible, because the idioplasm is situated in the nucleus and not in the cell-body, and because a nucleus can only divide into two exactly equal halves by the indirect method of division, which takes place, as we must believe, in these cases. 'It might indeed be supposed,' says Strasburger, 'that during nuclear division certain molecular groups remain unchanged in the nuclear substance which is in other respects transformed, and that these groups are uniformly distributed through the whole organism; but we cannot imagine that their transmission could only be effected along certain lines.'

I do not think that Strasburger's objections can be maintained. I base this opinion on my previous criticism upon the assumed equality of the two daughter-nuclei formed by indirect division. I do not see any reason why the two halves must always possess the same structure, although they may be of equal size and weight. I am surprised that Strasburger should admit the possibility that the germ-plasm, which, as I think, is mixed with the idioplasm of the somatic cells, may remain unchanged in its passage through the body; for if this writer be correct in maintaining that the changes of nuclear substance in ontogeny are effected by the nutritive influence of the cell-body (cytoplasm), it follows that the whole nuclear substance of a cell must be changed at every division, and that no unchanged part can remain. We can only imagine that one part of a nucleus may undergo change while the other part remains unchanged, if we hold that the necessary transformations of nuclear substance are effected, by purely internal causes, viz. that they follow from the constitution of the nucleoplasm. But that one part may remain unchanged, and that such persistence does, as a matter of fact, occur is shown by the cases above described, in which the germ-cells separate very early from the developing egg-cell. Thus in the egg of Diptera, the two nuclei which are first separated by division from the segmentation nucleus, form the sexual cells, and this proves that they receive the germ-plasm of the segmentation nucleus unchanged. But during or before the separation of these two nuclei, the remaining part of the segmentation nucleus must have become changed in nature, or else it would continue to form 'pole-cells' at a later period instead of forming somatic cells. Although in many cases the cell-bodies of such early embryonic cells fail to exhibit any visible differences, the idioplasm of their nuclei must undoubtedly differ, or else they could not develope in different directions. It seems to me not only possible, but in every way probable, that the bodies of such early embryonic cells are equal in reality as well as in appearance; for, although the idioplasm of the nucleus determines the character of the cell-body, and although every differentiation of the latter depends upon a certain structure of its nucleoplasm, it does not necessarily follow that the converse proposition is true, viz. that each change in the structure of the nucleoplasm must effect a change in the cell-body. Just as rain is impossible without clouds, but every cloud does not necessarily produce rain, so growth is impossible without chemical change, but chemical processes of every kind and degree need not produce growth. In the same manner every kind of change in the molecular structure of the nucleoplasm need not exercise a transforming influence on the cytoplasm, and we can easily imagine that a long series of changes in the nucleoplasm may appear only in the kind and energy of the nuclear divisions which take place, the cell-substance remaining unchanged, as far as its molecular and chemical structure is concerned. This suggestion is in accordance with the fact that during the first period of embryonic development in animals, the cell-bodies do not exhibit any visible differences, or only such as are very slight; although exceptional instances occur, especially among the lower animals. But even these latter (e.g. the difference in appearance of the cells of the ectoderm and endoderm in sponges and Coelenterata) perhaps depend more largely upon a different admixture of nutritive substances than upon any marked difference in the cytoplasm itself. It is obvious that, in the construction of the embryo, the amount of cell-material must be first of all increased, and that it is only at a later period that the material must be differentiated so as to possess various qualities, according to the principle of division of labour. Facts of this kind are also opposed to Strasburger's view, that the cause of changes in the nucleoplasm does not lie within this substance itself but within the cell-body.

I believe I have shown that theoretically hardly any objections can be raised against the view that the nuclear substance of somatic cells may contain unchanged germ-plasm, or that this germ-plasm may be transmitted along certain lines. It is true that we might imagine *a priori* that all somatic nuclei contain a small amount of unchanged germ-plasm. In Hydroids such an assumption cannot be made, because only certain cells in a certain succession possess the power of developing into germ-cells; but it might well be imagined that in some organisms it would be a great advantage if every part possessed the power of growing up into the whole organism and of producing sexual cells under appropriate circumstances. Such cases might exist if it were possible for all somatic nuclei to contain a minute fraction of unchanged germ-plasm. For this reason, Strasburger's other objection against my theory also fails to hold; viz. that certain plants can be propagated by pieces of rhizomes, roots, or even by means of leaves, and that plants produced in this manner may finally give rise to flowers, fruit and seeds, from which new plants arise. 'It is easy to grow new plants from the leaves of *Begonia* which have been cut off and merely laid upon moist sand, and yet in the normal course of ontogeny the molecules of germ-plasm would not have been compelled to pass through the leaf; and they ought therefore to be absent from its tissue. Since it is possible to raise from the leaf a plant which produces flower and fruit, it is perfectly certain that special cells containing the germ substance cannot exist in the plant.' But I think that this fact only proves, that in *Begonia* and similar plants, all the cells of the leaves or perhaps only certain cells contain a small amount of germ-plasm, and that consequently these plants are specially adapted for propagation by leaves. How is it then that all plants cannot be reproduced in this way? No one has ever grown a tree from the leaf of the lime or oak, or a flowering plant from the leaf of the tulip or convolvulus. It is insufficient to reply that, in the last-mentioned cases, the leaves are more strongly specialized, and have thus become unable to produce germ-substance; for the leaf-cells in these different plants have hardly undergone histological differentiation in different degrees. If, notwithstanding, the one can produce a flowering plant, while the others have not this power, it is of course clear that reasons other than the degree of histological differentiation must exist; and, according to my opinion, such a reason is to be found in the admixture of a minute quantity of unchanged germ-plasm with some of their nuclei.

In Sachs' excellent lectures on the physiology of plants, we read on page 723[140]—'In the true mosses almost any cell of the roots, leaves and shoot-axes, and even of the immature sporogonium, may grow out under favourable conditions, become rooted, form new shoots, and give rise to an independent living plant.' Since such plants produce germ-cells at a later period, we have here a case which requires the assumption that all or nearly all cells must contain germ-plasm.

The theory of the continuity of the germ-plasm seems to me to be still less disproved or even rendered improbable by the facts of the alternation of generations. If the germ-plasm may pass on from the egg into certain somatic cells of an individual, and if it can be further transmitted along certain lines, there is no difficulty in supposing that it may be transmitted through a second, third, or through any number of individuals produced from the former by budding. In fact, in the Hydroids, on which my theory of the continuity of the germ-plasm has been chiefly based, alternation of generations is the most important means of propagation.

II. The Significance of the Polar Bodies.

We have already seen that the specific nature of a cell depends upon the molecular structure of its nucleus; and it follows from this conclusion that my theory is further, and as I believe strongly, supported, by the phenomenon of the expulsion of polar bodies, which has remained inexplicable for so long a time.

For if the specific molecular structure of a cell-body is caused and determined by the structure of the nucleoplasm, every kind of cell which is histologically differentiated must have a specific nucleoplasm. But the egg-cell of most animals, at any rate during the period of growth, is by no means an indifferent cell of the most primitive type. At such a period its cell-body has to perform quite peculiar and specific functions; it has to secrete nutritive substances of a certain chemical nature and physical constitution, and to store up this food-material in such a manner that it may be at the disposal of the embryo during its development. In most cases the egg-cell also forms membranes which are often characteristic of particular species of animals. The growing egg-cell is therefore histologically differentiated: and in this respect resembles a somatic cell. It may perhaps be compared to a gland-cell, which does not expel its secretion, but deposits it within its own substance[14]. To perform such specific functions it requires a specific cell-body, and the latter depends upon a specific nucleus. It therefore follows that the growing egg-cell must possess nucleoplasm of specific molecular structure, which directs the above-mentioned secretory functions of the cell. The nucleoplasm of histologically differentiated cells may be called histogenetic nucleoplasm, and the growing egg-cell must contain such a substance, and even a certain specific modification of it. This nucleoplasm cannot possibly be the same as that which, at a later period, causes embryonic development. Such development can only be produced by true germ-plasm of immensely complex constitution, such as I have previously attempted to describe. It therefore follows that the nucleus of the egg-cell contains two kinds of nucleoplasm:—germ-plasm and a peculiar modification of histogenetic nucleoplasm, which may be called *ovogenetic nucleoplasm*. This substance must greatly preponderate in the young egg-cell, for, as we have already seen, it controls the growth of the latter. The germ-plasm, on the other hand, can only be present in minute quantity at first, but it must undergo considerable increase during the growth of the cell. But in order that the germ-plasm may control the cell-body, or, in other words, in order that embryonic development may begin, the still preponderating ovogenetic nucleoplasm must be removed from the cell. This removal takes place in the same manner as that in which differing nuclear substances are separated during the ontogeny of the embryo: viz. by nuclear division, leading to cell-division. The expulsion of the polar bodies is nothing more than the removal of ovogenetic nucleoplasm from the egg-cell. That the ovogenetic nucleoplasm continues to greatly preponderate in the nucleus up to the very last, may be concluded from the fact that two successive divisions of the latter and the expulsion of two polar bodies appear to be the rule. If in this way a small part of the cell-body is expelled from the egg, the extrusion must in all probability be considered as an inevitable loss, without which the removal of the ovogenetic nucleoplasm cannot be effected.

This is my theory of the significance of polar bodies, and I do not intend to contrast it, *in extenso*, with the theories propounded by others; for such theories are well known and differ essentially from my own. All writers agree in supposing that something which would be an obstacle to embryonic development is removed from the egg; but opinions differ as to the nature of this substance and the precise reasons for its removal[142]. Some observers (e. g. Minot[143], van Beneden, and Balfour) regard the nucleus as hermaphrodite, and assume that in the polar bodies the male element is expelled in order to render the egg capable of fertilization. Others speak of a rejuvenescence of the nucleus, others again believe that the quantity of nuclear substance must be reduced in order to become equal to that of the generally minute sperm-nucleus, and that the proportions for nuclear conjugation are in this way adjusted.

The first view seems to me to be disproved by the fact that male as well as female qualities are transmitted by the egg-cell, while the sperm-cell also transmits female qualities. The germ-plasm of the nucleus of the egg cannot therefore be considered as female, and that of the sperm-nucleus cannot be considered as male: both are sexually indifferent. The last view has been recently formulated by Strasburger, who holds that the quantity of the idioplasm contained in the germ-nucleus must be reduced by one half, and that a whole nucleus is again produced by conjugation with the sperm-nucleus. Although I believe that the fundamental idea underlying this hypothesis is perfectly correct, viz. that the influence of each nucleus is as largely dependent upon its quantity as upon its quality, I must raise the objection that the decrease in quantity is not the explanation of the expulsion of polar bodies. The quantity of idioplasm contained in the germ-nucleus is, as a matter of fact, not reduced by one-half but by three-fourths, for two divisions take place one after the other. Thus by conjugation with the sperm-nucleus, which we may assume to be of the same size as the germ-nucleus, a nucleus is produced which can only contain half as much idioplasm as was present in the original germ-nucleus, before division. Strasburger's view leaves unexplained the question why the size of the germ-nucleus, before the expulsion of polar bodies, was thus twice as large; and even if we neglect the theory of ovogenetic nucleoplasm and assume that this larger nucleus was entirely made up of germ-plasm, it must be asked why the egg did not commence segmentation earlier. The theory which explains the sperm-cell as the vitalizing principle which starts embryonic development, like the spark which kindles the gunpowder, would indeed answer this question in a very simple manner. But Strasburger does not accept this theory, and holds, as I do, a very different view, which will be explained later on.

If, on the other hand, we assume that the germ-nucleus contains two different kinds of nucleoplasm, the question is answered quite satisfactorily. In treating of parthenogenesis, further on, I shall mention a fact which seems to me to furnish a real proof of the validity of this explanation; and, if we accept this fact for the present, it will be clear that the simple explanation now offered of phenomena which are otherwise so difficult to understand, would also largely support the theory of the continuity of the germ-plasm. Such an explanation would, above all, very clearly demonstrate the co-existence of two nucleoplasms with different qualities in one and the same nucleus. My theory must stand or fall with this explanation, for if the latter were disproved, the continuity of the germ-plasm could not be assumed in any instance, not even in the simplest cases, where, as in Diptera, the germ-cells are the first-formed products of embryonic development. For even in these insects the egg possesses a specific histological character which must depend upon a specifically differentiated nucleus. If then two kinds of nucleoplasm are not present, we must assume that in such cases the germ-plasm of the newly formed germ-cells, which has passed on unchanged from the segmentation nucleus, is at once transformed entirely into ovogenetic nucleoplasm, and must be re-transformed into germ-plasm at a later period when the egg is fully mature. We could not in any way understand why such a re-transformation requires the expulsion of part of the nuclear substance.

At all events, my explanation is simpler than all others, in that it only assumes a single transformation of part of the germ-plasm, and not the later re-transformation of ovogenetic nucleoplasm into germ-plasm, which is so improbable. The ovogenetic nucleoplasm must possess entirely different qualities from the germ-plasm; and, above all, it does not readily lead to division, and thus we can better understand the fact, in itself so remarkable, that egg-cells do not increase in number by division, when they have assumed their specific structure, and are controlled by the ovogenetic nucleoplasm. The tendency to nuclear division, and consequently to cell-division, is not produced until changes have to a certain extent taken place in the mutual relation between the two kinds of nucleoplasm contained in the germ-nucleus. This change is coincident with the attainment of maximum size by the body of the egg-cell. Strasburger, supported by his observations on *Spirogyra*, concludes that the stimulus towards cell-division emanates from the cell-body; but the so-called centres of attraction at the poles of the nuclear spindle obviously arise under the influence of the nucleus itself, even if we admit that they are entirely made up of cytoplasm. But this point has not been decided upon, and we may presume that the so-called 'Polkörperchen' of the spindle (Fol) are derived from the nucleus, although they are placed outside the nuclear membrane[144]. Many points connected with this subject are still in a state of uncertainty, and we must abstain from general conclusions until it has been possible to demonstrate clearly the precise nature of certain phenomena attending indirect nuclear division, which still remain obscure in spite of the efforts of so many excellent observers. We cannot even form a decided opinion as to whether the chromatin or the achromatin of the nuclear thread is the real idioplasm. But although these points are not yet thoroughly understood, we are justified in maintaining that the cell enters upon division under the influence of certain conditions of the nucleus, although the latter are invisible until cell-division has already commenced.

I now pass on to examine my hypothesis as to the significance of the formation of polar bodies, in the light of those ascertained facts which bear upon it.

If the expulsion of the polar bodies means the removal of the ovogenetic nucleoplasm after the histological differentiation of the egg-cell is complete, we must expect to find polar bodies in all species except those in which the egg-cell has remained in a primitive undifferentiated condition, if indeed such species exist. Wherever the egg-cell assumes the character of a specialized cell, e.g. in the attainment of a particular size or constitution, in the admixture of food-yolk, or the formation of membranes, it must also contain ovogenetic nucleoplasm, which must ultimately be removed if the germ-plasm is to gain control over the egg-cell. It does not signify at all, in this respect, whether the egg is or is not destined for fertilization.

If we examine the Metazoa in regard to this question, we find that polar bodies have not yet been discovered in sponges[145], but this negative evidence is no proof that they are really absent. In all probability, no one has ever seriously endeavoured to find them, and there are perhaps difficulties in the way of the proofs of their existence, because the egg-cell lies free for a long time and even moves actively in the tissue of the mesogloea. We might expect that the formation of polar bodies takes place here, as in all other instances, when the egg becomes mature, that is, at a time when the eggs are already closely enveloped in the sponge tissue. At all events the eggs of sponges, as far as they are known, attain a specific nature, in the possession of a peculiar cell-body, frequently containing food-yolk, and of the nucleus which is characteristic of all animal eggs during the process of growth. Hence we cannot doubt the presence of a specific ovogenetic nucleoplasm, and must therefore also believe that it is ultimately removed in the polar bodies.

In other Coelenterata, in worms, echinoderms, and in molluscs polar bodies have been described, as well as in certain Crustacea, viz. in *Balanus* by Hoek and in *Cetochilus septentrionale* by Grobben. The latter instance appears to be quite trustworthy, but there is some doubt as to the former and also as regards *Moina* (a Daphnid), in which Grobben found a body, which he considered to be a polar body, on the upper pole of an egg which was just entering upon segmentation. In insects polar bodies have not been described up to the present time[146], and only in a few cases in Vertebrata, as in *Petromyzon* by Kupffer and Benecke.

It must be left to the future to decide whether the expulsion of polar bodies occurs in those large groups of animals in which they have not been hitherto discovered. The fact, however, that they have not been so discovered cannot be urged as an objection to my theory, for we do not know *a priori* whether the removal of the ovogenetic nucleoplasm has not been effected in the course of phylogeny in some other and less conspicuous manner. The cell-body of the polar globules is so minute in many eggs that it was a long time before the cellular nature of these structures was recognized[147]; and it is possible that their minute size may point to the fact that a phyletic process of reduction has taken place, to the end that the egg may be deprived of as little material as possible. It is at all events proved that in all Metazoan groups the nucleus undergoes changes during the maturation of the egg, which are entirely similar to those which lead to the formation of polar bodies in those eggs which possess them. In the former instances it is possible that nature has taken a shortened route to gain the same end.

It would be an important objection if it could be shown that no process corresponding to the expulsion of polar bodies takes place in the male germ-cells, for it is obvious that here also we should, according to my theory, expect such a process to occur. The great majority of sperm-cells differ so widely in character from the ordinary indifferent (i. e. undifferentiated) cells, that they are evidently histologically differentiated in a very high degree, and hence the sperm-cells, like the yolk-forming germ-cells, must possess a specific nuclear substance. The majority of sperm-cells therefore resemble the somatic cells in that they have a specific histological structure, but their characteristic form has nothing to do with their fertilizing power, viz. with their power of being the bearers of germ-plasm. Important as this structure is, in order to render it possible that the egg-cell may be approached and penetrated, it has nothing to do with the property of the sperm-cell to transmit the qualities of the species and of the individual to the following generation. The nuclear substance which causes such a cell to assume the appearance of a thread, or a stellate form (in Crustacea), or a boomerang form (present in certain Daphnids), or a conical bullet shape (Nematodes), cannot possibly be the same nuclear substance as that which, after conjugation with the egg-cell, contains in its molecular structure the tendency to build up a new Metazoon of the same kind as that by which it was produced. We must, therefore, conclude that the sperm-cell also contains two kinds of nucleoplasm, namely, germ-plasm and spermogenetic nucleoplasm.

It is true that we cannot say *a priori* whether the influence exercised on the sperm-cell by the spermogenetic nucleoplasm might not be eliminated by some means other than its removal from the cell. It is conceivable, for instance, that this substance may be expelled from the nucleus, but may remain in the cell-body, where it is in some way rendered powerless. We do not yet really know anything of the essential conditions of nuclear division, and it is quite impossible to bring forward any facts in support of my previous suggestion. The germ-plasm is supposed to be present in the nucleus of the growing egg-cell in smaller quantity than the ovogenetic nucleoplasm, and the germ-plasm gradually increases in quantity: thus when the egg has attained its maximum size, the opposition between the two different kinds of nucleoplasm becomes so marked, in consequence of the alteration in their quantitative relations, that their separation, viz. nuclear division, results. But although we are not able to distinguish, by any visible characteristics, the different kinds of nucleoplasm which may be united in one nuclear thread, the assumption that the influence of each kind bears a direct proportion to its quantity is the most obvious and natural one. The tendency of the germ-plasm contained in the nucleus cannot make itself felt so long as an excess of ovogenetic nucleoplasm is also present. We may imagine that the effects of the two different kinds of nucleoplasm are combined to produce a resultant effect; but when the two influences exerted upon the cell are nearly opposed, only the stronger can make itself felt, and in such a case the latter must exceed the former in quantity, because part of it is as it were neutralized by the other nucleoplasm working in an opposite direction. This metaphorical representation may give us a clue to explain the fact that the ovogenetic nucleoplasm comes to exceed the germ-plasm in quantity. For obviously these two kinds of nucleoplasm exert opposite tendencies in at least one respect. The germ-plasm tends to effect the division of the cell into the two first segmentation spheres; the ovogenetic nucleoplasm, on the other hand, possesses a tendency towards the growth of the cell-body without division. Hence the germ-plasm cannot make itself felt, and is not able to expel the ovogenetic nucleoplasm until it has reached such a relative size as enables it to successfully oppose the latter.

Applying these ideas to the sperm-cells we must see whether the expulsion of part of the nuclear substance, viz. of the spermogenetic nucleoplasm, corresponding to the ovogenetic nucleoplasm, takes place in them also.

As far as we can judge from thoroughly substantiated observations such phenomena are indeed found in many cases, although they appear to be different from those occurring in the egg-cell, and cannot receive quite so certain an interpretation.

The attempt to prove that a process similar to the expulsion of polar bodies takes place in the formation of sperm-cells has already been made by those observers who regard such expulsion as the removal of the male element from the egg, thus leading to sexual differentiation; for such a theory also requires the removal of part of the nuclear substance from the maturing sperm-cell. Thus, according to E. van Beneden and Ch. Julin, the cells which, in *Ascaris*, produce the spermatogonia (mother-cells of the sperm-cells), expel certain elements from their nuclear plate, a phenomenon which has not been hitherto observed in any other animal, and even in this instance has only been inferred and not directly observed. Moreover the sperm-cells have not attained their specific form (conical bullet-shaped) at the time when this expulsion takes place from the spermatogonia, and we should expect that the spermogenetic nucleoplasm would not be removed until it has completed its work, viz. not until the specific shape of the sperm-cell has been attained. We might rather suppose that phenomena explicable in this way are to be witnessed in those sperm-blastophores (mother-cells of sperm-cells) which, as has been known for a long time, are not employed in the formation of the nuclei of sperm-cells, but for the greater part remain at the base of the latter and perish after their maturation and separation. In this case an influence might be exerted by these nuclei upon the specific form of the sperm-cells, for the former arise and develope in the form of bundles of spermatozoa in the interior of the mother-cell.

It has been already shown in many groups of animals that parts of the sperm-mother-cells[148] perish, without developing into sperm-cells, as in Selachians, in the frog, in many worms and snails, and also in mammals (Blomfield). But the attention of observers has been directed to that part of the cell-body which is not used in the formation of sperm-cells, rather than to the nucleus; and the proof that part of the nucleus also perishes is still wanting in many of these cases. Fresh investigation must decide whether the nucleus of the sperm-mother-cell perishes as a general rule, and whether part of the nucleus is rendered powerless in some other way, where such mother-cells do not exist. Perhaps the paranucleus (Nebenkern) of the sperm-cell, first described by La Valette St. George, and afterwards found in many animals of very different groups, is the analogue of the polar body. It is true that this so-called paranucleus is now considered as a condensed part of the cell-body, but we must remember that it has been hitherto a question whether the head of the spermatozoon is formed from the nucleus of the cell or from the paranucleus; and that the observers who held the former view were in consequence obliged to regard the paranucleus as a product of the cell-body. But according to the most recent investigations of Fol[149], Roule[150], Balbiani[151], and Will[152], upon the formation of the follicular epithelium in the ovary of different groups, it is not improbable that parts of the nucleus may become detached without passing through the process of karyokinesis. Thus it is very possible that the paranucleus may be a product of the main nucleus and not a condensed part of the cell-body. This view is supported by its behaviour with staining reagents, while the other view, that it arises from the cell-substance, is not based upon direct observation. Consequently future investigation must decide whether the paranucleus is to be considered as the spermogenetic nucleoplasm expelled from the nucleus. But even if this question is answered in the affirmative, we should still have to explain why this nuclear substance, remaining in the cell-body, does not continue to exercise any control over the latter.

Strasburger has recently enumerated a large number of cases from different groups of plants, in which the maturation of both male and female germ-cells is accompanied by phenomena similar to the expulsion of polar bodies. In this respect the phenomena occurring in the pollen-grains of Phanerogams bear an astonishing resemblance to the maturation of the animal egg. For instance, in the larch, the sperm-mother-cell divides three times in succession, the products of division being very unequal on each occasion; and exactly as in the case of polar bodies, the three small so-called vegetative cells shrink rapidly after separation, and have no further physiological value.

According to Strasburger, the so-called 'ventral canal-cell,' which, in mosses, ferns, and Conifers, separates from the female germ-cell, reminds us, in every way, of the polar bodies of animal eggs. Furthermore, the spermatozoids in the mosses and vascular cryptogams throw off a small vesicle before performing their functions[153]. On the other hand the equivalents of 'polar bodies' (the 'ventral canal-cells') are said to be absent in the Cycads, although these are so nearly allied to Conifers. Furthermore, 'no phenomenon occurs in the oospheres (ova) of Angiosperms which can be compared to the formation of polar bodies.' Strasburger therefore concludes that the separation of certain parts from the germ-cells is not in all cases necessary for maturation, and that such phenomena are not fundamental, like those of fertilization, which must always take place along the same morphological lines. He further concludes that the former phenomena are only necessary in the case of the germ-cells of certain organisms, in order to bring the nuclei destined for the sexual act into the physiological condition necessary for its due performance.

I am unwilling to abandon the idea that the expulsion of the histogenetic parts of the nuclear substance, during the maturation of germ-cells, is also a general phenomenon in plants; for the process appears to be fundamental, while the argument that it has not been proved to occur universally is only of doubtful value. The embryo-sac of Angiosperms is such a complex structure that it seems to me to be possible (as it does to Strasburger) that 'processes which precede the formation of the egg-cell have borne relation to the sexual differentiation of the nucleus of the egg.' Besides, it is possible that the vegetable egg-cell may, in certain cases, possess so simple a structure and so small a degree of histological specialization, that it would not be necessary for it to contain any specific histogenetic nucleoplasm: thus it would consist entirely of germ-plasm from the first. In such cases, of course, its maturation would not be accompanied by the expulsion of somatic nucleoplasm.

I have hitherto abstained from discussing the question as to whether the process of the formation of polar bodies may require an interpretation which is entirely different from that which I have given it, whether it may receive a purely morphological interpretation.

In former times it could only be regarded as of purely phyletic significance: it could only be looked upon as the last remnant of a process which formerly possessed some meaning, but which is now devoid of any physiological importance. We are indeed compelled to admit that a process does occur in connexion with the true polar bodies of animal eggs, which we cannot explain on physiological grounds; I mean the division of the polar bodies after they have been expelled from the egg. In many animals the two polar bodies divide again after their expulsion, so as to form four bodies, which distinctly possess the structure of cells, as Trinchese observed in the case of gastropods. But, in the first place, this second division does not always take place, and, secondly, it is very improbable that a process which occurs during the first stage of ontogeny, or more properly speaking, before the commencement of ontogeny, and which is, therefore, a remnant of some excessively ancient phyletic stage, would have been retained up to the present day unless it possessed some very important physiological significance. We may safely maintain that it would have disappeared long ago if it had been without any physiological importance. Relying on our knowledge of the slow and gradual, although certain, disappearance, in the course of phylogeny, of organs which have lost their functions, and of processes which have become meaningless, we are compelled to regard the process of the formation of polar bodies as of high physiological importance. But this view does not exclude the possibility that the process possessed a morphological meaning also, and I believe that we are quite justified in attempting (as Bütschli[154] has recently done) to discover what this morphological meaning may have been.

Should it be finally proved that the expulsion of polar bodies is nothing more than the removal of histogenetic nucleoplasm from the germ-cell, the opinion (which is so intimately connected with the theory of the continuity of the germ-plasm) that a re-transformation of specialised idioplasm into germ-plasm cannot occur, would be still further confirmed; for we do not find that any part of an organism is thrown away simply because it is useless: organs that have lost their functions are re-absorbed, and their material is thus employed to assist in building up the organism.

III. On the Nature of Parthenogenesis.

It is well known that the formation of polar bodies has been repeatedly connected with the sexuality of germ-cells, and that it has been employed to explain the phenomena of parthenogenesis. I may now, perhaps, be allowed to develope the views as to the nature of parthenogenesis at which I have arrived under the influence of my explanation of polar bodies.

The theory of parthenogenesis adopted by Minot and Balfour is distinguished by its simplicity and clearness, among all other interpretations which had been hitherto offered. Indeed, their explanation follows naturally and almost as a matter of course, if the assumption made by these observers be correct, that the polar body is the male part of the hermaphrodite egg-cell. An egg which has lost its male part cannot develope into an embryo until it has received a new male part in fertilization. On the other hand, an egg which does not expel its male part may develope without fertilization, and thus we are led to the obvious conclusion that parthenogenesis is based upon the non-expulsion of polar bodies. Balfour distinctly states 'that the function of forming polar cells has been acquired by the ovum for the express purpose of preventing parthenogenesis[155].'

It is obvious that I cannot share this opinion, for I regard the expulsion of polar bodies as merely the removal of the ovogenetic nucleoplasm, on which depended the development of the specific histological structure of the egg-cell. I must assume that the phenomena of maturation in the parthenogenetic egg and in the sexual egg are precisely identical, and that in both, the ovogenetic nucleoplasm must in some way be removed before embryonic development can begin.

Unfortunately the actual proof of this assumption is not so complete as might be desired. In the first place, we are as yet uncertain whether polar bodies are or are not expelled by parthenogenetic eggs[156]; for in no single instance has such expulsion been established beyond doubt. It is true that this deficiency does not afford any support to the explanation of Minot and Balfour, for in all cases in which polar bodies have not been found in parthenogenetic eggs, these structures are also absent from the eggs which require fertilization in the same species. But although the expulsion of polar bodies in parthenogenesis has not yet been proved to occur, we must assume it to be nearly certain that the phenomena of maturation, whether connected or unconnected with the expulsion of polar bodies, are the same in the eggs which develope parthenogenetically and in those which are capable of fertilization, in one and the same species. This conclusion depends, above all, upon the phenomena of reproduction in bees, in which, as a matter of fact, the same egg may be fertilized or may develope parthenogenetically, as I shall have occasion to describe in greater detail at a later period.

Hence when we see that the eggs of many animals are capable of developing without fertilization, while in other animals such development is impossible, the difference between the two kinds of eggs must rest upon something more than the mode of transformation of the nucleus of the germ-cell into the first segmentation nucleus. There are, indeed, facts which distinctly point to the conclusion that the difference is based upon quantitative and not qualitative relations. A large number of insects are exceptionally reproduced by the parthenogenetic method, e. g. in Lepidoptera. Such development does not take place in all the eggs laid by an unfertilized female, but only in part, and generally a small fraction of the whole, while the rest die. But among the latter there are some which enter upon embryonic development without being able to complete it, and the stage at which development may cease also varies. It is also known that the eggs of higher animals may pass through the first stages of segmentation without having been fertilized. This was shown to be the case in the egg of the frog by Leuckart[157], in that of the fowl by Oellacher[158], and even in the egg of mammals by Hensen[159].

Hence in such cases it is not the impulse to development, but the power to complete it, which is absent. We know that force is always bound up with matter, and it seems to me that such instances are best explained by the supposition that too small an amount of that form of matter is present, which, by its controlling agency, effects the building-up of the embryo by the transformation of mere nutritive material. This substance is the germ-plasm of the segmentation nucleus, and I have assumed above that it is altered in the course of ontogeny by changes which arise from within, so that when sufficient nourishment is afforded by the cell-body, each succeeding stage necessarily results from the preceding one. I believe that changes arise in the constitution of the nucleoplasm at each cell-division which takes place during the building-up of the embryo, changes which either correspond or differ in the two halves of each nucleus. If, for the present, we neglect the minute amount of unchanged germ-plasm which is reserved for the formation of the germ-cells, it is clear that a great many different stages in the development of somatic nucleoplasm are thus formed, which may be denominated as stages 1, 2, 3, 4, &c., up to n. In each of these stages the cells differ more as development proceeds, and as the number by which the stage is denominated becomes higher. Thus, for instance, the two first segmentation spheres would represent the first stage of somatic nucleoplasm, a stage which may be considered as but slightly different in its molecular structure from the nucleoplasm of the segmentation nucleus; the four first segmentation spheres would represent the second stage; the succeeding eight spheres the third, and so on. It is clear that at each successive stage the molecular structure of the nucleoplasm must be further removed from that of the germ-plasm, and that, at the same time, the cells of each successive stage must also diverge more widely among themselves in the molecular structure of their nucleoplasm. Early in development each cell must possess its own peculiar nucleoplasm, for the further course of development is peculiar to each cell. It is only in the later stages that equivalent or nearly equivalent cells are formed in large numbers, cells in which we must also suppose the existence of equivalent nucleoplasm.

If we may assume that a certain amount of germ-plasm must be contained in the segmentation nucleus in order to complete the whole process of the ontogenetic differentiation of this substance; if we may further assume that the quantity of germ-plasm in the segmentation nucleus varies in different cases; then we should be able to understand why one egg can only develope after fertilization, while another can begin its development without fertilization, but cannot finish it, and why a third is even able to complete its development. We should also understand why one egg only passes through the first stages of segmentation and is then arrested, while another reaches a few more stages in advance, and a third developes so far that the embryo is nearly completely formed. These differences would depend upon the extent to which the germ-plasm, originally present in the egg, was sufficient for the development of the latter; development will be arrested as soon as the nucleoplasm is no longer capable of producing the succeeding stage, and is thus unable to enter upon the following nuclear division.

From a general point of view such a theory would explain many difficulties, and it would render possible an explanation of the phyletic origin of parthenogenesis, and an adequate understanding of the strange and often apparently abrupt and arbitrary manner of its occurrence. In my works on *Daphnidae* I have already laid especial stress upon the proposition that parthenogenesis in insects and Crustacea certainly cannot be an ancestral condition which has been transmitted by heredity, but that it has been derived from a sexual condition. In what other way can we explain the fact that parthenogenesis is present in certain species or genera, but absent in others closely allied to them; or the fact that males are entirely wanting in species of which the females possess a complete apparatus for fertilization? I will not repeat all the arguments with which I attempted to support this conclusion[160]. Such a conclusion may be almost certainly accepted for the *Daphnidae*, because parthenogenesis does not occur in their still living ancestors, the Phyllopods, and especially the *Estheridae*. In *Daphnidae* the cause and object of the phyletic development of parthenogenesis may be traced more clearly than in any other group of animals. In *Daphnidae* we can accept the conclusion with greater certainty than in all other groups, except perhaps the *Aphidae*, that parthenogenesis is extremely advantageous to species in certain conditions of life; and that it has only been adopted when, and as far as, it has been beneficial; and further, that at least in this group parthenogenesis became possible, and was adopted, in each species as soon as it became useful. Such a result can be easily understood if it is only the presence of more or less germ-plasm which decides whether an egg is, or is not, capable of development without fertilization.

If we now examine the foundations of this hypothesis we shall find that we may at once accept one of its assumptions, viz. that fluctuations occur in the quantity of germ-plasm in the segmentation nucleus; for there can never be absolute equality in any single part of different individuals. As soon therefore as these fluctuations become so great that parthenogenesis is produced, it may become, by the operation of natural selection, the chief mode of reproduction of the species or of certain generations of the species. In order to place this theory upon a firm basis, we have simply to decide whether the quantity of germ-plasm contained in the segmentation nucleus is the factor which determines development; although for the present it will be sufficient if we can render this view to some extent probable, and show that it is not in contradiction with established facts.

At first sight this hypothesis seems to encounter serious difficulties. It will be objected that neither the beginning nor the end of embryonic development can possibly depend upon the quantity of nucleoplasm in the segmentation nucleus, since the amount may be continually increased by growth; for it is well known that during embryonic development the nuclear substance increases with astonishing rapidity. By an approximate calculation I found[161] that, in the egg of a *Cynips*, the quantity of nuclear substance present at the time when the blastoderm was about to be formed, and when there were twenty-six nuclei, was even then seven times as great as the quantity which had been contained in the segmentation nucleus. How then can we imagine that embryonic development would ever be arrested from want of nuclear substance, and if such deficiency really acted as an arresting force, how then could development begin at all? We might suppose that when germ-plasm is present in sufficient quantity to start segmentation, it must also be sufficient to complete the development; for it grows continuously, and must presumably always possess a power equal to that which it possessed at the beginning, and which was just sufficient to start the process of segmentation. If at each ontogenetic stage, the quantity of nucleoplasm is just sufficient to produce the following stage, we might well imagine that the whole ontogeny would necessarily be completed.

The flaw in this argument lies in the erroneous assumption that the growth of nuclear substance is, when the quality of the nucleus and the conditions of nutrition are equal, unlimited and uncontrolled. The intensity of growth must depend upon the quantity of nuclear substance with which growth and the phenomena of segmentation commenced. There must be an optimum quantity of nucleoplasm with which the growth of the nucleus proceeds most favourably and rapidly, and this optimum will be represented in the normal size of the segmentation nucleus. Such a size is just sufficient to produce, in a certain time and under certain external conditions, the nuclear substance necessary for the construction of the embryo, and to start the long series of cell-divisions. When the segmentation nucleus is smaller, but large enough to enter upon segmentation, the nuclei of the two first embryonic cells will fall rather more below the normal size, because the growth of the segmentation nucleus during and after division will be less rapid on account of its unusually small size. The succeeding generations of nuclei will depart more and more from the normal size in each respective stage, because they do not pass into a resting-stage during embryonic development, but divide again immediately after their formation. Hence nuclear growth would become less vigorous as the nuclei fell more and more below the optimum size, and at last a moment would arrive when they would be unable to divide, or would be at least unable to control the cell-body in such a manner as to lead to its division.

The first event of importance for embryonic development is the maturation of the egg, i. e. the transformation of the nucleus of the germ-cell into a nuclear spindle and the removal of the ovogenetic nucleoplasm by the separation of polar bodies, or by some analogous process. There must be some cause for this separation, and I have already tried to show that it may lie in the quantitative relations which obtain between the two kinds of nucleoplasm contained in the nucleus of the egg. I have suggested that the germ-plasm, at first small in quantity, undergoes a gradual increase, so that it can finally oppose the ovogenetic nucleoplasm. I will not further elaborate this suggestion, for the ascertained facts are insufficient for the purpose. But the appearances witnessed in nuclear division indicate that there are opposing forces, and that such a contest is the motive cause of division; and Roux[162] may be right in referring the opposition to electrical forces. However this may be, it is perfectly certain that the development of this opposition is based upon internal conditions arising during growth in the nucleus itself. The quantity of nuclear thread cannot by itself determine whether the nucleus can or cannot enter upon division; if so, it would be impossible for two divisions to follow each other in rapid succession, as is actually the case in the separation of the two polar bodies, and also in their subsequent division. In addition to the effects of quantity, the internal conditions of the nucleus must also play an important part in these phenomena. Quantity alone does not necessarily produce nuclear division, or the nucleus of the egg would divide long before maturation is complete, for it contains much more nucleoplasm than the female pronucleus, which remains in the egg after the expulsion of the polar bodies, and which is in most cases incapable of further division. But the fact that segmentation begins immediately after the conjugation of male and female pronuclei, also shows that quantity is an essential requisite. The effect of fertilization has been represented as analogous to that of the spark which kindles the gunpowder. In the latter case an explosion ensues, in the former segmentation begins. Even now, many authorities are inclined to refer the polar repulsion manifested in the nuclear division which immediately follows fertilization, to the antagonism between male and female elements. But, according to the important discoveries of Flemming and van Beneden, the polar repulsion in each nuclear division is not based on the antagonism between male and female loops, but depends upon the antagonism and mutual repulsion between the two halves of the same loop. The loops of the father and those of the mother remain together and divide together throughout the whole ontogeny.

What can be the explanation of the fact that nuclear division follows immediately after fertilization, but that without fertilization it does not occur in most cases? There is only one possible explanation, viz. the fact that the quantity of the nucleus has been suddenly doubled, as the result of conjugation. The difference between the male and female pronuclei cannot serve as an explanation, even though the nature of this difference is entirely unknown, because polar repulsion is not developed between the male and female halves of the nucleus, but within each male and each female half. We are thus forced to conclude that increase in the quantity of the nucleus affords an impulse for division, the disposition towards it being already present. It seems to me that this view does not encounter any theoretical difficulties, and that it is an entirely feasible hypothesis to suppose that, besides the internal conditions of the nucleus, its quantitative relation to the cell-body must be taken into especial account. It is imaginable, or perhaps even probable, that the nucleus enters upon division as soon as its idioplasm has attained a certain strength, quite apart from the supposition that certain internal conditions are necessary for this end. As above stated, such conditions may be present, but division may not occur because the right quantitative relation between nucleus and cell-body, or between the different kinds of nuclear idioplasm, has not been established. I imagine that such a quantitative deficiency exists in an egg, which, after the expulsion of the ovogenetic nucleoplasm in the polar bodies, requires fertilization in order to begin segmentation. The fact that the polar bodies were expelled proves that the quantity of the nucleus was sufficient to cause division, while afterwards it was no longer sufficient to produce such a result.

This suggestion will be made still clearer by an example. In *Ascaris megalocephala* the nuclear substance of the female pronucleus forms two loops, and the male pronucleus does the same; hence the segmentation nucleus contains four loops, and this is also the case with the first segmentation spheres. If we suppose that in embryonic development, the first nuclear division requires such an amount of nuclear substance as is necessary for the formation of four loops,—it follows that an egg, which can only form two or three loops from its nuclear reticulum, would not be able to develope parthenogenetically, and that not even the first division would take place. If we further suppose that, while four loops are sufficient to start nuclear division, these loops must be of a certain size and quantity in order to complete the whole ontogeny (in a certain species), it follows that eggs possessing a reticulum which contains barely enough nuclear substance to divide into four segments, would be able to produce the first division and perhaps also the second and third, or some later division, but that at a certain point during ontogeny, the nuclear substance would become insufficient, and development would be arrested. This will occur in eggs which enter upon development without fertilization, but are arrested before its completion. One might compare this retardation leading to the final arrest of development, to a railway train which is intended to meet a number of other trains at various junctions, and which can only travel slowly because of some defect in the engine. It will be a little behind time at the first junction, but it may just catch the train, and it may also catch the second or even the third; but it will be later at each successive junction, and will finally arrive too late for a certain train; and after that it will miss all the trains at the remaining junctions. The nuclear substance grows continuously during development, but the rate at which it increases depends upon the nutritive conditions together with its initial quantity. The nutritive changes during the development of an egg depend upon the quantity of the cell-body which was present at the outset, and which cannot be increased. If the quantity of the nuclear substance is rather too small at the beginning, it will become more and more insufficient in succeeding stages, as its growth becomes less vigorous, and differs more from the standard it would have reached if the original quantity had been normal. Consequently it will gradually fall more and more short of the normal quantity, like the train which arrives later and later at each successive junction, because its engine, although with the full pressure of steam, is unable to attain the normal speed.

It will be objected that four loops cannot be necessary for nuclear division in *Ascaris*, since such division takes place in the formation of the polar bodies, resulting in the appearance of the female pronucleus with only two loops. But this fact only shows that the quantity of nuclear substance necessary for the formation of four loops is not necessary for all nuclear divisions; it does not disprove the assumption that such a quantity is required for the division of the segmentation nucleus. In addition to these considerations we must not leave the substance of the cell-body altogether out of account, for, although it is not the bearer of the tendencies of heredity, it must be necessary for every change undergone by the nucleus, and it surely also possesses the power of influencing changes to a large extent. There must be some reason for the fact that in all animal eggs with which we are acquainted, the nucleus moves to the surface of the egg at the time of maturation, and there passes through its well-known transformation. It is obvious that it is there subjected to different influences from those which would have acted upon it in the centre of the cell-body, and it is clear that such an unequal cell-division as takes place in the separation of the polar bodies could not occur if the nucleus remained in the centre of the egg.

This explanation of the necessity for fertilization does not exclude the possibility, that, under certain circumstances, the substance of the egg-nucleus may be larger, so that it is capable of forming four loops. Eggs which thus possess sufficient nucleoplasm, viz. germ-plasm, for the formation of the requisite four loops of normal size, (namely, of the size which would have been produced by fertilization), can and must develope by the parthenogenetic method.

Of course the assumption that four loops must be formed has only been made for the sake of illustration. We do not yet know whether there are always exactly four loops in the segmentation nucleus[163]. I may add that, although the details by which these considerations are illustrated are based on arbitrary assumptions, the fundamental view that the development of the egg depends, *ceteris paribus*, upon the quantity of nuclear substance, is certainly right, and follows as a necessary conclusion from the ascertained facts. It is not unlikely that such a view may receive direct proof in the results of future investigations. Such proof might for instance be forthcoming if we were to ascertain, in the same species, the number of loops present in the segmentation nucleus of fertilization, as compared with those present in the segmentation nucleus of parthenogenesis.

The reproductive process in bees will perhaps be used as an argument against my theory. In these insects, the same egg will develope into a female or male individual, according as fertilization has or has not taken place, respectively. Hence, one and the same egg is capable of fertilization, and also of parthenogenetic development, if it does not receive a spermatozoon. It is in the power of the queen-bee to produce male or female individuals: by an act of will she decides whether the egg she is laying is to be fertilized or unfertilized. She 'knows beforehand'[164] whether an egg will develope into a male or a female animal, and deposits the latter kind in the cells of queens and workers, the former in the cells of drones. It has been shown by the discoveries of Leuckart and von Siebold that all the eggs are capable of developing into male individuals, and that they are only transformed into 'female eggs' by fertilization. This fact seems to be incompatible with my theory as to the cause of parthenogenesis, for if the same egg, possessing exactly the same contents, and above all the same segmentation nucleus, may develope sexually or parthenogenetically, it appears that the power of parthenogenetic development must depend on some factor other than the quantity of germ-plasm.

Although this appears to be the case, I believe that my theory encounters no real difficulty. I have no doubt whatever, that the same egg may develope with or without fertilization. From a careful study of the numerous excellent investigations upon this point which have been conducted in a particularly striking manner by Bessels[165] (in addition to the observers quoted above), I have come to the conclusion that the fact is absolutely certain. It must be candidly admitted that the *same* egg will develope into a drone when not fertilized, or into a worker or queen when fertilized. One of Bessels' experiments is sufficient to prove this assertion. He cut off the wings of a young queen and thus rendered her incapable of taking 'the nuptial flight.' He then observed that all the eggs which she laid developed into male individuals. This experiment was made in order to prove that drones are produced by unfertilized eggs; but it also proves that the assertion mentioned above is correct, for the eggs which ripen first and are therefore first laid, would have been fertilized had the queen been impregnated. The supposition that, at certain times, the queen produces eggs requiring fertilization, while at other times her eggs develope parthenogenetically, is quite excluded by this experiment; for it follows from it, that the eggs must all be of precisely the same kind, and that there is no difference between the eggs which require fertilization and those which do not.

But does it therefore follow that the quantity of germ-plasm in the segmentation nucleus is not the factor which determines the beginning of embryonic development? I believe not. It can be very well imagined that the nucleus of the egg, having expelled the ovogenetic nucleoplasm, may be increased to the size requisite for the segmentation nucleus in one of two ways: either by conjugation with a sperm-nucleus, or by simply growing to double its size. There is nothing improbable in this latter assumption, and one is even inclined to inquire why such growth does not take place in all unfertilized eggs. The true answer to this question must be that nature generally pursues the sexual method of reproduction, and that the only way in which the general occurrence of parthenogenesis could be prevented, was by the production of eggs which remained sterile unless they were fertilized. This was effected by a loss of the capability of growth on the part of the egg-nucleus after it had expelled the ovogenetic nucleoplasm.

The case of the bee proves in a very striking manner that the difference between eggs which require fertilization, and those which do not, is not produced until after the maturation of the egg, and the removal of the ovogenetic nucleoplasm. The increase in the quantity of the germ-plasm cannot have taken place at any earlier period, or else the nucleus of the egg would always start embryonic development by itself, and the egg would probably be incapable of fertilization. For the relation between egg-nucleus and sperm-nucleus is obviously based upon the fact that each of them is insufficient by itself, and requires completion. If such completion had taken place at an early stage the egg-nucleus would either cease to exercise any attractive force upon the sperm-nucleus, or else conjugation would be effected, as in Fol's interesting experiments upon fertilization by many spermatozoa; and, as in these experiments, malformation of the embryo would result. In *Daphnidae* I believe I have shown[166] that the summer-eggs are not only developed parthenogenetically, but also that they are never fertilized; and the explanation of this incapacity for fertilization may perhaps be found in the fact that their segmentation nucleus is already formed.

We may therefore conclude that, in bees, the nucleus of the egg, formed during maturation, may either conjugate with the sperm-nucleus, or else if no spermatozoon reaches the egg may, under the stimulus of internal causes, grow to double its size, thus attaining the dimensions of the segmentation nucleus. For our present purpose we may leave out of consideration the fact that in the latter case the individual produced is a male, and in the former case a female.

It is clear that such an increase in the germ-plasm must depend, to a certain extent, upon the nutrition of the nucleus, and thus indirectly upon the body of the egg-cell; but the increase must chiefly depend upon internal nuclear conditions, viz. upon the capability of growth. We must further assume that the latter condition plays the chief part in the process, for everywhere in the organic world the limit of growth depends upon the internal conditions of the growing body, and can only be altered to a small extent by differences of nutrition. The phyletic acquisition of the capability of parthenogenetic development must therefore depend upon an alteration in the capability of growth possessed by the nucleus of the egg.

This theory of parthenogenesis most nearly approaches Strasburger's views upon the subject, for he also explains the non-occurrence of parthenogenetic development by the insufficient quantity of nucleoplasm remaining in the egg after the expulsion of polar bodies. The former theory differs however in that the occurrence of parthenogenesis is supposed to be only due to an increase of this nucleoplasm to the normal size of the segmentation nucleus. Strasburger assumes that 'specially favourable conditions of nutrition counteract the deficiency of nuclear idioplasm,' while it seems to me that nutrition must be considered as only of secondary importance. Thus in bees, as above stated, the same egg may develope parthenogenetically or after fertilization, the nucleus being subject to the same conditions of nutrition in both cases. Strasburger[167] considers that parthenogenesis may be interpreted by one of three possible explanations. First, he suggests that especially favourable nutrition may lead to the completion of the nuclear idioplasm. But if this assumption be made, we must ask why a part of the idioplasm should be previously expelled, when immediately afterwards the presence of an equal amount becomes necessary. Such a view can only be explained by the above-made assumption that the expelled nucleoplasm has a different constitution from that possessed by the nucleoplasm which is afterwards formed. It is true that we do not yet certainly know whether a polar body is expelled in eggs in which parthenogenesis occurs, but we do know that the egg of the bee passes through the same stages of maturation whether it is to be fertilized or not. I can hardly accept Strasburger's second suggestion, 'that under some favourable conditions of nutrition half [or perhaps better, a quarter] of the idioplasm of the egg-nucleus is sufficient to start the processes of development in the cyto-idioplasm.' Finally, his third suggestion, 'that the cyto-idioplasm, nourished by its surroundings and thus increased in quantity, compels the nucleus of the egg to enter upon division,' presupposes that the cell-body gives the impulse for nuclear division, a supposition which up to the present time remains at least unproved. The ascertained facts appear to me to indicate rather that the cell-body serves only as a medium for the nutrition of the nucleus, and Fol's recently mentioned observations, which have been especially quoted by Strasburger in support of his theories, seem to me to rather confirm my conclusions. If supernumerary sperm-nuclei penetrate into the egg, they may, under the nutritive influence of the cell-body, become centres of attraction, and may take the first step towards nuclear and cell-division by forming amphiasters. Such nuclei cannot control the whole cell-body and force it to divide, but each one of them, having grown to a certain size at the expense of the cell-body, makes its influence felt over a certain area. Strasburger is quite right in considering this process as a 'partial parthenogenesis.' Such partial parthenogenesis presumably occurs in all egg-nuclei, but the latter cannot attain to complete parthenogenesis when, as in Fol's supernumerary sperm-nuclei, their powers of assimilation are insufficient to enable them to reach the requisite size. As before stated, the cell-body does not force the nucleus to divide, but *vice versa*. It would, moreover, be quite erroneous to suppose that parthenogenetic eggs must contain a larger amount of nutritive material in order to facilitate the growth of the nucleus. The parthenogenetic eggs of certain *Daphnidae* (*Bythotrephes*, *Polyphemus*) are very much smaller than the winter-eggs, which require fertilization, in the same species. It is also an error for Strasburger to conclude that 'it has been established with certainty that favourable conditions of nutrition cause parthenogenetic development in *Daphnidae*, while unfavourable conditions cause the formation of eggs requiring fertilization.' It is true that Carl Düsing[168], in his notable work upon the origin of sex, has attempted, in a most ingenious manner, to prove, from my observations and experiments on the reproduction of *Daphnidae*, 'that winter or summer-eggs are formed according to the nutritive condition of the ovary.' I do not, however, believe that he has succeeded in this attempt, and at all events it is quite clear that the validity of such conclusions is not fully established. I have observed that the maturing eggs break up in the ovaries and are absorbed in those *Daphnidae* (*Sida*) which are starved because sufficient food cannot be provided in captivity. Hence such animals live, as it were, at the expense of their descendants; but it would be quite erroneous to conclude with Düsing, from the similarity which such disappearing egg-follicles bear to the groups of germ-cells which normally break up in the formation of winter-eggs, that with a less degree of starvation winter-eggs would have been formed. Düsing further quotes my incidental remark that the formation of resting-eggs in *Daphnia* has been especially frequent in aquaria 'which had been for some time neglected, and in which it was found that a great increase in the number of individuals had taken place.' He is entirely wrong in concluding that there was any want of food in these neglected aquaria; and if I had foreseen that such conclusions would have been drawn, I might have easily guarded against them by adding that in these very aquaria an undisturbed growth of different algae was flourishing, so that there could have been no deficiency, but, on the contrary, a great abundance of nutritive material. I may add that since that time I have conducted some experiments directly bearing upon this question, by bringing virgin females as near to the verge of starvation as possible, but in no case did they enter upon sexual reproduction[169].

An author must have been to some extent misled by preconceived ideas when he is unable to see that the manner in which the two kinds of eggs are respectively formed, directly excludes the possibility of the origin of sexual eggs from the effects of deficient or poor nutrition. The resting eggs, which require fertilization, are always larger, and require for their formation far more nutritive material, than the parthenogenetic summer-eggs. In *Moina*, for instance, forty large food-cells are necessary for the formation of a resting egg, while a summer-egg only requires three. And Düsing is aware of these facts, and quotes them. How can the formation of resting eggs depend upon the effects of poor nutrition when food is most abundant at the very time of their formation? In all those species which inhabit lakes, sexual reproduction occurs towards the autumn, and in such cases the resting eggs are true winter-eggs, destined to preserve the species during the winter. But at no time of the year is the food of the *Daphnidae* so abundant as in September and October, and frequently even until late in November (in South Germany). At this period of the year, the water is filled with flakes of animal and vegetable matter in a state of partial decomposition, thus affording abundant food for many species. It also swarms with a large number of species of Crustacea, Radiolaria, and Infusoria; and thus such Daphnids as the *Polyphemidae* are also well provided for. Hence there is no deficiency in the supply of food. Any one who has used a fine net in our fresh waters at this time of the year must have been at first astonished at the enormous abundance of the lower forms of animal life; and he must have been much more astonished if he has been able to compare such results with the scanty population of the same localities in spring. But it is during the spring and summer that these very *Daphnidae* reproduce themselves parthenogenetically. I am far from believing that my experiments on *Daphnidae* are exhaustive and final, and I have stated this in my published writings on the subject; but it seems to me that I have established the fact that direct influences, whether of food or of temperature, acting upon single individuals, do not determine the kind of eggs which are to be produced; but that such a decisive influence is to be found in the indirect conditions of life, and especially in the average frequency of the recurrence of adverse circumstances which kill whole colonies at once, such as the winter cold, or the drying-up of small ponds in summer. It is unnecessary for me to controvert Düsing in detail, as I have already taken this course in the case of Herbert Spencer[170], who had also formed the hypothesis that diminished nutrition causes sexual reproduction.

One of my observations seems, indeed, to support such a view, but only when it is considered as an isolated example. I refer to the behaviour of the genus *Moina*. Females of this genus which possess sexual eggs in their ovaries, and which would have continued to produce such eggs if males had been present, enter in the absence of the latter upon the formation of parthenogenetic summer-eggs, that is, if the sexual eggs have not all been extruded, but have been re-absorbed in the ovary. At first sight, indeed, such a result appears to indicate that the increase in nutrition, produced by the breaking-up of the large winter-egg in the ovary, determines the formation of parthenogenetic eggs. This apparent conclusion seems to be further confirmed by the following fact. The transition from sexual to parthenogenetic reproduction only occurs in one species of *Moina* (*M. rectirostris*), but in this species it occurs always and without exception, while in the other species which I have investigated (*M. paradoxa*), winter-eggs, when once formed, are always laid, and such females can never produce summer-eggs. But in spite of this fact, Düsing is mistaken when he explains the continuous formation of sexual eggs in the latter species as due to the absence of any great increase in the amount of nutrition, such as would have followed if the egg had broken up in the ovary. In many other *Daphnidae* which have come under my notice, the females frequently enter again upon the formation of parthenogenetic summer-eggs, after having laid fertilized resting eggs, upon one or more occasions. This is the case, for instance, in all the species of *Daphnia* with which I am acquainted, and such a fact at once proves that the abnormal increase in nutrition produced by the absorption of winter-eggs cannot be the cause of the succeeding parthenogenesis. It also supports the proof that a high or low nutritive condition of the whole animal can have nothing to do with the kind of eggs which are produced, for in the above-quoted instance, the nutrition has remained the same throughout, or at all events has not been increased. It is erroneous to always look for the explanation of the mode of egg-formation in the direct action of external causes. Of course there must be direct causes which determine that one germ shall become a winter-egg, and another a summer-egg; but such causes do not lie outside the animal, and have nothing to do with the nutritive condition of the ovary: they are to be found in those conditions which we are not at present able to analyze further, and which we must, in the meantime, call the specific constitution of the species. In the young males of *Daphnidae* the testes have precisely the same appearance as the ovaries of the young females[171], but the former will, nevertheless, produce sperm-cells and not ova. In such cases the sex of the young individual can always be identified by the form of the first antenna and of the first thoracic appendage, both of which are always clawed in the male. But who can point to the direct causes which determine that the sexual cells shall become sperm-cells in this case, and not egg-cells? Does the determining cause depend on the conditions of nutrition? Or, again, in the females, can the state of nutrition determine that the third out of a group of four germ-cells shall become an egg-cell, and that the others shall break up to serve as its food?

It is, I think, clear that these are obvious instances of the general conclusion that the direct causes determining the direction of development in each case are not to be looked for in external conditions, but in the constitution of the organs concerned.

We arrive at a like conclusion when we consider the quality of the eggs which are produced. The constitution of one species of *Moina* contains the cause which determines that each individual shall produce winter-eggs only, or summer-eggs only; while in another species the transition from the formation of sexual eggs to the formation of summer-eggs can take place, but only when the winter-egg remains unfertilized. The latter case appears to me to be notably a special adaptation, in this and other species, to the deficiency of males, which is apt to occur. At all events, it is obvious that it is an advantage that an unfertilized sexual egg shall not be lost to the organism. The re-absorption of the winter-egg is an arrangement which, without being the cause, is favourable to the production of summer-eggs.

This subject is by no means a simple one, as is proved by the behaviour of the small group of *Daphnidae*. Thus in some species, the winter-eggs are produced by purely sexual females, which never enter upon parthenogenesis; in others, the sexual females may take the latter course, but only when males are absent; in others, again, they regularly enter upon parthenogenesis. In my work on *Daphnidae*, I have attempted to show that their behaviour in this respect is associated with the various external conditions under which the different species live; and also that the ultimate occurrence of the sexual period, and finally the whole cyclical alternation of sexual and parthenogenetic reproduction, depend upon adaptation to certain external conditions of life.

With the aid of my hypothesis that the egg-nucleus is composed of ovogenetic nucleoplasm and germ-plasm, I can now attempt to give an approximate explanation of the nature and origin of the direct causes which determine the production, at one time of parthenogenetic summer-eggs, and at another time of winter-eggs, requiring fertilization. But in such an explanation I should also wish to include a consideration of the causes which determine the formation of the nutritive cells of the egg and of the sperm-cells to which I have alluded above.

I believe that the direct cause which determines why the apparently identical cells of the young testis and ovary in the *Daphnidae* develope in such different directions, is to be found in the fact, that their nuclei possess different histogenetic nucleoplasms, while, if we neglect individual differences, the germ-plasm remains precisely the same. In the sperm-cells the histogenetic nucleoplasm is spermogenetic, in the egg-cells it is ovogenetic. This must be conceded if our fundamental view is correct, that the specific nature of the cell-body is determined by the nature of its nucleus.

Similarly, the germ-cells of female *Daphnidae*, which at first do not exhibit the smallest differences, must really differ in that their nuclei must contain different kinds of nucleoplasm, which are present in different proportions. Germ-cells which are to produce a finely granular, brick-red, winter yolk (*Moina rectirostris*) must possess an ovogenetic nucleoplasm of a somewhat different molecular structure from those germ-cells which have only to form a few large blue fat-globules, as in the summer-eggs of the same species. It is further probable that different proportions obtain between germ-plasm and ovogenetic nucleoplasm, in these two kinds of germ-cells; and it would be a very simple explanation of the otherwise obscure part played by the food-cells, if we were to suppose that they do not contain any germ-plasm at all, and on this account do not enter upon embryonic development, but are arrested after growing to a certain size. Such an explanation, however, would not by itself show why they subsequently undergo gradual solution in the surrounding fluids. But since we know that egg-cells also begin to undergo solution as soon as the parent Daphnid is poorly nourished, we can hardly help also referring the solution of the food-cells to insufficient nourishment, occurring as soon as the egg-cell, after the attainment of a certain size, exercises a superior power of assimilation. But hitherto we could not in any way understand why the third out of a group of germ-cells should always gain this superior power and become an egg-cell. If it could be shown that its position is more highly favoured in respect of nutrition, we could understand why it outstrips the other three in development, and thus prevents them from further growth. But nothing of the kind can be shown to occur with any degree of probability, as I have previously mentioned in my works on the subject. At that time, having no better explanation, I adopted the view in question, although only as a provisional interpretation. It was not possible for me to seek in the substance of those four apparently identical cells for the cause of their different development; but now I am justified in offering the supposition that during the division of a primitive germ-cell into two, and afterwards into four germ-cells, an unequal division of the nucleoplasms takes place, in that one of the four cells receives germ-plasm as well as ovogenetic nucleoplasm, while the other three receive the latter alone. Similarly, the fact that the second cell of the group may occasionally become an egg is also intelligible, although this fact remained quite inexplicable by my former interpretation. The fact that true egg-cells, or even the whole ovary with all its germ-cells, may break up and become absorbed when the animal has been starved for a certain period of time, seems to me to be no objection to our present view, any more than the fact that an Infusorian may die from starvation would be an objection to the supposition of the immortality of unicellular organisms. The growth of an organism is not only arrested by its constitution, but also by absolute want of food; but it would be very foolish to explain the differences in size of the various species of animals as results of the different conditions of nutrition to which they were subject. Just as a sparrow, however highly nourished, could never attain the size or form of an eagle, so a germ-cell destined to become a summer-egg could never attain the size, form, or colour of a winter-egg. It is by internal constitutional causes that the course of development is determined in both these cases; and in the latter, the cause can hardly be anything more than the different constitution of the nucleoplasms.

All these considerations depend upon the supposition that the egg-nucleus contains two kinds of idioplasm, viz. germ-plasm and ovogenetic nucleoplasm. I have not hitherto brought forward any direct evidence in favour of this assumption, but I believe that such proofs can be obtained.

It is well known that there are certain eggs in which the polar bodies are not expelled until after the entrance of spermatozoa. Brooks[172] has already made use of this fact as evidence against Minot's and Balfour's theory; for he quite rightly concludes that if the polar bodies really possess the significance of male cells, we cannot understand why such eggs are unable to develope without fertilization, when they still possess the male half of the nucleus necessary for development. But such eggs (e.g. that of the oyster) do not develope, but always die if they remain unfertilized.

This argument can only be met by a new hypothesis, the construction of which I must leave to the defenders of the above-mentioned theory. But the observation in question seems to me to furnish at the same time a proof of the co-existence of two different nucleoplasms in the egg-nucleus. If the nucleoplasm of the polar bodies was also germ-plasm, we could not understand why such eggs are unable to develope parthenogenetically, for at least as much germ-plasm is contained in the unfertilized egg as would have been present after fertilization.

The only objection which can be raised against this conclusion depends upon the supposition that the nucleoplasm of the sperm-cell is qualitatively different from that of the egg-cell. I have already dealt with this view, but I should wish to refer to it again rather more in detail. Some years ago I expressed the opinion[173] that the physiological values of the sperm-cell and of the egg-cell must be identical; that they stand in the ratio of 1 : 1. But Valaoritis[174] has brought forward the objection that if we consider the function of a cell as the measure of its physiological value, it is only necessary to point to the respective functions of ovum and spermatozoon in order to show that their physiological values must be different. 'The egg-cell alone, by passing more or less completely through the phyletic stages of the female parent, developes into a similar organism; and although the presence of the spermatozoon is in most cases required in order to render possible such a result, the cases of parthenogenesis prove nevertheless that the egg can do without this stimulus.' This objection appeared to be fully justified as long as fertilization was looked upon as the 'vitalization of the germ,' and so long as the sperm-cell was considered as merely 'the spark that kindles the gunpowder,' and further so long as the germ-substance was believed to be contained in the cell-body. But now we can hardly give to the body of the egg-cell a higher significance than that of the common nutritive soil of the two nuclei which conjugate in fertilization. But these two nuclei 'are not different in nature,' as Strasburger says, and as I fully believe. They cannot differ in kind, for they both consist of germ-plasm belonging to the same species of animal or plant; and there cannot be any deeper contrast between them such as would correspond to the differences between mature individuals. They cannot, from their essential nature, exercise any special attraction upon each other, and when we see that sperm-cell and egg-cell do nevertheless attract each other, as has been shown in both plants and animals, such a property must have been secondarily acquired, and has no other significance than to favour the union of sexual cells—an arrangement which may be compared to the vibrating flagellum of the spermatozoon or the micropyle of the egg, but which is not fundamental, and is not based upon the molecular structure of the germ-plasm. In lower plants, Pfeffer has proved that certain chemical stimuli emanate from the egg and attract the spermatozoid; and according to Strasburger, the synergidae in the upper part of the embryo-sac of Phanerogams secrete a substance which is capable of directing the growth of the pollen-tube towards the egg-cell. In animals it is only known as yet that spermatozoa and ova do attract each other, so that the former find the latter and bore their way through its membranes. It has also been shown that the substance of the egg-body moves towards the penetrating spermatozoon ('*cones d'exsudation*' in *Asteridae*: Fol); and that it sometimes enters upon convulsive movements (*Petromyzon*). Here therefore a mutual stimulation and attraction must exist; and perhaps we must also assume that there is an attraction between the two conjugating nuclei, for we cannot readily understand how the cytoplasm alone could direct the one to the other, as Strasburger supposes. According to Strasburger's hypothesis, we must suppose that part of the specific cytoplasm of the sperm-cell continues to surround the nucleus after it has penetrated into the body of the egg. But however this may be, the assumed attraction between the conjugating nuclei certainly cannot depend upon the molecular structure of their germ-plasm, which is the same in both, but it must be due to some accessory circumstance. If it were possible to introduce the female pronucleus of an egg into another egg of the same species, immediately after the transformation of the nucleus of the latter into the female pronucleus, it is very probable that the two nuclei would conjugate just as if a fertilizing sperm-nucleus had penetrated. If this were so, the direct proof that egg-nucleus and sperm-nucleus are identical would be furnished. Unfortunately the practical difficulties are so great that it is hardly possible that the experiment can ever be made; but such want of experimental proof is partially compensated for by the fact, ascertained by Berthold, that in certain Algae (*Ectocarpus* and *Scytosiphon*) there is not only a female, but also a male parthenogenesis; for he shows that in these species the male germ-cells may sometimes develope into plants, which however are very weakly[175]. Furthermore the process of conjugation may be considered as a proof that this view as to the secondary importance of sexual differentiation is the true one. At the present time there can hardly be any hesitation in accepting the view that conjugation is the sexual reproduction of unicellular organisms. In these the two conjugating cells are almost always identical in appearance, and there is no evidence in favour of the assumption that they are not also identical in molecular structure, at least so far as one individual of the same species may be identical with another. But there are also forms in which the conjugating cells are distinctly differentiated into male and female, and these are connected with the former by a gradual transition: thus in *Pandorina*, a genus of *Volvocineae*, we are unable to make out any differences between the conjugating cells, while large egg-cells and minute sperm-cells exist in the closely allied *Volvox*. If we must suppose that the conjugation of two entirely identical Infusoria has the same physiological effect as the union of two sexual cells in higher animals and plants, we cannot escape the conclusion that the process is essentially the same throughout: and that therefore the differences, which are perhaps already indicated in *Pandorina* and are very distinct in *Volvox* and in all higher organisms, have nothing to do with the nature of the process, but are of quite secondary importance. If we further take into account the extremely different constitution of the two kinds of sexual cells in size, appearance, membranes, motile power, and finally in number, no doubt remains that these differences are only adaptations which secure the meeting of the two kinds of conjugating cells: that in each species they are adaptations to the peculiar conditions under which fertilization takes place.

NOTE.

It is of considerable importance for the proper appreciation of the views advanced in the present essay, to ascertain whether a polar body is or is not expelled from eggs which develope parthenogenetically. I wish therefore to briefly state that I have recently succeeded in proving the formation of a polar body of distinctly cellular structure in the summer-eggs of *Daphnidae*. I propose to publish a more detailed account in a future paper.

A. W.

June 22, 1885.

Footnotes for Essay IV.

94. Häckel, 'Ueber die Wellenzeugung der Lebenstheilchen etc.,' Berlin, 1876.

95. Darwin, 'The Variation of Animals and Plants under Domestication,' vol. ii. 1875, chap. xxvii. pp. 344-399.

96. His, 'Unsre Körperform etc.,' Leipzig, 1875.

97. Brooks, 'The Law of Heredity,' Baltimore, 1883.

98. Galton's experiments on transfusion in Rabbits have in the mean time really proved that Darwin's gemmules do not exist. Roth indeed states that Darwin has never maintained that his gemmules make use of the circulation as a medium, but while on the one hand it cannot be shown why they should fail to take the favourable opportunities afforded by such a medium, inasmuch as they are said to be constantly circulating through the body; so on the other hand we

cannot understand how the gemmules could contrive to avoid the circulation. Darwin has acted very wisely in avoiding any explanation of the exact course in which his gemmules circulate. He offered his hypothesis as a formal and not as a real explanation.

Professor Meldola points out to me that Darwin did not admit that Galton's experiments disproved pangenesis ('Nature,' April 27, 1871, p. 502), and Galton also admitted this in the next number of 'Nature' (May 4, 1871, p. 5).—A. W. 1889.

99. Weismann, 'Ueber die Vererbung.' Jena, 1883; translated in the present volume as the second essay 'On Heredity.'

100. E. Roth, 'Die Thatsachen der Vererbung.' 2. Aufl., Berlin, 1885, p. 14.

101. Jäger, 'Lehrbuch der allgemeinen Zoologie,' Bd. II. Leipzig, 1878.

102. M. Nussbaum, 'Die Differenzirung des Geschlechts im Thierreich,' Arch. f. Mikrosk. Anat., Bd. XVIII. 1880.

103. I have since learnt that Professor Rauber of Dorpat also expressed similar views in 1880; and Professor Herdman of Liverpool informs me that Mr. Francis Galton had brought forward in 1876 a theory of heredity of which the fundamental idea in some ways approached that of the continuity of the germ-plasm ('Journal of the Anthropological Institute,' vol. v; London, 1876).—A. W., 1888.

[A less complete theory was brought forward by Galton at an earlier date, in 1872 (see Proc. Roy. Soc. No. 136, p. 394). In this paper he proposed the idea that heredity chiefly depends upon the development of the offspring from elements directly derived from the fertilized ovum which had produced the parent. Galton speaks of the fact that 'each individual may properly be conceived as consisting of two parts, one of which is latent and only known to us by its effects on his posterity, while the other is patent, and constitutes the person manifest to our senses. The adjacent and, in a broad sense, separate lines of growth in which the patent and latent elements are situated, diverge from a common group and converge to a common contribution, because they were both evolved out of elements contained in a structureless ovum, and they, jointly, contribute the elements which form the structureless ova of their offspring.' The following diagram shows clearly 'that the span of each of the links in the general chain of heredity extends from one structureless stage to another, and not from person to person:—

Structureless elements {...Adult Father... } structureless elements

in Father   {...Latent in Father...} in Offspring.'

Again Galton states—'Out of the structureless ovum the embryonic elements are taken ... and these are developed (*a*) into the visible adult individual; on the other hand ..., after the embryonic elements have been segregated, the large residue is developed (*b*) into the latent elements contained in the adult individual.' The above quoted sentences and diagram indicate that Galton does not derive the whole of the hereditary tendencies from the latent elements, but that he believes some effect is also produced by the patent elements. When however he contrasts the relative power of these two influences, he attaches comparatively little importance to the patent elements. Thus if any character be fixed upon, Galton states that it 'may be conceived (1) as purely personal, without the concurrence of any latent equivalents, (2) as personal but conjoined with latent equivalents, and (3) as existent wholly in a latent form.' He argues that the hereditary power in the first case is exceedingly feeble, because 'the effects of the use and disuse of limbs, and those of habit, are transmitted to posterity in only a very slight degree.' He also argues that many instances of the supposed transmission of personal characters are really due to latent equivalents. 'The personal manifestation is, on the average, though it need not be so in every case, a certain proof of the existence of latent elements.' Having argued that the strength of the latter in heredity is further supported by the facts of reversion, Galton considers it is safe to conclude 'that the contribution from the patent elements is very much less than from the latent ones.' In the later development of his theory, Galton adheres to the conception of 'gemmules' and accepts Darwin's views, although 'with considerable modification.' Together with pangenesis itself, Galton's theory must be looked upon as *preformational*, and so far it is in opposition to Weismann's theory which is *epigenetic*. See Appendix IV. to the next Essay (V.), pp. 316-319.—E. B. P.]

104. Nägeli, 'Mechanisch-physiologische Theorie der Abstammungslehre.' München u. Leipzig, 1884.

105. O. Hertwig, 'Beiträge zur Kenntniss der Bildung, Befruchtung und Theilung des thierischen Eies.' Leipzig, 1876.

106. Fol, 'Recherches sur la fécondation, etc.' Genève, 1879.

107. Kölliker formerly stated, and has again repeated in his most recent publication, that the spermatozoa ('Samenfäden') are mere nuclei. At the same time he recognizes the existence of sperm-cells in certain species. But proofs of the former assertion ought to be much stronger in order to be sufficient to support so improbable a hypothesis as that the elements of fertilization may possess a varying morphological value. Compare Zeitschr. f. wiss. Zool., Bd. XLII.

108. F. M. Balfour, 'Comparative Embryology,' vol. i. p. 69.

109. Arch. f. mikr. Anat., Bd. 23. p. 182, 1884.

110. Born, 'Biologische Untersuchungen,' I, Arch. Mikr. Anat., Bd. XXIV.

111. Roux, 'Beiträge zum Entwicklungsmechanismus des Embryo,' 1884.

112. O. Hertwig, 'Welchen Einfluss übt die Schwerkraft,' etc. Jena, 1884.

113. [Our present knowledge of the development of vegetable ova (including the position of the parts of the embryo) is also in favour of the view that it is not influenced by external causes, such as gravitation and light. It takes place in a manner characteristic of the genus or species, and essentially depends on other causes which are fixed by heredity, see Heinricher 'Beeinflusst das

Licht die Organanlage am Farnembryo?' in Mittheilungen aus dem Botanischen Institute zu Graz, II. Jena, 1888.—S. S.]

114. E. van Beneden, 'Recherches sur la maturation de l'œuf,' etc., 1883.

115. M. Nussbaum, 'Ueber die Veränderung der Geschlechtsprodukte bis zur Eifurchung,' Arch. Mikr. Anat., 1884.

116. Eduard Strasburger, 'Neue Untersuchungen über den Befruchtungsvorgang bei den Phanerogamen als Grundlage für eine Theorie der Zeugung.' Jena, 1884.

[It is now generally admitted that, in the Vascular Cryptogams, as also in Mosses and Liverworts, the bodies of the spermatozoids are formed by the nuclei of the cells from which they arise. Only the cilia which they possess, and which obviously merely serve as locomotive organs, are said to arise from the surrounding cytoplasm. It is therefore in these plants also the nucleus of the male cell which effects the fertilization of the ovum. See Göbel, 'Outlines of Classification and Special Morphology,' translated by H. E. F. Garnsey, edited by I. B. Balfour, Oxford, 1887, p. 203, and Douglas H. Campbell, 'Zur Entwicklungsgeschichte der Spermatozoiden,' in Berichte d. deutschen bot. Gesellschaft, vol. v (1887), p. 120.—S. S.]

117. O. Hertwig, 'Das Problem der Befruchtung und der Isotropie des Eies.' Jena, 1885.

118. This opinion was first expressed in my lecture, 'Ueber die Dauer des Lebens,' Jena, 1882, translated as the first essay in the present volume.

119. M. Nussbaum, 'Sitzungber. der Niederrheinischen Gesellschaft fur Natur- und Heilkunde.' Dec. 15, 1884.

120. A. Gruber, 'Biologisches Centralblatt,' Bd. IV. No. 23, and V. No. 5.

121. According to the observations of Nussbaum and van Beneden, the egg of *Ascaris* departs from the ordinary type, but I think that the latter observer goes too far when he concludes from the form of the nuclear spindle (of which the two halves are inclined to each other at an angle) that we have before us a process entirely different from that of ordinary nuclear division.

122. Trinchese, 'I primi momenti dell' evoluzione nei molluschi,' Atti Acad. Lyncei (3) vii. 1879, Roma.

123. M. Nussbaum, 'Archiv für Mikroskopische Anatomie,' Bd. XVIII und XXIII.

124. Valaoritis, 'Die Genesis des Thier-Eies.' Leipzig, 1882.

125. Kölliker, 'Die Bedeutung der Zellkerne,' etc.; Zeitschr. f. wiss. Zool. Bd. XLII.

126. 'Compt. rend.' Tom. LIV. p. 150.

127. 'Entwicklung der Dipteren.' Leipzig, 1864.

128. 'Zeitschr. f. wiss. Zool.' Bd. XVI. p. 389 (1866).

129. 'Compt. rend.' Nov. 13, 1882.

130. Grobben, 'Arbeiten d. Wien. Zool. Instituts,' Bd. II. p. 203.

131. Bütschli, 'Zeitschrift f. wiss. Zool.' Bd. XXIII. p. 409.

132. 'Science,' vol. iv. No. 90, 1884.

133. Among unicellular organisms, encysted individuals are often called germs. They sometimes differ from the adult organism in their smaller size and simpler structure (*Gregarinidae*), but they represent the same morphological stage of individuality.

134. Compare Bütschli in Bronn's 'Klassen und Ordnungen des Thierreichs,' Bd. I. p. 777.

135. Gustav Jäger, 'Lehrbuch der Allgemeinen Zoologie,' Leipzig, 1878; II. Abtheilung. Probably on account of the extravagant and superficial speculations of the author, the valuable ideas contained in his book have been generally overlooked. It is only lately that I have become aware of Jäger's above-mentioned hypothesis. M. Nussbaum seems to have also arrived at the same conclusion quite independently of Jäger. The latter has not attempted to work out his hypothesis with any degree of completeness. The above-mentioned observations are followed immediately by quite valueless considerations, as, for instance, that the ontogenetic and phyletic groups are in concentric ratio! The author might as well speak of a quadrangular or triangular ratio!

136. [Facts of the same kind are also known in the Vascular Cryptogams, Muscineae, Characeae, Florideae, etc.—S. S.]

137. Weismann, 'Die Entstehung der Sexualzellen bei den Hydromedusen.' Jena, 1883.

138. [I adopt this term, suggested by E. Ray Lankester and G. C. Bourne, as the name of the supporting lamina of Coelenterata. See 'Quart. Journ. Microsc. Sci.' Jan. 1887, p. 28.—E. B. P.]

139. Dr. Clemens Hartlaub, 'Ueber die Entstehung der Sexualzellen bei Obelia.' Freiburg, Inaugural Dissertation: see also 'Zeitschrift für wissenschaftliche Zoologie.' Bd. XLI. 1884.

140. English translation, by H. Marshall Ward. Oxford, 1887, Clarendon Press.

141. [Such gland-cells are known in both animals and plants. See W. Gardiner and Tokutaro Ito, On the structure of the mucilage-secreting cells of *Blechnum occidentale* L., and *Osmunda regalis* L., 'Annals of Botany,' vol. i. p. 49.—S. S.]

142. Thus in 1877 Bütschli thought that 'the chief significance of the formation of polar bodies lies in the removal of part of the nucleus of the egg, whether this removal is effected by simple expulsion or by the budding of the egg-cell.' 'Entwicklungsgeschichtliche Beiträge;' Zeitschrift für wissenschaftliche Zoologie, Bd. XXIX. p. 237, footnote.

143. C. S. Minot, 'Account, etc.;' Proc. Boston Soc. Nat. Hist. vol. xix. p. 165, 1877.

144. E. van Beneden and Boveri have recently, quite independently of each other, made a more exact study of these 'Polkörperchen' ('Centrosoma,' Boveri). They show that nuclear division starts from these bodies, although the mode of origin of the latter is not yet quite clear.—A. W., 1888.

145. The existence of polar bodies in sponges has been recently proved by Fiedler: Zool., Anzeiger., Nov. 28, 1887.—A. W., 1888.

146. They have now been observed in many species, so that their general occurrence in insects is tolerably certain. Compare bibliography given in Weismann and Ischikawa, 'Weitere Untersuchungen zum Zahlengesetz der Richtungskörper,' 'Zoolog. Jahrbücher,' vol. iii. 1888, p. 593.—A. W., 1888.

147. Van Beneden, even in his last work, considers these bodies to have only the value of nuclei; l. c., p. 394.

148. I purposely abstain from using a more precise term, for the complicated terminology employed in spermatogenesis hardly contributes anything to the elucidation of the phenomena themselves. Why do we not simply speak of sperm-cells and spermatoblasts, and distinguish the latter by numbers when they occur in successive generations of different form? Moreover, all the names which have been suggested for successive stages of development, can only be applied to the special group of animals upon which the observations have been made. Hence great confusion results from the use of such terms as spermatoblasts, spermatogonia, spermatomeres, spermatocysts, spermatocytes, spermatogemmae, etc.

149. Fol, 'Sur l'origine des cellules du follicule et de l'ovule chez les Ascidies.' Compt. rend., 28 mai, 1883.

150. Roule, 'La structure de l'ovaire et la formation des œufs chez les Phallusiadées.' Ibid., 9 avril, 1883.

151. Balbiani, 'Sur l'origine des cellules du follicule et du noyau vitellin de l'œuf chez les Géophiles.' Zool. Anzeiger, 1883, Nos. 155, 156.

152. Will, 'Ueber die Entstehung des Dotters und der Epithelzellen bei den Amphibien und Insecten.' Ibid., 1884, Nos. 167, 168.

153. [It is almost certain that this vesicle is not derived from the nucleus, but from the cytoplasm of the sperm-mother-cell. See Douglas H. Campbell, 'Zur Entwicklungsgeschichte der

Spermatozoiden' in Berichte der deutschen botanischen Gesellschaft, vol. v, 1887, p. 122.—S. S.]

154. Bütschli, 'Gedanken über die morphologische Bedeutung der sogenannten Richtungskörperchen,' Biolog. Centralblatt, Bd. VI. p. 5, 1884.

155. F. M. Balfour, 'Comparative Embryology,' vol. i. p. 63.

156. The formation of a polar body in parthenogenetic eggs has now been proved: see note at the end of this Essay; see also Essay VI.—A. W., 1888.

157. R. Leuckart,—article 'Zeugung,' in R. Wagner's 'Handwörterbuch der Physiologie,' 1853, Bd. IV. p. 958. Similar observations were made by Max Schultze. These observations appear however to be erroneous, for Pflüger has since shown that the eggs of frogs never develope if the necessary precautions are taken to prevent the access of any spermatozoa to the water.—A. W., 1888.

158. Oellacher, 'Die Veränderungen des unbefruchteten Keims des Hühncheneies. 'Zeitschrift für wissenschaftliche Zoologie,' Bd. XXII. p. 181. 1872.

159. Hensen, 'Centralblatt,' 1869, No. 26.

160. Weismann, 'Beiträge zur Naturgeschichte der Daphnoiden,' Leipzig, 1876-79, Abhandlung VII, and 'Zeitschrift für wissenschaftliche Zoologie,' Bd. XXXIII.

161. Weismann, 'Beiträge zur Kenntniss der ersten Entwicklungsvorgänge im Insectenei,' Bonn, 1882, p. 106.

162. W. Roux, 'Ueber die Bedeutung der Kerntheilungsfiguren.' Leipzig, 1883.

163. We now know that the number of loops varies considerably in different species, even when they belong to the same group of animals (e.g. Nematodes).—A.W., 1888.

164. This expression is used by bee-keepers, for instance by the well-known Baron Berlepsch. Of course, it would be more accurate to say that the queen, seeing the cell of a drone, is stimulated to lay an unfertilized egg, and that, on the other hand, she is stimulated to lay a fertilized egg when she sees the cell of a worker, or that of a queen.

165. E. Bessels, 'Die Landois'sche Theorie widerlegt durch das Experiment.' Zeitschrift für wissenschaftliche Zoologie, Bd. XVIII. p. 124. 1868.

166. 'Daphniden,' Abhandlung, vi. p. 324.

167. l. c., p. 150.

168. Carl Düsing, 'Die Regulirung des Geschlechtsverhältnisses.' Jena. 1884.

169. I intend to publish these experiments elsewhere in connexion with other observations.

170. Weismann, 'Daphniden,' Abhandlung, VII. p. 329; Herbert Spencer, 'The Principles of Biology,' 1864, vol. i. pp. 229, 230.

171. The same fact has since been ascertained in species belonging to several groups of animal.

172. Brooks, 'The Law of Heredity.' Baltimore, 1883, p. 73.

173. 'Zeitschrift für wissenschaftliche Zoologie,' Bd. XXXIII. p. 107. 1873.

174. Valaoritis, l. c., p. 6.

175. I quote from Falkenberg, in Schenk's 'Handbuch der Botanik,' Bd. II. p. 219. He further states that these are the only instances hitherto known in which undoubted male cells have proved to be capable of further development when they have been unable to exercise their powers of fertilization. It must be added that the two kinds of germ-cells do not differ in appearance, but only in behaviour; the female germ-cells becoming fixed, and withdrawing one of their two flagella, while the male cells continue to swarm. But even this slight degree of differentiation requires the supposition of internal molecular differentiation.

V.

THE SIGNIFICANCE OF SEXUAL REPRODUCTION

IN THE THEORY OF NATURAL SELECTION.

1886.

SIGNIFICANCE OF SEXUAL REPRODUCTION, ETC.

PREFACE.

The greater part of the present essay was delivered at the first general meeting of the Association of German Naturalists, at Strassburg, on September 18th, 1885, and is printed in the Proceedings of the fifty-eighth meeting of that Society.

The form of a lecture has been retained in the present publication, but its contents have been extended in many ways. Besides many small and a few large additions to the text, I have added six appendices in order to treat of certain subjects more fully than was possible in the lecture itself, in which I was often obliged to be content with mere hints and suggestions. This appears to be all the more necessary because it is impossible to suppose that many views and ideas upon which the lecture was based would be well known to all readers, although they have been described in my former papers. It was above all necessary to deal with the class of acquired characters, which, as it seems to me, is easily confounded, especially by the medical profession, with the much broader class of new characters generally. Only those new characters can be called 'acquired' which owe their origin to external influences, and the term 'acquired' must be denied to those which depend upon the mysterious relationship between the different hereditary tendencies which meet in the fertilized ovum. These latter are not 'acquired' but inherited, although the ancestors did not possess them as such, but only as it were the elements of which they are composed. Such new characters as these do not at present admit of an exact analysis: we have to be satisfied with the undoubted fact of their occurrence. The transmission or non-transmission of acquired characters must be of the highest importance for a theory of heredity, and therefore for the true appreciation of the causes which lead to the transformation of species. Any one who believes, as I do, that acquired characters are not transmitted, will be compelled to assume that the process of natural selection has had a far larger share in the transformation of species than has been as yet accorded to it; for if such characters are not transmitted, the modifying influence of external circumstances in many cases remains restricted to the individual, and cannot have any part in producing transformation. We shall also be compelled to abandon the ideas as to the origin of individual variability which have been hitherto accepted, and shall be obliged to look for a new source of this phenomenon, upon which the processes of selection entirely depend.

In the following pages I have attempted to suggest such a source.

A. W.

Freiburg I. Br.,

November 22, 1885.

SIGNIFICANCE OF SEXUAL REPRODUCTION, etc.

CONTENTS.

APPENDICES.

causes

V.

THE SIGNIFICANCE OF SEXUAL REPRODUCTION IN THE THEORY OF NATURAL SELECTION.

During the quarter of a century which has elapsed since Biology began to occupy itself again with general problems, at least one main fact has been made clear by the united labours of numerous men of science, viz. the fact that the Theory of Descent, the idea of development in the organic world, is the only conception as to the origin of the latter, which is scientifically tenable. It is not only that, in the light of this theory, numerous facts receive for the first time a meaning and significance; it is not only that, under its influence, all the ascertained facts can be harmoniously grouped together; but in some departments it has already yielded the highest results which can be expected from any theory, it has rendered possible the prediction of facts, not indeed with the absolute certainty of calculation, but still with a high degree of probability. It has been predicted that man, who, in the adult state, only possesses twelve pairs of ribs, would be found to have thirteen or fourteen in the embryonic state: it has been predicted that, at this early period in his existence, he would possess the insignificant remnant of a very small bone in the wrist, the so-called *os centrale*, which must have existed in the adult condition of his extremely remote ancestors. Both predictions have been fulfilled, just as the planet Neptune was discovered after its existence had been predicted from the disturbances induced in the orbit of Uranus.

That existing species have not arisen independently, but have been derived from other and mostly extinct species, and that on the whole this development has taken place in the direction of greater complexity, may be maintained with the same degree of certainty as that with which astronomy asserts that the earth moves round the sun; for a conclusion may be arrived at as safely by other methods as by mathematical calculation.

If I make this assertion so unhesitatingly, I do not make it in the belief that I am bringing forward anything new nor because I think that any opposition will be encountered, but simply because I wish to begin by pointing out the firm ground on which we stand, before considering the numerous problems which still remain unsolved. Such problems appear as soon as we pass from the facts of the case to their explanation; as soon as we pass from the statement 'The organic world has arisen by development,' to the question 'But how has this been effected, by the action of what forces, by what means, and under what circumstances?'

In attempting to answer these questions we are very far from dealing with certainties; and opinions are still conflicting. But the answer lies in the domain of future investigation, that unknown country which we have to explore.

It is true that this country is not entirely unknown, and if I am not mistaken, Charles Darwin, who in our time has been the first to revive the long-dormant theory of descent, has already given a sketch, which may well serve as a basis for the complete map of the domain; although perhaps many details will be added, and many others taken away. In the principle of natural selection, Darwin has indicated the route by which we must enter this unknown land.

But this opinion is not universal, and only recently Carl Nägeli[176], the famous botanist, has expressed decided doubts as to the general applicability of the principle of natural selection. According to Nägeli, the co-operation of the external conditions of life with the known forces of the organism, viz. heredity and variability, are insufficient to explain the regular course of development pursued by the organic world. He considers that natural selection is at best an auxiliary principle, which accepts or rejects existing characters, but which is unable to create anything new: he believes that the causes of transformation reside within the organism alone. Nägeli further assumes that organisms contain forces which cause periodical transformation of the species, and he imagines that the organic world, as a whole, has arisen in a manner similar to that in which a single individual arises.

Just as a seed produces a certain plant because it possesses a certain constitution, and just as, in this process, certain conditions must be favourable (light, warmth, moisture, &c.) in order that development may take place, although they do not determine the kind or the manner of development; so, in precisely the same way, the tree of the whole organic world has grown up from the first and lowest forms of life on our planet, under a necessity arising from within, and on the whole independently of external influences. According to Nägeli, the cause which compels every form of living substance to change, from time to time, in the course of its secular growth, and which moulds it afresh into new species, must lie within the organic substance itself, and must depend upon its molecular structure.

It is with sincere admiration and real pleasure that we read the exposition in which Nägeli gives, as it were, the result of all his researches which bear upon the great question of the development of the organic world. But although we derive true enjoyment from the contemplation of the elaborate and ingeniously wrought-out theoretical conception,—which like a beautiful building or a work of art is complete in itself,—and although we must be convinced that its rise has depended upon the progress of knowledge, and that by its means we shall eventually reach a fuller knowledge; it is nevertheless true that we cannot accept the author's fundamental hypothesis. I at least believe that I am not alone in this respect, and that but few zoologists will be found who can adopt the hypothesis which forms the foundation of Nägeli's theory.

It is not my intention at present to justify my own widely different views, but the subject of this lecture compels me to briefly explain my position in relation to Nägeli, and to give some of the reasons why I cannot accept his theory of an active force of transformation arising and working within the organism; and I must also explain the reasons which induce me to adhere to the theory of natural selection.

The supposition of such a phyletic force of transformation (see Appendix I, p.298) possesses, in my opinion, the greatest defect that any theory can have,—it does not explain the phenomena. I do not mean to imply that it is incapable of rendering certain subordinate phenomena intelligible, but that it leaves a larger number of facts entirely unexplained. It does not afford any explanation of the purposefulness seen in organisms: and this is just the main problem which the organic world offers for our solution. That species are, from time to time, transformed into new ones might perhaps be understood by means of an internal transforming force, but that they are so changed as to become better adapted to the new conditions under which they have to live, is left entirely unintelligible by this theory. For we certainly cannot accept as an explanation Nägeli's statement that organisms possess the power of being transformed in an adaptive manner simply by the action of an external stimulus (see Appendix II, p. 300).

In addition to this fundamental defect, we must also note that there are absolutely no proofs in support of the foundation of this theory, viz. of the existence of an internal transforming force.

Nägeli has very ingeniously worked out his conception of idioplasm, and this conception is certainly an important acquisition and one that will last, although without the special meaning given to it by its author. But is this special meaning anything more than pure hypothesis? Can we say more than this of the ingenious description of the minute molecular structure of the hypothetical basis of life? Could not idioplasm be built up in a manner entirely different from that which Nägeli supposes? And can conclusions drawn from its supposed structure be brought forward to prove anything? The only proof that idioplasm must necessarily change, in the course of time, as the result of its own structure, is to be found in the fact that Nägeli has so constructed it; and no one will doubt that the structure of idioplasm might have been so conceived as to render any transformation from within itself entirely impossible.

But even if it is theoretically possible to imagine that idioplasm possesses such a structure that it changes in a certain manner, as the result of mere growth, we should not be justified in thus assuming the existence of a new and totally unknown principle until it had been proved that known forces are insufficient for the explanation of the observed phenomena.

Can any one assert that this proof has been forthcoming? It has been again and again pointed out that the phyletic development of the vegetable kingdom proceeds with regularity and according to law, as we see in the preponderance and constancy of so-called purely 'morphological' characters in plants. The formation of natural groups in the animal and vegetable kingdoms compels us to admit that organic evolution has frequently proceeded for longer or shorter periods along certain developmental lines. But we are not on this account compelled to adopt the supposition of unknown internal forces which have determined such lines of development.

Many years ago I attempted to prove[177] that the constitution or physical nature of an organism must exercise a restricting influence upon its capacity for variation. A given species cannot change into any other species, which may be thought of. A beetle could not be transformed into a vertebrate animal: it could not even become a grasshopper or a butterfly; but it could change into a new species of beetle, although only at first into a species of the same genus. Every new species must have been directly continuous with the old one from which it arose, and this fact alone implies that phyletic development must necessarily follow certain lines.

I can fully understand how it is that a botanist has more inclination than a zoologist to take refuge in internal developmental forces. The relation of form to function, the adaptation of the organism to the internal and external conditions of life, is less prominent in plants than in animals; and it is even true that a large amount of observation and ingenuity is often necessary in order to make out any adaptation at all. The temptation to accept the view that everything depends upon internal directing causes is therefore all the greater. Nägeli indeed looks at the subject from the opposite point of view, and considers that the true underlying cause of transformation is in animals obscured by adaptation, but is more apparent in plants[178]. Sufficient justification for this opinion cannot, however, be furnished by the fact that in plants many characters have not been as yet explained by adaptation. We should do well to remember the extent to which the number of so-called 'morphological' characters in plants has been lessened during the last twenty years. What a flood of light was thrown upon the forms and colours of flowers, so often curious and apparently arbitrary, when Sprengel's long-neglected discovery was extended and duly appreciated as the result of Darwin's investigations, and when the subject was further advanced by Hermann Müller's admirable researches! Even the venation of leaves, which was formerly considered to be entirely without significance, has been shown to possess a high biological value by the ingenious investigations of J. Sachs (see Appendix III, p. 308). We have not yet reached the limits of investigation, and no reason can be assigned for the belief that we shall not some day receive an explanation of characters which are now unintelligible[179].

It is obvious that the zoologist cannot lay too much stress upon the intimate connexion between form and function, a connexion which extends to the minutest details: it is almost impossible to insist too much upon the perfect manner in which adaptation to certain conditions of life is carried out in the animal body. In the animal body we find nothing without a meaning, nothing which might be otherwise; each organ, even each cell or part of a cell is, as it were, tuned for the special part it has to perform in relation to the surroundings.

It is true that we are as yet unable to explain the adaptive character of every structure in any single species, but whenever we succeed in making out the significance of a structure, it always proves to be a fresh example of adaptation. Any one who has attempted to study the structure of a species in detail, and to account for the relation of its parts to the functions of the whole, will be altogether inclined to believe with me that everything depends upon adaptation. There is no part of the body of an individual or of any of its ancestors, not even the minutest and most insignificant part, which has arisen in any other way than under the influence of the conditions of life; and the parts of the body conform to these conditions, as the channel of a river is shaped by the stream which flows over it.

These are indeed only convictions, not real proofs; for we are not yet sufficiently intimately acquainted with any species to be able to recognize the nature and meaning of all the details of its structure, in all their relations: and we are still less able to trace the ancestral history in each case, and to make out the origin of those structures of which the presence in the descendants depends primarily upon heredity. But already a fair advance towards the attainment of inductive proof has been made; for the number of adaptations which have been established is now very large and is increasing every day. If, however, we anticipate the results of future researches, and admit that an organism only consists of adaptations, based upon an ancestral constitution, it is obvious that nothing remains to be explained by a phyletic force, even though the latter be presented to us in the refined form of Nägeli's self-changing idioplasm.

It will perhaps be useful to illustrate my views by a familiar example. I choose the well-known group of the whales. These animals are placental mammals, which, probably in secondary times, arose from terrestrial Mammalia, by adaptation to an aquatic life.

Everything that is characteristic of these animals and distinguishes them from other mammals depends upon this adaptation. Their fore-limbs have been transformed into rigid paddles, only movable at the shoulder-joint; upon the back and the tail there are ridges with a form somewhat similar to the dorsal and caudal fins of fishes. The organ of hearing is without any external ear and without an air-containing external auditory meatus. The aerial vibrations do not pass, as in other mammals, from the external auditory passage to the tympanic cavity and thus to the nerve-terminations of the inner ear; but they reach the tympanic cavity by direct transmission through the bones of the skull, which possess a special structure and contain abundant air-cavities. This arrangement is obviously adapted for hearing in water. The nostrils also exhibit peculiarities, for they do not open near the mouth, but upon the forehead, so that the animal can breathe, even in a rough sea, as soon as it comes to the surface. In order to facilitate rapid movement in water, the whole body has become extended in length, and spindle-shaped, like the body of a fish. The hind limbs are absent in no other mammals, the fish-like *Sirenia* being alone excepted. In the whales, as in the *Sirenia*, these appendages have become useless, owing to the powerfully developed tail-fin; they are now rudimentary and consist of some small bones and muscles deeply buried in the body of the animal, which nevertheless, in certain species, still exhibit the original structure of the hind-limb. The hairy covering of other mammals has also disappeared, its place having been taken by a thick layer of fat beneath the skin, which affords a much better protection against cold. This fatty layer was also necessary in order to diminish the specific gravity of the animal, and to thus render it equal to that of sea-water. In the structure of the skull there are also a number of peculiarities, all of which are directly or indirectly connected with the conditions under which these animals live. In the whalebone whales, the enormous size of the face, the immense jaws, and wide mouth are very striking. Can it be suggested that this very characteristic appearance is entirely due to the guidance of some internal transforming force, or to some spontaneous modification of the idioplasm? Any such suggestion cannot be accepted, for it is easy to show that all these structural features depend upon adaptation to a peculiar mode of feeding. Functional teeth are absent, but rudimentary ones exist in the embryo as relics of an ancestral condition in which these organs were fully developed. Large plates of whalebone with finely divided ends are suspended vertically from the roof of the mouth. These whales feed upon small organisms, about an inch in length, which swim or float upon the water in countless numbers; and in order that they may subsist upon such minute animals, it is necessary to obtain them in immense numbers. This is achieved by means of the huge mouth which takes in a vast quantity of water at a single mouthful. The water then filters away through the plates of whalebone, while the organisms which form the whale's food remain stranded in the mouth. Is it necessary to add that the internal organs—so far as we understand the details of their functions, and so far as their structure differs from that of the corresponding organs in other Mammalia— have also been directly or indirectly modified by adaptation to an aquatic life? Thus all whales possess a very peculiar arrangement of the nasal passages and larynx, enabling them to breathe and swallow at the same time: the lungs are of enormous length, and thus cause the animal to assume a horizontal position in the water without the exercise of muscular effort: in consequence of this latter modification, the diaphragm extends in a nearly horizontal direction: there are moreover certain arrangements in the vascular system which enable the animal to remain under water for a considerable time, and so on.

And now, in reference to this special example, I will repeat the question which I have asked before:—'If everything that is characteristic of a group of animals depends upon adaptation, what remains to be explained by the operation of an internal developmental force?' What remains of a whale when we have taken away its adaptive characters? We are compelled to reply that nothing remains except the general plan of mammalian organization, which existed previously in the mammalian ancestors of the *Cetacea*. But if everything which stamps these animals as whales has arisen by adaptation, it follows that the internal developmental force cannot have had any share in the origin of this group.

And yet this very force is said to be the main factor in the transformation of species, and Nägeli unhesitatingly asserts that both the animal and vegetable kingdoms would have become very much as they now are, if there had been no adaptation to new conditions, and no such thing as competition in the struggle for existence[180].

But even if we admit that such an assumption affords some explanation, instead of being the renunciation of all attempts at explanation; if we admit that an organism, the characteristic peculiarities of which entirely depend upon adaptation, has been formed by an internal developmental force; we should still be unable to explain how it happens that such an organism, suited to certain conditions of life, and unable to exist under other conditions, appeared at that very place on the earth's surface, and at that very time in the earth's history, which offered the conditions appropriate for its existence. As I have previously argued, the believers in an internal developmental force are compelled to invent an auxiliary hypothesis, a kind of 'pre-established harmony' which explains how it is that changes in the organic world advance step by step, parallel with changes in the crust of the earth and in other conditions of life; just as, according to Leibnitz, body and soul, although independent of each other, proceed along parallel courses, like two chronometers which keep perfect time. And even this supposition would not be sufficient, because the place must be taken into account as well as the time: thus the whales could not have existed if they had first appeared upon dry land. We know of countless instances in which a species is exclusively and precisely adapted to a certain localized area, and could not thrive anywhere else. We have only to remember the cases of mimicry in which one insect gains protection by resembling another, the cases of protective resemblance to the bark or the leaves of a certain species of plant, or the numerous marvellous adaptations of parasitic animals to certain parts of certain species of hosts.

A mimetic species cannot have appeared at any place other than that in which it exists: it cannot have arisen through an internal developmental force. But if single species, or even whole orders like the *Cetacea*, have arisen independently of any such force, then we may safely assert that the existence of the supposed force is neither required by reason nor necessity.

Hence, abstaining from the invocation of unknown forces, we are justified in carrying on Darwin's attempt to explain the transformation of organisms by the action of known forces and known phenomena. I say 'carry on the attempt,' because I do not believe that our knowledge in this direction has ended with Darwin, and it seems to me that we have already arrived at ideas which are incompatible with certain important points in his general theory, and which therefore necessitate some modification of the latter.

The theory of natural selection explains the rise of new species by supposing that changes occur, from time to time, in those conditions of life to which an organism must adapt itself if it is to continue in existence. Thus a selective process is set up which ensures that only those out of the existing variations are preserved, which correspond in the highest degree to the changed conditions of life. By continued selection in the same direction the deviations from the type, although at first very insignificant, are accumulated and increased until they become specific differences.

I should wish to assert more definitely than Darwin has done, that alterations in the conditions of life, together with changes in the organism itself, must have advanced very gradually and by the smallest steps, in such a way that, at each period in the whole process of transformation, the species has remained sufficiently adapted to the surrounding conditions. An abrupt transformation of a species is inconceivable, because it would render the species incapable of existence. If the whole organization of an animal depends upon adaptation, if the animal body is, as it were, an extremely complex combination of new and old adaptations, it would be a highly remarkable coincidence if, after any sudden alteration occurring simultaneously in many parts of the body, all these parts were changed in such a manner that they again formed a whole which exactly corresponded to the altered external conditions. Those who assume the existence of such a sudden transformation overlook the fact that everything in the animal body is exactly calculated to maintain the existence of the species, and that it is just sufficient for this purpose; and they forget that the minutest change in the least important organ may be enough to render the species incapable of existence.

It may perhaps be objected that the case is different in plants, as is proved by the American weeds which have spread all over Europe, or the European plants which have become naturalized in Australia. Reference might also be made to the plants which inhabited the plains during the glacial epoch, and which at its close migrated to the Alpine mountains and to the far north, and which have remained unaltered under the apparently diverse conditions of life to which they have been subjected for so long a time. Similar instances may also be found among animals. The rabbit, which was brought by sailors to the Atlantic island of Porto Santo, has bred abundantly and remains unchanged in this locality; the European frogs, which were introduced into Madeira, have increased immensely and have become almost a plague; and the European sparrow now thrives in Australia quite as well as with us. But these instances do not prove that adaptation to external conditions of life is not of primary importance; they do not prove that an organism which is adapted to a certain environment will, when unmodified, remain capable of existence amid new surroundings. They only prove that the above-mentioned species found in those countries the same conditions of life as at home, or at least that they met with conditions to which their organization could be subjected without the necessity for modification. Not every new environment includes such changed conditions as will be effective in modifying every species of plant or animal. The rabbit of Porto Santo certainly feeds on herbs different from those which form the food of its relations in Europe, but such a change does not mean an effective alteration in the conditions under which this species lives, for the herbs in both localities are equally well suited to the needs of the animal.

But if we suppose that the wild rabbit, occurring in Europe, were to suddenly lose but a trifle of its wariness, its keen sight, its fine sense of hearing or of smell, or were to suddenly acquire a colour different from that which it now possesses, it would become incapable of existence as a species, and would soon die out. The same result would probably occur if any of its internal organs, such as the lungs or the liver, were suddenly modified. Perhaps single individuals would still remain capable of existence under these circumstances, but the whole species would suffer a certain decline from the maximum development of its powers of resistance, and would thus become extinct. The sudden transformation of a species appears to me to be inconceivable from a physiological point of view, at any rate in animals.

Hence the transformation of a species can only take place by the smallest steps, and must depend upon the accumulation of those differences which characterise individuals, or, as we call them, 'individual differences.' There is no doubt that these differences are always present, and thus, at first sight, it appears to be simply a matter of course that they will afford the material by means of which natural selection produces new forms of life. But the case is not so simple as it appeared to be until recently; that is if I am right in believing that in all animals and plants which are reproduced by true germs, only those characters which were potentially present in the germ of the parent can be transmitted to the succeeding generation.

I believe that heredity depends upon the fact that a small portion of the effective substance of the germ, the germ-plasm, remains unchanged during the development of the ovum into an organism, and that this part of the germ-plasm serves as a foundation from which the germ-cells of the new organism are produced[181]. There is therefore continuity of the germ-plasm from one generation to another. One might represent the germ-plasm by the metaphor of a long creeping root-stock from which plants arise at intervals, these latter representing the individuals of successive generations.

Hence it follows that the transmission of acquired characters is an impossibility, for if the germ-plasm is not formed anew in each individual but is derived from that which preceded it, its structure, and above all its molecular constitution, cannot depend upon the individual in which it happens to occur, but such an individual only forms, as it were, the nutritive soil at the expense of which the germ-plasm grows, while the latter possessed its characteristic structure from the beginning, viz. before the commencement of growth.

But the tendencies of heredity, of which the germ-plasm is the bearer, depend upon this very molecular structure, and hence only those characters can be transmitted through successive generations which have been previously inherited, viz. those characters which were potentially contained in the structure of the germ-plasm. It also follows that those other characters which have been acquired by the influence of special external conditions, during the life-time of the parent, cannot be transmitted at all.

The opposite view has, up to the present time, been maintained, and it has been assumed, as a matter of course, that acquired characters can be transmitted; furthermore, extremely complicated and artificial theories have been constructed in order to explain how it may be possible for changes produced by the action of external influences, in the course of a life-time, to be communicated to the germ and thus to become hereditary. But no single fact is known which really proves that acquired characters can be transmitted, for the ascertained facts which seem to point to the transmission of artificially produced diseases cannot be considered as a proof; and as long as such proof is wanting we have no right to make this supposition, unless compelled to do so by the impossibility of suggesting a mode in which the transformation of species can take place without its aid. (See Appendix IV, p. 310.)

It is obvious that the unconscious conviction that we need the aid of acquired characters has hitherto securely maintained the assumed axiom of the transmission of such features. It was believed that we could not do without such an axiom in order to explain the transformation of species; and this was believed not only by those who hold that the direct action of external influences plays an important part in the process, but also by those who hold that the operation of natural selection is the main factor.

Individual variability forms the most important foundation of the theory of natural selection: without it the latter could not exist, for this alone can furnish the minute differences by the accumulation of which new forms are said to arise in the course of generations. But how can such hereditary individual characters exist if the changes wrought by the action of external influences, during the life of an individual, cannot be transmitted? We are clearly compelled to find some other source of hereditary individual differences, or the theory of natural selection would collapse, as it certainly would if hereditary individual variations did not exist. If, on the other hand, acquired differences are transmitted, this would prove that there must be something wrong in the theory of the continuity of the germ-plasm, as above described, and in the non-transmission of acquired characters which results from this theory. But I believe that it is possible to suggest that the origin of hereditary individual characters takes place in a manner quite different from any which has been as yet brought forward. To explain this origin is the task which I am about to undertake in the following pages.

The origin of individual variability has been hitherto represented somewhat as follows. The phenomena of heredity lead to the conclusion that each organism is capable of producing germs, from which, theoretically at least, exact copies of the parent may arise. In reality this is never the case, because each organism possesses the power of reacting on the different external influences with which it is brought into contact, a power without which it could neither develope nor exist. Each organism reacting in a different way must be to some extent changed. Favourable nutrition makes such an organism strong and large; unfavourable nutrition renders it small and weak, and what is true of the whole organism may also be said of its parts. Now it is obvious that even the children of the same mother meet with influences different in kind and degree, from the very beginning of their existence, so that they must necessarily become unlike, even if we suppose them to have been derived from absolutely identical germs, with precisely the same hereditary tendencies.

In this manner individual differences are believed to have been introduced. But if acquired characters are not transmitted the whole chain of argument collapses, for none of those changes which are caused by the conditions of nutrition acting upon single parts of the whole organism, including the results of training and of the use or disuse of single organs,—none of these changes can furnish hereditary differences, nor can they be transmitted to succeeding generations. They are, as it were, only transient characters as far as the species is concerned.

The children of accomplished pianists do not inherit the art of playing the piano; they have to learn it in the same laborious manner as that by which their parents acquired it; they do not inherit anything except that which their parents also possessed when children, viz. manual dexterity and a good ear. Furthermore, language is not transmitted to our children, although it has been practised not only by ourselves but by an almost endless line of ancestors. Only recently, facts have again been worked up and brought together, which show that children of highly civilized nations have no trace of a language when they have grown up in a wild condition and in complete isolation[182]. The power of speech is an acquired or transient character: it is not inherited, and cannot be transmitted: it disappears with the organism which manifests it. Not only do similar phenomena occur in the vegetable kingdom, but they present themselves in an especially striking manner.

When Nägeli[183] introduced Alpine plants, taken from their natural habitat, into the botanical garden at Munich, many of the species were so greatly altered that they could hardly be recognized: for instance, the small Alpine hawk-weeds became large and thickly branching, and they blossomed freely. But if such plants, or even their descendants, were removed to a poor gravelly soil the new characters entirely disappeared, and the plants were re-transformed into the original Alpine form. The re-transformation was always complete, even when the species had been cultivated in rich garden soil for several generations.

Similar experiments with identical results were made twenty years ago by Alexis Jordan[184], who chiefly made use of *Draba verna* in his researches. These experiments furnish very strong proofs, because they were originally undertaken without the bias which may be given by a theory. Jordan only intended to decide experimentally whether the numerous forms of the plant, as it occurs wild in different habitats, are mere varieties or true species. He found that the different forms do not pass into one another, and are in all cases re-transformed after they have been altered by cultivation in a soil different from that in which they usually grow, and he therefore assumed that they were true species. All these experiments therefore confirm the conclusion that external influences may alter the individual, but that the changes produced are not transmitted to the germs, and are never hereditary.

Nägeli indeed asserts that innate individual differences do not exist in plants. The differences which we find, for instance, between two beeches or oaks, are always, according to him, modifications produced by the influence of varying local conditions. But it is obvious that Nägeli goes too far in this respect, although it may be conceded that innate individual differences in plants are much more difficult to distinguish from those which are acquired, than in animals.

There is no doubt about the occurrence of innate and hereditary individual characters in animals, and we may find an especially interesting illustration in the case of man. The human eye can with practice appreciate the most minute differences between individual men, and especially differences of feature. Every one knows that peculiarities of feature persist in certain families through a long series of generations. I need hardly remind the reader of the broad forehead of the Julii, the projecting chin of the Hapsburgs, or the curved nose of the Bourbons. Hence every one can see that hereditary individual characters do unquestionably exist in man. The same conclusion may be affirmed with equal certainty for all our domestic animals, and I do not see any reason why there should be any doubt about its application to other animals and to plants.

But now the question arises,—How can we explain the presence of such characters consistently with a belief in the continuity of the germ-plasm, a theory which implies the rejection of the supposition that acquired characters can become hereditary? How can the individuals of any species come to possess various characters which are undoubtedly hereditary, if all changes which are due to the influence of external conditions are transient and disappear with the individual in which they arose? Why is it that individuals are distinguished by innate characters, as well as by those which I have previously called transient, and how can deep-seated hereditary characters arise at all, if they are not produced by the external influences to which the individual is exposed?

In the first place it may be argued that external influences may not only act on the mature individual, or during its development, but that they may also act at a still earlier period upon the germ-cell from which it arises. It may be imagined that such influences of different kinds might produce corresponding minute alterations in the molecular structure of the germ-plasm, and as the latter is, according to our supposition, transmitted from one generation to another, it follows that such changes would be hereditary.

Without altogether denying that such influences may directly modify the germ-cells, I nevertheless believe that they have no share in the production of hereditary *individual* characters.

The germ-plasm or idioplasm of the germ-cell (if this latter term be preferred) certainly possesses an exceedingly complex minute structure, but it is nevertheless a substance of extreme stability, for it absorbs nourishment and grows enormously without the least change in its complex molecular structure. With Nägeli we may indeed safely affirm so much, although we are unable to acquire any direct knowledge as to the constitution of germ-plasm. When we know that many species have persisted unchanged for thousands of years, we have before us the proof that their germ-plasm has preserved exactly the same molecular structure during the whole period. I may remind the reader that many of the embalmed bodies of the sacred Egyptian animals must be four thousand years old, and that the species are identical with those now existing in the same locality. Now, since the quantity of germ-plasm contained in a single germ-cell must be very minute, and since only a very small fraction can remain unchanged when the germ-cell develops into an organism, it follows that an enormous growth of this small fraction must take place in every individual, for it must be remembered that each individual produces thousands of germ-cells. It is therefore not too much to say that, during a period of four thousand years, the growth of the germ-plasm in the Egyptian ibis or crocodile must have been quite stupendous. But in the animals and plants which inhabit the Alps and the far north, we have instances of species which have remained unchanged for a much longer period, viz. for the time which has elapsed between the close of the glacial epoch and the present day. In such organisms the growth of the germ-plasm must therefore have been still greater.

If nevertheless the molecular structure of the germ-plasm has remained precisely the same, this substance cannot be readily modifiable, and there is very little chance of the smallest changes being produced in its molecular structure, by the operation of those minute transient variations in nutrition to which the germ-cells, together with every other part of the organism, are exposed. The rate of growth of the germ-plasm will certainly vary, but its structure is unlikely to be affected for the above-mentioned reasons, and also because the influences are mostly changeable, and occur sometimes in one and sometimes in another direction.

Hereditary individual differences must therefore be derived from some other source.

I believe that such a source is to be looked for in the form of reproduction by which the great majority of existing organisms are propagated: viz. in sexual, or, as Häckel calls it, amphigonic reproduction.

It is well known that this process consists in the coalescence of two distinct germ-cells, or perhaps only of their nuclei. These germ-cells contain the germ-substance, the germ-plasm, and this again, owing to its specific molecular structure, is the bearer of the hereditary tendencies of the organism from which the germ-cell has been derived. Thus in amphigonic reproduction two groups of hereditary tendencies are as it were combined. I regard this combination as the cause of hereditary individual characters, and I believe that the production of such characters is the true significance of amphigonic reproduction. The object of this process is to create those individual differences which form the material out of which natural selection produces new species.

At first sight this conclusion appears to be very startling and almost incredible, because we are on the contrary inclined to believe that the continued combination of existing differences, which is implied by the very existence of amphigonic reproduction, cannot lead to their intensification, but rather to their diminution and gradual obliteration. Indeed the opinion has already been expressed that deviations from the specific type are rapidly destroyed by the operation of sexual reproduction. Such an opinion may be true with regard to specific characters, because the deviations from a specific type occur in such rare cases that they cannot hold their ground against the large number of normal individuals. But the case is different with those minute differences which are characteristic of individuals, because every individual possesses them, although of a different kind and degree. The extinction of such differences could only take place if a few individuals constituted a whole species; but the number of individuals which together represent a species is not only very large but generally incalculable. Cross-breeding between all individuals is impossible, and hence the obliteration of individual differences is also impossible.

In order to explain the effects of sexual reproduction, we will first of all consider what happens in monogonic or unisexual reproduction, which actually occurs in parthenogenetic organisms. Let us imagine an individual producing germ-cells, each of which may by itself develope into a new individual. If we then suppose a species to be made up of individuals which are absolutely identical, it follows that their descendants must also remain identical through any number of generations, if we neglect the transient non-transmissible peculiarities caused by differences of food and other external conditions.

Although the individuals of such a species might be actually different, they would be potentially identical: in the mature state they might differ, but they must have been identical in origin. The germs of all of them must contain exactly the same hereditary tendencies, and if it were possible for their development to take place under exactly the same conditions, identical individuals would be produced.

Let us now assume that the individuals of such a species, reproducing itself by the monogonic process and therefore without cross-breeding, differ, not only in transient but also in hereditary characters. If this were the case, each individual would produce descendants possessing the same hereditary differences which were characteristic of itself; and thus from each individual a series of generations would emanate, the single individuals of which would be potentially identical with each other and with their first ancestor. Hence the same individual differences would be repeated again and again, in each succeeding generation, and even if all the descendants lived to reproduce themselves, there would be at last just as many groups of potentially identical individuals as there were single individuals at the beginning.

Similar cases actually occur in many species in which sexual reproduction has been entirely replaced by the parthenogenetic method, as in many species of *Cynips* and in certain lower Crustacea. But all these differ from our hypothetical case in one important respect; it is always impossible for all the descendants to reach maturity and reproduce themselves. The vast majority of the descendants generally perish at an early stage, and only about as many remain to continue the species as reached maturity in the preceding generation.

We have now to consider whether such a species can be subject to the operation of natural selection. Let us take the case of an insect living among green leaves, and possessing a green colour as a protection against discovery by its enemies. We will assume that the hereditary individual differences consist of various shades of green. Let us further suppose that the sudden extinction of its food-plant compelled this species to seek another plant with a somewhat different shade of green. It is clear that such an insect would not be completely adapted to the new environment. It would therefore be compelled, metaphorically speaking, to endeavour to bring its colour into closer harmony with that of the new food-plant, or else the increased chances of detection given to its enemies would lead to its slow but certain extinction.

It is obvious that such a species would be altogether unable to produce the required adaptation, for *ex hypothesi*, its hereditary variations remain the same, one generation after another. If therefore the required shade of green was not previously present, as one of the original individual differences, it could not be produced at any time. If, however, we suppose that such a colour existed previously in certain individuals, it follows that those with other shades of green would be gradually exterminated, while the former would alone survive. But this process would not be an adaptation in the sense used in the theory of natural selection. It would indeed be a process of selection, but it could form no more than the beginning of that process which we call natural selection. If the latter could only bring existing characters into prominence, it would not be worth much consideration, for it could never produce a new species. A species never includes, from the beginning, individuals which deviate from the specific type as widely as the individuals of the most nearly allied species deviate from it. And it would be still less possible to explain, on such a principle, the origin of the whole organic world; for, if so, all existing species would have been included as variations of the first species. Natural selection must be able to do infinitely more than this, if it is to be of any importance as a principle of development. It must be able to accumulate minute existing differences in the required direction, and thus to create new characters. In our example it ought to be able, after preserving those individuals with a colour nearest to the required shade, to lead their descendants onward through successive stages towards a complete harmony of colour.

But such a result is quite unattainable with the asexual method of reproduction: in other words, natural selection, in the true meaning of the term, viz. a process which could produce new characters in the manner above described, is an impossibility in a species propagated by asexual reproduction.

If it could be shown that a purely parthenogenetic species had become transformed into a new one, such an observation would prove the existence of some force of transformation other than selective processes, for the new species could not have been produced by these latter. As already explained, the only selection which would be possible for such a species, would lead to the survival of one group of individuals and to the extinction of all others. Thus in our example that group of individuals would alone survive, the ancestors of which originally possessed the appropriate colour. But if one group alone survived, it follows that all hereditary individual differences would have disappeared from the species, for the members of such a single group are identical with one another and with their original ancestors. We thus reach the conclusion that monogonic reproduction can never cause hereditary individual variability, but that, on the other hand, it is very likely to lead to its entire suppression.

But the case is very different with sexual reproduction. When once individual differences have begun to appear in a species propagated by this process, uniformity among its individuals can never again be reached. So far from this being the case, the differences must even be increased in the course of generations, not indeed in intensity, but in number, for new combinations of the individual characters will continually arise.

Again, assuming the existence of a number of individuals which differ from one another by a few hereditary individual characters, it follows that no individual of the second generation can be identical with any other. They must all differ, not only actually but also potentially, for their differences exist at the very beginning of development, and do not solely depend upon the accidental conditions under which they live. Moreover, no one of the descendants can be identical with any of the ancestors, for each of the former unites within itself the hereditary tendencies of two parents, and its organism is therefore, as it were, a compromise between two developmental tendencies. Similarly in the third generation, the hereditary tendencies of two individuals of the second generation enter into combination. But since the germ-plasm of the latter is not simple, but composed of two individually distinct kinds of germ-plasm, it follows that an individual of the third generation is a compromise between four different hereditary tendencies. In the fourth generation, eight; in the fifth, sixteen; in the sixth, thirty-two different hereditary tendencies must come together, and each of them will make itself more or less felt in some part of the future organism. Thus by the sixth generation a large number of varied combinations of ancestral individual characters will appear, combinations which have never existed before and which can never exist again.

We do not know the number of generations over which the specific hereditary tendencies of the first generation can make themselves felt. Many facts seem to indicate however that the number is large, and it is at all events greater than six. When we remember that, in the tenth generation, a single germ contains 1024 different germ-plasms, with their inherent hereditary tendencies, it is quite clear that continued sexual reproduction can never lead to the re-appearance of exactly the same combination, but that new ones must always arise.

New combinations are all the more probable because the different idioplasms composing the germ-plasm in the germ-cells of any individual are present in different degrees of intensity at different times of its life; in other words, the intensity of the component idioplasms is a function of time. This conclusion follows from the fact that children of the same parents are never exactly identical. In one child the characters of the father may predominate, in another those of the mother, in another again those of either grand-parent or great-grand-parent.

We are thus led to the conclusion that even in a few sexually produced generations a large number of well-marked individuals must arise: and this would even be true of generations springing from our hypothetical species, assumed to be without ancestors, and characterised by few individual differences. But of course organisms which reproduce themselves sexually are never without ancestors, and if these latter were also propagated by the sexual method, it follows that each generation of every sexual species is in the stage which we have previously assumed for the tenth or some much later generation of the hypothetical species. In other words, each individual contains a maximum of hereditary tendencies and an infinite variety of possible individual characters (see Appendix VI, p. 326).

In this manner we can explain the origin of hereditary individual variability as it is known in man and the higher animals, and as it is required for the theory which explains the transformation of species by means of natural selection.

Before proceeding further, I must attempt to answer a question which obviously suggests itself. For the sake of argument, I have assumed the existence of a first generation, of which the individuals were already characterised by individual differences. Can we find any explanation of these latter, or are we compelled to take them for granted, without any attempt to enquire into their origin? If we abandon this enquiry, we can never achieve a complete solution of the problems of heredity and variability. We have, it is true, shown that hereditary differences, when they have once appeared, would, through sexual reproduction, undergo development into the diverse forms which actually exist; but this conclusion affords us no explanation of the source whence such differences have been derived. If the external conditions acting directly upon an organism can only produce transient (viz. non-hereditary) differences in the latter, and if, on the other hand, the external influences which act upon the germ-cell can only produce a change in its molecular structure after operating over very long periods, it seems that we have exhausted all the possible sources of hereditary differences without reaching any satisfactory explanation.

I believe, however, that an explanation can be given. The origin of hereditary individual variability cannot indeed be found in the higher organisms—the Metazoa and Metaphyta; but it is to be sought for in the lowest—the unicellular organisms. In these latter the distinction between body-cell and germ-cell does not exist. Such organisms are reproduced by division, and if therefore any one of them becomes changed in the course of its life by some external influence, and thus receives an individual character, the method of reproduction ensures that the acquired peculiarity will be transmitted to its descendants. If, for instance, a Protozoon, by constantly struggling against the mechanical influence of currents in water, were to gain a somewhat denser and more resistent protoplasm, or were to acquire the power of adhering more strongly than the other individuals of its species, the peculiarity in question would be directly continued on into its two descendants, for the latter are at first nothing more than the two halves of the former. It therefore follows that every modification which appears in the course of its life, every individual character, however it may have arisen, must necessarily be directly transmitted to the two offspring of a unicellular organism.

The pianist, whom I have already used as an illustration, may by practice develope the muscles of his fingers so as to ensure the highest dexterity and power; but such an effect would be entirely transient, for it depends upon a modification in local nutrition which would be unable to cause any change in the molecular structure of the germ-cells, and could not therefore produce any effect upon the offspring. And even if we admit that some change might be caused in the germ-cells, the chances would be infinity to nothing against the production of the appropriate effect, viz. such a change as would lead to the development in the child of the acquired characters of the parent.

In the lowest unicellular organisms, however, the case is entirely different. Here parent and offspring are still, in a certain sense, one and the same thing: the child is a part, and usually half, of the parent. If therefore the individuals of a unicellular species are acted upon by any of the various external influences, it is inevitable that hereditary individual differences will arise in them; and as a matter of fact it is indisputable that changes are thus produced in these organisms, and that the resulting characters are transmitted. It has been directly observed that individual differences do occur in unicellular organisms,—differences in size, colour, form, and the number or arrangement of cilia. It must be admitted that we have not hitherto paid sufficient attention to this point, and moreover our best microscopes are only very rough means of observation when we come to deal with such minute organisms. Nevertheless we cannot doubt that the individuals of the same species are not absolutely identical.

We are thus driven to the conclusion that the ultimate origin of hereditary individual differences lies in the direct action of external influences upon the organism. Hereditary variability cannot however arise in this way at every stage of organic development, as biologists have hitherto been inclined to believe. It can only arise in the lowest unicellular organisms; and when once individual difference had been attained by these, it necessarily passed over into the higher organisms when they first appeared. Sexual reproduction coming into existence at the same time, the hereditary differences were increased and multiplied, and arranged in ever-changing combinations.

Sexual reproduction can also increase the differences between individuals, because constant cross-breeding must necessarily and repeatedly lead to a combination of forces which tend in the same direction, and which may determine the constitution of any part of the body. If, for instance, the same part of the body is strongly developed in both parents, the experience of breeders tells us that the part in question is likely to be even more strongly developed in the offspring; and that weakly developed parts will in the same manner tend to become still weaker. Amphigonic reproduction therefore ensures that every character which is subject to individual fluctuation must appear in many individuals with a strengthened degree of development, in many others with a development which is less than normal, while in a still larger number of individuals the average development will be reached. Such differences afford the material by means of which natural selection is able to increase or weaken each character according to the needs of the species. By the removal of the less well-adapted individuals, natural selection increases the chance of beneficial cross-breeding in the subsequent generations.

Every one must admit that, if a species came into existence having only a small number of individual differences which appeared in the different parts of different individuals, the number of differences would increase with each sexually produced generation, until all the parts in which the variations occurred had received a peculiar character in all individuals.

Moreover sexual reproduction not only adds to the number of existing differences, but it also brings them into new combinations, and this latter consequence is as important as the former.

The former consequence can hardly make itself felt in any existing species, because in them every part already possesses its peculiar character in all individuals. The second consequence is, however, more important, viz. the production of new combinations of individual characters by sexual reproduction; for, as Darwin has already pointed out, we must imagine that not only are single characters changed in the process of breeding, but that probably several, and perhaps very many characters, are simultaneously modified. No two species, however nearly allied, differ from each other in but a single character. Even our eyesight, which has by no means reached the highest pitch of development, can always detect several, and often very many points of difference; and if we possessed the powers necessary for making an absolutely accurate comparison, we should probably find that everything is different in two nearly allied species.

It is true that a great number of these differences depend upon correlation, but others must depend upon simultaneous primary changes.

A large butterfly (*Kallima paralecta*), found in the East Indian forests, has often been described in its position of rest as almost exactly resembling a withered leaf; the resemblance in colour being aided by the markings which imitate the venation of a leaf. These markings are composed of two parts, the upper of which is on the fore-wings, while the lower one is on the hind wings. The butterfly when at rest must therefore keep the wings in such a position that the two parts of each marking exactly correspond, for otherwise the character would be valueless; and as a matter of fact the wings are held in the appropriate position, although the butterfly is of course unconscious of what it is doing. Hence a mechanism must exist in the insect's brain which compels it to assume this attitude, and it is clear that the mechanism cannot have been developed before the peculiar manner of holding the wings became advantageous to the butterfly, viz. before the similarity to a leaf had made its first appearance. Conversely, this latter resemblance could not develope before the butterfly had gained the habit of holding its wings in the appropriate position. Both characters must therefore have come into existence simultaneously, and must have undergone increase side by side: the marking progressing from an imperfect to a very close similarity, while the position of the wings gradually approached the attitude which was exactly appropriate. The development of certain minute structural elements of the central nervous system, and the appropriate distribution of colouring matter on the wings, must have taken place simultaneously, and only those individuals have been selected to continue the species which possessed the favourable variations in both these directions.

It is, however, obvious that sexual reproduction will readily afford such combinations of required characters, for by its means the most diverse features are continually united in the same individual, and this seems to me to be one of its most important results.

I do not know what meaning can be attributed to sexual reproduction other than the creation of hereditary individual characters to form the material upon which natural selection may work. Sexual reproduction is so universal in all classes of multicellular organisms, and nature deviates so rarely from it, that it must necessarily be of pre-eminent importance. If it be true that new species are produced by processes of selection, it follows that the development of the whole organic world depends on these processes, and the part that amphigony has to play in nature, by rendering selection possible among multicellular organisms, is not only important, but of the very highest imaginable importance.

But when I maintain that the meaning of sexual reproduction is to render possible the transformation of the higher organisms by means of natural selection, such a statement is not equivalent to the assertion that sexual reproduction originally came into existence in order to achieve this end. The effects which are now produced by sexual reproduction did not constitute the causes which led to its first appearance. Sexual reproduction came into existence before it could lead to hereditary individual variability. Its first appearance must therefore have had some other cause; but the nature of this cause can hardly be determined with any degree of certainty or precision from the facts with which we are at present acquainted. The general solution of the problem will, however, be found to lie in the conjugation of unicellular organisms, which forms the precursor of true sexual reproduction. The coalescence of two unicellular individuals which represents the simplest and therefore probably the most primitive form of conjugation, must have some directly beneficial effect upon the species in which it occurs.

Various assumptions may be made as to the nature of these beneficial effects, and it will be useful to consider in detail some of those suggestions which have been brought forward. Eminent biologists, such as Victor Hensen[185] and Edouard van Beneden[186], believe that conjugation, and indeed sexual reproduction generally, must be considered as 'a rejuvenescence of life.' Bütschli also accepts this view, at any rate as regards conjugation. These authorities imagine that the wonderful phenomena of life, of which the underlying cause is still an unsolved problem, cannot be continued indefinitely by the action of forces arising from within itself, that the clock-work would be stopped after a longer or shorter time, that the reproduction of purely asexual organisms would cease, just as the life of the individual finally comes to an end, or as a spinning wheel comes to rest in consequence of friction, and requires a renewed impetus if its motion is to continue. In order that reproduction may continue without interruption, these writers believe that a rejuvenescence of the living substance is necessary, that the clock-work of reproduction must be wound up afresh; and they recognize such a rejuvenescence in sexual reproduction and in conjugation, or in other words in the fusion of two cells, whether in the form of germ-cells or of two unicellular organisms.

Edouard van Beneden expresses this idea in the following words:—'Il semble que la faculté que possèdent les cellules, de se multiplier par division soit limitée: il arrive un moment où elles ne sont plus capables de se diviser ultérieurement, à moins qu'elles ne subissent le phénomène du rajeunissement par le fait de la fécondation. Chez les animaux et les plantes les seules cellules capables d'être rajeunies sont les œufs; les seules capables de rajeunir sont les spermatocytes. Toutes les autres parties de l'individu sont vouées à la mort. La fécondation est la condition de la continuité de la vie. Par elle le générateur échappe à la mort' (l. c., p. 405). Victor Hensen thinks it possible that the germ and its products are prevented from dying by means of normal fertilization: he says that the law which states that every egg must be fertilized, was formulated before the discovery of parthenogenesis and cannot now be maintained, but that we are nevertheless compelled to assume that even the most completely parthenogenetic species requires fertilization after many generations (l. c., p. 236).

If the theory of rejuvenescence be thoroughly examined, it will be found to be nothing more than the expression of the fact that sexual reproduction persists without any ascertainable limit. From the fact of its general occurrence, the conclusion is, however, drawn that asexual reproduction could not persist indefinitely as the only mode of reproduction in any species of animal. But proofs in support of this opinion are wanting, and it is very probable that it would never have been advanced if it had been possible to explain the general occurrence of sexual reproduction in any other way,—if we had been able to ascribe any other significance to this pre-eminently important process.

But quite apart from the fact that it is impossible to bring forward any proofs, the theory of rejuvenescence seems to me to be unsatisfactory in other ways. The whole conception of rejuvenescence, although very ingenious, has something uncertain about it, and can hardly be brought into accordance with the usual conception of life as based upon physical and mechanical forces. How can any one imagine that an Infusorian, which by continued division had lost its power of reproduction, could regain this power by forming a new individual, after fusion with another Infusorian, which had similarly become incapable of division? Twice nothing cannot make one. If indeed we could assume that each animal contained half the power necessary for reproduction, then both together would certainly form an efficient whole; but it is hardly possible to apply the term rejuvenescence to a process which is simply an addition, such as would be attained under other circumstances by mere growth; neglecting, for the present, that factor which, in my opinion, is of the utmost importance in conjugation,—the fusion of two hereditary tendencies. If rejuvenescence possesses any significance at all, it must be this,—that by its means a force, which did not previously exist in the conjugating individuals, is called into activity. Such a force would, however, owe its existence to latent energy stored up in each single animal during the period of asexual reproduction, and such latent forces would necessarily be of different natures, and of such a constitution that their union at the moment of conjugation would give rise to the active force of reproduction.

The process might perhaps be compared to the flight of two rockets, which by the combustion of some explosive substance (such as nitro-glycerine) stored up within themselves are impelled in such a direction that they would meet at the end of their course, when all the nitro-glycerine had been completely exhausted. The movement would then come to an end, unless the explosive material could have been meanwhile renewed. Now suppose that such a renewal were achieved by the formation of nitric acid in one of the rockets and glycerine in the other, so that when they came into contact nitro-glycerine would be formed afresh equal in quantity and in distribution on both the rockets to that which was originally present. In this way the movement would be renewed again and again with the same velocity, and might continue for ever.

Rejuvenescence can be rendered intelligible in theory by some such metaphor, but considerable difficulties are encountered in the rigid application of the metaphor to the facts of the case. In the first place, how is it possible that the motive force can be exhausted by continual division, while one of its components is being formed afresh in the same body and during the same time? When thoroughly examined the loss of the power of division is seen to follow from the loss of the powers of assimilation, nutrition, and growth. How is it possible that such a power can be weakened and finally entirely lost while one of its components is accumulated?

I believe that, instead of accepting such daring assumptions, it is better to be satisfied with the simple conception of living matter possessing as attributes the powers of unlimited assimilation and capacity for reproduction. With such a theory the mere form of reproduction, whether sexual or asexual, will have no influence upon the duration of the capacity: for force and matter undergo simultaneous increase, and are inseparably connected in this as in all other instances. This theory does not, however, exclude the possible occurrence of circumstances under which such an association is no longer necessary.

I could only consent to adopt the hypothesis of rejuvenescence, if it were rendered absolutely certain that reproduction by division could never under any circumstances persist indefinitely. But this cannot be proved with any greater certainty than the converse proposition, and hence, as far as direct proof is concerned, the facts are equally uncertain on both sides. The hypothesis of rejuvenescence is, however, opposed by the fact of parthenogenesis; for if fertilization possesses in any way the meaning of rejuvenescence, and depends upon the union of two different forms of force and of matter, which thus produce the power of reproduction, it follows that we cannot understand how it happens that the same power of reproduction may be sometimes produced from one form of matter, alone and unaided. Logically speaking, parthenogenesis should be as impossible as that either nitric acid or glycerine should separately produce the effect of nitro-glycerine. The supposition has indeed been made that in the case of parthenogenesis, one fertilization is sufficient for a whole series of generations, but this supposition is not only incapable of proof, but it is contradicted by the fact that certain eggs which may develope parthenogenetically are also capable of fertilization. If, in this case, the power of reproduction were sufficient for development, how is it that the egg is also capable of fertilization; and if the power were insufficient, how is it that the egg can develope parthenogenetically? And yet one and the same egg (in the bee) can develope into a new individual, with or without fertilization. We cannot escape this dilemma by making the further assumption, which is also incapable of proof, that a smaller amount of reproductive force is required for the development of a male individual than for the development of a female. It is true that the unfertilized eggs of the bee produce male individuals, while the fertilized ones develope into females, but in certain other species the converse association holds good, while in others, again, fertilization bears no relation to the sex of the offspring.

Although the mere fact that parthenogenesis occurs at all is, in my opinion, sufficient to disprove the theory of rejuvenescence, it is well to remember that parthenogenesis is now the only method of reproduction in many species (although we do not know the period of time over which these conditions have extended), and is nevertheless unattended by any perceptible decrease in fertility.

From all these considerations we may draw the conclusion that the process of rejuvenescence, as described above, cannot be accepted either as the existing or the original meaning of conjugation, and the question naturally arises as to what other significance this latter process can have possessed at its first beginning.

Rolph[187] has expressed the opinion that conjugation is a form of nutrition, so that the two conjugating individuals, as it were, devour each other. Cienkowsky[188] also regards conjugation as merely 'accelerated' assimilation. There is, however, not only an essential difference but a direct contrast between the processes of conjugation and nutrition. With regard to Cienkowsky's view, Hensen[189] has well said that 'coalescence in itself cannot be an accelerated nutrition, because even if we admit that both individuals are in want of nourishment, it is impossible that the need can be supplied by this process, unless one of them perishes and is really devoured.' In order that an animal may serve as the food of another, it must perish and must be brought into a fluid form, and finally it must be assimilated. In the case before us, however, two protoplasmic bodies are placed side by side and coalesce, without either of them passing into the liquid state. Two idioplasms unite, together with all the hereditary tendencies contained in them; but although it is certain that nutrition in the proper sense of the word cannot take place, because neither of the animals receives an addition of liquid food by the coalescence, yet the consequence of this process must be in one respect similar to that of nutrition and growth:—the mass of the body and the quantity of the forces contained in it undergo simultaneous increase. It is not inconceivable that effects are by this means rendered possible, which under the peculiar circumstances leading to conjugation, could not have been otherwise produced.

I believe that this is at any rate the direction in which we shall have to seek for the first meaning of conjugation and for its phyletic origin. This first result and meaning of conjugation may be provisionally expressed in the following formula:—conjugation originally signified a strengthening of the organism in relation to reproduction, which happened when from some external cause, such as want of oxygen, warmth, or food, the growth of the individual to the extent necessary for reproduction could not take place.

This explanation must not be regarded as equivalent to that afforded by the theory of rejuvenescence; for the latter process is said to be necessary for the continuance of reproduction, and ought therefore to occur periodically quite independently of external circumstances; while according to my theory, conjugation at first only occurred under unfavourable conditions, and assisted the species to overcome such difficulties.

But whatever the original meaning of conjugation may have been, it seems to have become already subordinated in the higher Protozoa, as is indicated by the changes in the course taken by this process. The higher Protozoa when conjugating do not as a rule coalesce completely and permanently[190] in the manner followed by the lower Protozoa, and it seems to me possible, or even probable, that in the former the process has already gained the full significance of sexual reproduction, and is to be looked upon as a source of variability.

Whether this be so or not, I believe it is certain that sexual reproduction could not have been entirely abandoned at any period since the time when the Metazoa and Metaphyta first arose; for they derived this form of reproduction from their unicellular ancestors.

We know that organs and characters which have persisted through a long series of generations are transmitted with extreme tenacity, even when they have ceased to be of any direct use to their immediate possessors. The rudimentary organs in various animals, and not least in man, afford very strong proofs of the soundness of this conclusion. Another example has only recently been discovered in the sixth finger, which has been shown to exist in the human embryo[191], a part which has only been present in a rudimentary form ever since the origin of the Amphibia[192]. Superfluous organs become rudimentary very slowly, and enormous periods must elapse before they completely disappear, while the older a character is, the more firmly it becomes rooted in the organism. What I have above called the physical constitution of a species is based upon these facts, and upon them depend the *tout ensemble* of inherited characters, which are adapted to one another and woven together into a harmonious whole. It is this specific nature of an organism which causes it to respond to external influences in a manner different from that followed by any other organism, which prevents it from changing in any way except along certain definite lines of variation, although these may be very numerous. Furthermore these facts ensure that characters cannot be taken at random from the constitution of a species and others substituted for them. Such a variation as a mammal wanting the firm axis of the backbone is an impossibility, not only because the backbone is necessary as a support to the body, but chiefly because this structure has been inherited from times immemorial, and has become so impressed upon the mammalian organization that any variation so great as to threaten its very existence cannot now take place. The view here set forth of the origin of hereditary variability by amphigonic reproduction, makes it clear that an organism is in a state of continual oscillation only upon the surface, so to speak, while the fundamental parts of its constitution, which have been inherited from extremely remote periods, remain unaffected.

Thus sexual reproduction itself did not cease after it had existed in the form of conjugation through innumerable generations of the vast numbers of species which have been included under the Protozoa; it did not cease even when its original physiological significance had lost its importance, either completely or in part. This process, however, had come to possess a new significance which ensured its continuance, in the enormous advantage conferred on a species by the power of adapting itself to new conditions of life, a power which could only be preserved by means of this method of reproduction. The formation of new species which among the lower Protozoa could be achieved without amphigony, could only be attained by means of this process in the Metazoa and Metaphyta. It was only in this way that hereditary individual differences could arise and persist. It was impossible for amphigony to disappear, for each species in which it was preserved was necessarily superior to those which had lost it, and must have replaced them in the course of time; for the former alone could adapt itself to the ever-changing conditions of life, and the longer sexual reproduction endured, the more firmly was it necessarily impressed upon the constitution of the species, and the more difficult its disappearance became.

Sexual reproduction has nevertheless been lost in some cases, although only at first in certain generations. Thus in the *Aphidae* and in many lower Crustacea, generations with parthenogenetic reproduction alternate with others which reproduce themselves by the sexual method. But in most cases it is clear that this partial loss of amphigony conferred considerable advantages upon the species by giving increased capabilities for the maintenance of existence. By means of partial parthenogenesis a much more rapid increase in the number of individuals could be attained in a given time, and this fact is of the highest importance for the peculiar circumstances under which these species exist. A species of Crustacean which inhabits rapidly drying pools, and developes from winter-eggs which have remained dried up in the mud, has, as a rule, only a very short time in which to secure the existence of succeeding generations. The few sexual eggs which have escaped the attacks of numerous enemies develope immediately after the first shower of rain; the animals attain their full size in a few days and reproduce themselves as virgin females. Their descendants are propagated in the same manner, and thus in a short time almost incredible numbers of individuals are formed, until finally the sexual eggs are again produced. If now the pool dries up again, the existence of the colony is secured, for the number of animals which produce sexual eggs is very large, and the eggs themselves are of course far more numerous, so that in spite of the destructive agencies to which they are subjected, there will be every chance of the survival of a sufficient number to produce a new generation at a later period. Here, therefore, sexual reproduction has not been abandoned accidentally or from any internal cause, but as an adaptation to certain definite necessities imposed upon the organism by its surroundings.

It is, however, well known that there are certain instances in which sexual reproduction has been altogether lost, and in which parthenogenesis is the only form of propagation. In the animal kingdom, such a condition chiefly occurs in species of which the closely-allied forms exhibit the above-mentioned alternation between parthenogenesis and amphigony, viz. in many *Cynipidae* and *Aphidae*, and also in certain freshwater and marine Crustacea. We may imagine that these parthenogenetic species have arisen from forms with alternating methods of reproduction, by the disappearance of the sexual phase.

In any particular case, it may be difficult to point out the motive by which this change has been determined; but it is most probable that the same conditions which originally caused the intercalation of a parthenogenetic stage have been efficient in causing the gradual disappearance of the sexual stage. If a species of Crustacean, with the above-described alternating method of reproduction (heterogeny), were killed off by its enemies on a larger scale than before, it is obvious that the threatened extinction of the species could be checked by the attainment of a correspondingly greater degree of fertility. Such increased fertility might well be produced by pure parthenogenesis (see Appendix V, p. <u>323</u>), by means of which the number of egg-producing individuals in all the previous sexual generations would be doubled.

In a certain sense, this would be the last and most extreme method by means of which a species might secure continued existence, for it is a method for which it would have to pay very dearly at a later period. If my theory as to the causes of hereditary individual variability be correct, it follows that all species with purely parthenogenetic reproduction are sure to die out; not, indeed, because of any failure in meeting the existing conditions of life, but because they are incapable of transforming themselves into new species, or, in fact, of adapting themselves to any new conditions. Such species can no longer be subject to the process of natural selection, because, with the disappearance of sexual reproduction, they have also lost the power of combining and increasing those hereditary individual characters which they possess.

All the facts with which we are acquainted confirm this conclusion, for whole groups of purely parthenogenetic species or genera are never met with, as would certainly be the case if parthenogenesis had been the only method of reproduction through a successional series of species. We always find it in isolated instances, and under conditions which compel the conclusion that it has become predominant in the species in question, and has not been transmitted from any preceding species.

There still remains a very different class of facts which, so far as we can judge, are in accordance with my theory as to the significance of sexual reproduction, and which may be quoted in its support. I refer to the condition of functionless organs in species with parthenogenetic reproduction.

Under the supposition that acquired characters cannot be transmitted—and this forms the foundation of the views here set forth—organs which are of no further use cannot become rudimentary in the direct and simple manner in which it has been hitherto imagined that degeneration takes place. It is true that an organ which does not perform any function exhibits a marked decrease of strength and perfection in the individual which possesses it, but such acquired degradation is not transmitted to its descendants, and we must therefore look for some other explanation of the firmly established fact that organs do become rudimentary through a series of generations. In seeking this explanation, we shall have to start from the supposition that new forms are not only created by natural selection, but are also preserved by its means. In order that any part of the body of an individual of any species may be kept at the maximum degree of development, it is necessary that all individuals possessing it in a less perfect form must be prevented from propagation—they must succumb in the struggle for existence. I will illustrate this by a special instance. In species which, like the birds of prey[193], depend for food upon the acuteness of their vision, all individuals with relatively weak eyesight must be exterminated, because they will fail in the competition for food. Such birds will perish before they have reproduced themselves, and their imperfect vision is not further transmitted. In this way the keen eyesight of birds of prey is kept up to its maximum.

But as soon as an organ becomes useless, the continued selection of individuals in which it is best developed must cease, and a process which I have termed *panmixia* takes place. When this process is in operation, not only those individuals with the best-developed organs have the chance of reproducing themselves, but also those individuals in which the organs are less well-developed. Hence follows a mixture of all possible degrees of perfection, which must in the course of time result in the deterioration of the average development of the organ. Thus a species which has retired into dark caverns must necessarily come to gradually possess less developed powers of vision; for defects in the structure of the eyes, which occur in consequence of individual variability, are not eliminated by natural selection, but may be transmitted and fixed in the descendants[194]. This result is all the more likely to happen, inasmuch as other organs which are of importance for the life of the species will gain what the functionless organ loses in size and nutrition. As at each stage of retrogressive transformation individual fluctuations always occur, a continued decline from the original degree of development will inevitably, although very slowly, take place, until the last remnant finally disappears. How inconceivably slowly this process goes on is shown by the numerous cases of rudimentary organs: by the above-mentioned embryonic sixth finger of man, or by the hind limbs of whales buried beneath the surface of the body, or by their embryonic tooth-germs. I believe that the very slowness with which functionless organs gradually disappear, agrees much better with my theory than with the one which has been hitherto held. The result of the disuse of an organ is considerable, even in the course of a single individual life, and if only a small fraction of such a result were transmitted to the descendants, the organ would be necessarily reduced to a minimum, in a hundred or at any rate in a thousand generations. But how many millions of generations may have elapsed since e. g. the teeth of the whalebone whales became useless, and were replaced by whalebone! We do not know the actual number of years, but we know that the whole material of the tertiary rocks has been derived from the older strata, deposited in the sea, elevated, and has been itself largely removed by denudation, since that time.

Now if this theory as to the causes of deterioration in disused organs be correct, it follows that rudimentary organs can only occur in species with sexual reproduction, and that they cannot be formed in species which are exclusively reproduced by the parthenogenetic method: for, according to my theory, variability depends upon sexual reproduction, while the deterioration of an organ when disused, no less than its improvement when in use, depends upon variability. There are therefore two reasons which lead us to expect that organs which are no longer used will remain unreduced in species with asexual reproduction: first, because only a very slight degree of hereditary variability can be present, viz. such a degree as was transmitted from the time when sexual reproduction was first abandoned by the ancestors; and, secondly, because even these slight degrees of variability are not combined, or, in other words, because panmixia cannot occur.

And the facts seem to point in the direction required by the theory, for superfluous organs do not become rudimentary in parthenogenetic species. For example, as far as my experience goes, the *receptaculum seminis* does not deteriorate, although it is, of course, altogether functionless when parthenogenesis has become established. I do not attach much importance to the fact that the Psychids and Solenobias—(genera of Lepidoptera which Siebold and Leuckart have shown to include species with parthenogenetic reproduction)—still retain the complete female sexual apparatus, because colonies containing males still occasionally occur in these species. Although the majority of colonies are now purely female, the occasional appearance of males points to the fact that the unisexuality of the majority cannot have been of very long duration. The process of transformation of the species from a bisexual into a unisexual form, only composed of females, is obviously incomplete, and is still in process of development. The case is similar with several species of *Cynipidae*, which reproduce by the parthenogenetic method. In these cases the occurrence of a very small proportion of males is the general rule, and is not confined to single colonies. Thus Adler[195] counted 7 males and 664 females in the common *Cynips* of the rose.

In some Ostracodes, on the other hand, the males appear to be entirely wanting: at least, I have tried in vain for years to discover them in any locality or at any time of the year[196].

Cypris vidua and *Cypris reptans* are such species. Now, although the transformation of these formerly bisexual species into purely unisexual female species appears to be complete[197], yet the females still possess a large, pear-shaped *receptaculum seminis*, with its long spirally twisted duct, which is surrounded by a thick glandular layer. This is the more remarkable as the apparatus is very complicated in the Ostracodes, and retrogressive changes could be therefore easily detected. Furthermore among insects, in the genus *Chermes* the *receptaculum seminis* of the females has also remained unreduced, although the males appear to be entirely wanting, or at least have never been found, in spite of the united efforts of several acute observers[198]. The case is quite different in species which retain both sexual and parthenogenetic reproduction. Thus, the summer females of the *Aphidae* have lost the *receptaculum seminis*; and in these insects sexual reproduction has not ceased, but alternates regularly with parthenogenetic reproduction.

Certainly this proof of the truth of my theory as to the significance of sexual reproduction is far from settling the question: it only renders the theory highly probable. At present it is impossible to do more than this, because we do not yet possess a sufficient number of facts, for many of them could not have been sought for until after the theory had been suggested. We are here concerned with complicated phenomena, into which we cannot acquire an immediate insight, but can only attain it gradually.

But, nevertheless, I hope to have shown that the theory of natural selection is by no means incompatible with the theory of 'the continuity of the germ-plasm;' and, further, that if we accept this latter theory, sexual reproduction appears in an entirely new light: it has received a meaning, and has to a certain extent become intelligible.

The time in which men believed that science could be advanced by the mere collection of facts has long passed away: we know that it is not necessary to accumulate a vast number of miscellaneous facts, or to make as it were a catalogue of them; but we know that it is necessary to establish facts which, when grouped together in the light of a theory, will enable us to acquire a certain degree of insight into some natural phenomenon. In order to direct our attention to those new facts which are of immediate importance, it is absolutely necessary to seek the aid of some general theory for the arrangement and grouping of those which we already possess. This has been my object in the present paper.

But it may be perhaps objected that these phenomena are far too complicated to be attacked at the present time, and that we ought to wait quietly until the simpler phenomena have been resolved into their components. It may be asked whether the trouble and labour involved in the attempt to solve such questions as heredity or the transformation of species are not likely to be wasted and useless.

It is true that we sometimes meet with such opinions, but I believe that they are based upon a misunderstanding of the method which mankind has always followed in the investigation of nature, and which must therefore be founded upon the necessary relations existing between mankind and nature.

Science has often been compared to an edifice which has been solidly built by laying stone upon stone, until it has gradually risen to greater height and perfection. This comparison holds good up to a certain point, but it leads us to easily overlook the fact that this metaphorical building does not at any point rest upon the ground, and that, at least up to the present time, it has remained floating in the air. Not a single branch of science, not even Physics itself, has commenced building from below; all branches have begun to build at greater or less heights in the air, and have then built downwards: and even Physics has not yet reached the ground, for it is still very uncertain as to the nature of matter and force. In no single group of phenomena can we begin with the investigation of ultimate causes, because at this very point our means of reasoning stop short. We cannot begin with ultimate phenomena and gradually lead up to those which are more complicated: we cannot proceed synthetically and deductively, building up the phenomena from below; but we must as a rule proceed analytically and inductively, proceeding from above downwards.

No one will dispute these statements, but they are often forgotten, as is proved by the above-mentioned objection. If we were only permitted to attack the more complicated phenomena after gaining a complete insight into the simpler ones, then all scientists would be physicists and chemists, and not until Physics and Chemistry were done with should we be permitted to proceed to the investigation of organic nature. Under these circumstances we ought not to possess now any scientific theory of medicine; for the study of pathological physiology could not be commenced until normal physiology was completely known and understood. Yet how great a debt is owing by normal to pathological physiology! This is an example which enforces the conclusion that it is not only permissible, but in the highest degree advantageous, for the different spheres of phenomena to be attacked simultaneously.

Furthermore, if we had been compelled to proceed from the simple to the complex, what would have become of the Theory of Descent, the influence of which has advanced our knowledge of Biology to an altogether immeasurable extent?

But in this often repeated criticism that we are not yet ready to attack such complicated phenomena as heredity, is hidden still another fallacy, for it is implied that facts become less certain in proportion to the complexity of their causes. But is it less certain that the egg of an eagle develops into an eagle, or that the peculiarities of the father and mother are transmitted to the child, than that a stone falls to the ground when its support is taken away? Again, is it not possible to draw a perfectly distinct and certain conclusion as to the relative quantity of the material basis of heredity, present in the germ-cells of either parent, from the fact that the father and mother possess an equal or nearly equal share in heredity? But it is really unnecessary to argue in this way: why should we do more than re-affirm that such a method of procedure in scientific investigation is the only way by which we can gradually penetrate the hidden depths of natural phenomena?

No! Biology is not obliged to wait until Physics and Chemistry are completely finished; nor have we to wait for the investigation of the phenomena of heredity until the physiology of the cell is complete. Instead of comparing the progress of science to a building, I should prefer to compare it to a mining operation, undertaken in order to open up a freely branching lode. Such a lode must not be attacked from one point alone, but from many points simultaneously. From some of these we should quickly reach the deep-seated parts of the lode, from others we should only reach its superficial parts; but from every point some knowledge of the complex *tout ensemble* of the lode would be gained. And the more numerous the points of attack, the more complete would be the knowledge acquired, for valuable insight will be obtained in every place where the work is carried on with discretion and perseverance.

But discretion is indispensable for a fruitful result; or, leaving our metaphor, facts must be connected together by theories, if science is to advance. Just as theories are valueless without a firm basis of facts, so the mere collection of facts, without relation and without coherence, is utterly valueless. Science is impossible without hypotheses and theories: they are the plummets with which we test the depth of the ocean of unknown phenomena, and thus determine the future course to be pursued on our voyage of discovery. They do not give us absolute knowledge, but they afford us as much insight as it is possible for us to gain at the present time. To go on investigating without the guidance of theories, is like attempting to walk in a thick mist without a track and without a compass. We should get somewhere under these circumstances, but chance alone would determine whether we should reach a stony desert of unintelligible facts or a system of roads leading in some useful direction; and in most cases chance would decide against us.

In this sense I trust that the sign-post or compass which I offer may be accepted. Even though it should be its fate to be replaced by a better one at a later period, it will have fulfilled its object if it enables science to advance for even a short distance.

Footnotes for Essay V.

176. C. Nägeli, 'Mechanisch-physiologische Theorie der Abstammungslehre.' München u. Leipzig, 1884.

177. 'Ueber die Berechtigung der Darwin'schen Theorie.' Leipzig, 1868, p. 27.

178. l. c., Preface, p. vi.

179. Since the above was written many other morphological peculiarities of plants have been rightly explained as adaptations. Compare, for instance, the investigations of Stahl on the means by which plants protect themselves against the attacks of snails and slugs (Jena, 1888).—A. W., 1888.

180. l. c., pp. 117, 286.

181. Compare the second and fourth of the preceding Essays, 'On Heredity' and 'The Continuity of the Germ-plasm as the Foundation of a Theory of Heredity.'

182. Compare Rauber, 'Homo sapiens ferus oder die Zustände der Verwilderten.' Leipzig, 1885.

183. 'Sitzungsberichte der baierischen Akademie der Wissenschaften,' vom 18 Nov. 1865. Compare also his 'Mechanisch-physiologische Theorie der Abstammungslehre,' p. 102, etc.

184. Jordan, 'Remarques sur le fait de l'existence en société des espèces végétales affines.' Lyon, 1873.

185. S. Hermann's 'Handbuch der Physiologie,' Theil II; 'Physiologie der Zeugung,' by V. Hensen.

186. E. van Beneden, 'Recherches sur la maturation de l'œuf, la fécondation et la division cellulaire.' Gand u. Leipzig, 1883, pp. 404 et seq.

187. Rolph, 'Biologische Probleme.' Leipzig, 1882.

188. Cienkowsky, 'Arch. f. mikr. Anat.,' ix. p. 47. 1873.

189. Hensen, 'Physiologie der Zeugung,' p. 139.

190. Coalescence takes place in the so-called bud-like conjugation of *Vorticellidae* and *Trichodinidae*, etc.

191. Compare (1) Bardeleben, 'Zur Entwicklung der Fusswurzel,' Sitzungsber. d. Jen. Gesellschaft, Jahrg. 1885, Feb. 6; also 'Verhandl. d. Naturforscherversammlung zu Strassburg,' 1885, p. 203; (2) G. Baur, 'Zur Morphologie des Carpus und Tarsus der Wirbelthiere,' Zool. Anzeiger, 1885, pp. 326, 486.

192. In frogs the sixth toe exists in the hind legs as a rudimentary prehallux. Compare Born, Morpholog. Jahrbuch, Bd. I, 1876.

193. I here make use of the same illustration which I employed in my first attempt to explain the effects of *panmixia*. Compare the second Essay 'On Heredity.'

194. [E. Ray Lankester has suggested (Encycl. Britann., art. 'Zoology,' pp. 818, 819) that the blindness of cave-dwelling and deep-sea animals is also due to the fact that 'those individuals with perfect eyes would follow the glimmer of light and eventually escape to the outer air or the shallower depths, leaving behind those with imperfect eyes to breed in the dark place. A natural selection would thus be effected.' Such a sifting process would certainly greatly quicken the rate of degeneration due to panmixia alone.—E. B. P.]

195. Adler, 'Zeitschrift f. wiss. Zool.,' Bd. XXXV, 1881.

196. Compare my paper, 'Parthenogenese bei den Ostracoden,' in 'Zool. Anzeiger,' 1880, p. 82. Purely negative evidence, unless on an immense scale, is quite rightly considered to be of no great value in most cases. But the condition of these animals renders the accumulation of such evidence unusually easy, because the presence of males in a colony of Ostracodes can be proved by a very simple indirect test. Thus if a colony contains any males the *receptacula seminis* of all mature females are filled with spermatozoa, and on the other hand we may be quite sure that males are absent, if after the examination of many mature females, no spermatozoa can be found in any of their *receptacula*.

197. We cannot, however, be absolutely certain of this, for it is conceivable that males may still occur in colonies other than those examined.

198. It has now been shown by Blochmann that males appear for a very short time towards the close of summer, as in the case of *Phylloxera*.—A. W., 1888.

APPENDICES.

Appendix I. Further considerations which oppose Nägeli's

explanation of transformation as due to internal causes[199].

When I describe Nägeli's theory of transformation as due to active causes lying within the organism, as a phyletic force of transformation, I do not mean to imply that it is one of those mysterious principles which, according to some writers, constitute the unconscious cause which directs the transformation of species. Nägeli's idioplasm, which changes from within itself, is conceived as a thoroughly scientific, mechanically operating principle. This cause is undoubtedly capable of theoretical conception: the only question is whether it has any real existence. According to Nägeli, the growing organic substance, the idioplasm, not only represents a *perpetuum mobile* rendered possible as long as its substance continually receives from without the matter and force which are necessary for continuous growth, but it also represents a *perpetuum variabile* due to the action of internal causes[200]. But this is just the doubtful point, viz., whether the structure of the idioplasm itself compels it to change gradually during the course of its growth, or whether it is not rather the external conditions which compel the ever slightly varying idioplasm to change in a certain direction by the summation of small differences. It has been shown above that we do not gain anything by adopting Nägeli's theory, because the main problem which organic nature offers for our solution, viz. adaptation, remains unsolved. Hence this theory does not explain the phenomena of nature, and I believe that there are also certain facts which are directly antagonistic to it.

If the idioplasm really possessed the power of spontaneous variability ascribed to it by Nägeli; if, as a result of its own growth, it were compelled to undergo gradual changes, and thus to produce new species, we should expect that the duration of species, genera, orders, &c. would be of approximately equal length respectively, at least in forms of equal structural complexity. The time required by the idioplasm to undergo such changes as would constitute transformation into a new species ought to be always the same at equal heights in the scale of organization, that is, with equal complexity in the molecular structure of the idioplasm. It appears to me to be a necessary consequence of Nägeli's theory that the causes of transformation lie solely in this molecular structure of the idioplasm. If nothing more than a certain amount of growth, and consequently a certain period of time during which the organism reproduces itself with a certain intensity, is required to produce a change in the idioplasm, then we must conclude that the alteration in the latter must take place when this certain amount of growth has been reached, or after this certain period has elapsed. In other words, the time during which a species exists—from its origin as a modification of some older species, until its own transformation into a new one—must be the same in species with the same degree of organization. But the facts are very far from supporting this consequence of Nägeli's theory. The duration of species is excessively variable: many arise and perish within the limits of a single geological formation, while others may be restricted to a very small part of a formation; others again may last through several formations. It must be admitted that we cannot estimate the exact position of extinct species in the scale of organization, and the differences may therefore depend upon differences of organization: or they may be explained by the supposition that certain species may have become incapable of transformation, and might, under favourable conditions, continue to exist for an indefinite period. But this reply would introduce a new hypothesis in direct antagonism to Nägeli's theory, which assumes that the variability of idioplasm takes place as the consequence of mere growth, and necessarily depends upon molecular structure. Nägeli himself asserts that the essential substance (idioplasm) of the descendants of the earliest forms of life is in a state of perpetual change, which would continue even if the series of successive generations were indefinitely prolonged[201]. Hence there can be no rest in the process of change which the idioplasm must undergo; and this is as true of each single species as it is of the organic world taken as a whole. We could, perhaps, find shelter in the insufficiency of our geological knowledge, but the number of ascertained facts is too great for this to be possible. Thus it is well known that the genus *Nautilus* has lasted from Silurian times, through all the three geological periods, up to the present day: while all its Silurian allies (*Orthoceras*, *Gomphoceras*, *Goniatites*, &c.) became extinct at a comparatively early period.

A keen and clever controversialist might still bring forward many objections against such an argument. I do not therefore place too much dependence upon the geological facts by themselves, as a disproof of the self-variability of Nägeli's idioplasm; for it must be admitted that the facts are not sufficiently complete for this purpose. For instance, in the case of *Nautilus* it might be argued that we do not know anything about the fossil Cephalopods of pre-Silurian times, and that it is therefore possible that the above-mentioned allies of *Nautilus* may have existed previously for as long a period as that through which *Nautilus* has lived in post-Silurian time. However this may be, it will be at least conceded that the geological facts do not lend any support to Nägeli's theory, for we can see no trace of even an approximately regular succession of forms.

Appendix II. Nägeli's explanation of adaptation[202].

In order to explain adaptation Nägeli assumes that, under certain circumstances, external influences may cause slight permanent changes in the idioplasm. If then such influences act continually in the same direction during long periods of time, the changes in the idioplasm may increase to a perceptible amount, i. e. to a degree which makes itself felt in visible external characters[203]. But such changes alone could not be considered as adaptations, for the essential character of an adaptation is that it must be a purposeful change. Nägeli, however, brings forward the fact that external stimuli often produce their chief effects at that very part of the organism to which the stimuli themselves were applied. 'If the results are detrimental, the organism attempts to defend itself against the stimulus: a confluence of nutrient fluid takes place towards the part upon which the stimulus has acted, and new tissues are formed which restore the integrity of the organism by replacing the lost structures as far as possible. Thus in plants the healthy tissues begin to grow actively around the seat of an injury, tending to close it up, and to afford protection by impenetrable layers of cork.' Purposeful reactions of this kind are certainly common in the organic world, occurring in animals as well as in plants. Thus in the human body an injury causes a rapid growth of the surrounding tissues, which leads to the closing-up of the wound; while in the Salamander even the amputated leg or tail is replaced by growth. An extreme example of these purposeful reactions is afforded by the tree-frog (*Hyla*), which is of a light-green colour when seated upon a light-green leaf, but becomes dark brown when transferred to dark surroundings. Hence this animal adapts itself to the colour of its environment, and thus gains protection from its enemies.

Admitting this capability on the part of organisms to react under certain stimuli in a purposeful manner, the question remains whether such a power is a primitive original quality belonging to the essential nature of each organism. The power of changing the colour of the skin in correspondence with that of the surroundings is not very common in the animal kingdom. In the frog this power depends upon a highly complex reflex mechanism. Certain chromatophores in the skin are connected with nerves[204] which pass to the brain and are there brought into relation, by means of nerve-cells, with the nervous centres of the organ of vision. The relation is of such a kind that strong light falling upon the retina constitutes a stimulus for the production of an impulse, which is conducted, along the previously mentioned motor nerves, from the brain to the chromatophores, thus determining the contraction of these latter and the consequent appearance of a light-coloured skin. When the strong stimulus (of light) ceases, the chromatophores expand again, and the skin becomes dark. That the chromatophores do not themselves react upon the direct stimulus of light was proved by Lister[205], who showed that blind frogs do not possess the power of altering their colour in correspondence with that of their environment. It is quite obvious that in this case we are not dealing with a primary, but with a secondarily produced character; and it has yet to be proved that all the purposeful reactions mentioned by Nägeli are not similarly secondary characters or adaptations, and thus very far from being primitive qualities of the organic substance of the forms in which they occur.

I do not by any means doubt that some of the reactions witnessed in organisms do not depend upon adaptation, but such reactions are not usually purposeful. Curiously enough, Nägeli mentions the formation of galls in plants among his instances of purposeful reactions under external stimuli. I think, however, that it can hardly be maintained that the galls are of any use to the plant: on the contrary, they may even be very injurious to it. The gall is only useful to the insect which it protects and supplies with food. The recent and most excellent investigations of Adler[206] and of Beyerinck[207] have shown that the puncture made by the *Cynips* in depositing its eggs is not the stimulus which produces the gall, as was formerly believed to be the case, but that such a stimulus is provided by the larva which developes from the egg. The presence of this small, actively moving, foreign body stimulates the tissue of the plant in a definite manner, always producing a result which is advantageous to the larva and not to the plant. It would be to the advantage of the latter if it killed the intruding larva, either enclosing it by woody tissue devoid of nourishment, or poisoning it by some acrid secretion, or simply crushing it by the active growth of the surrounding tissues. But nothing of the kind occurs: in fact an active growth of cells (forming the so-called 'Blastem' of Beyerinck) takes place around the embryo, while it is still enclosed in the egg-capsule; but the growth is not such as to crush the embryo, which remains free in the cavity, the so-called larval chamber, which is formed around it. It would be out of place to discuss here the question as to how we can conceive that the plant is thus compelled to produce a growth which is at any rate indifferent and may be injurious to it; and which, moreover, is exactly adapted to the needs of its insect-enemy. But it is at all events obvious that this cannot be an example of a self-protecting reaction under a stimulus, and that therefore an organism does not always respond to external stimuli in a manner useful to itself.

But even if we could accept the suggestion that the purposeful reaction of an organism under stimulation is a primary and not a secondarily produced character, such a principle would by no means suffice for the explanation of existing adaptations. Nägeli attempts to explain certain selected cases of adaptation as the direct results of external stimuli. He looks upon the thick hairy coat of mammals in arctic regions, and the winter covering of animals in temperate regions, as a direct reaction of the skin under the influence of cold. He considers that the horns, claws, and tusks of animals have arisen directly as reactions under stimuli applied to certain parts of the surface of the body in attack and defence[208]. This interpretation is similar to that offered by Lamarck at the beginning of this century. At first sight such a suggestion appears to be plausible, for the acquisition of a thick hairy covering by the mammals of temperate regions is actually contemporaneous with the cold season of the year. But the question arises as to whether the production of a larger number of hairs at the beginning of winter is not merely another instance of a secondary character, like the assumption of a green colour by the tree-frog under the stimulus exerted by strong light.

In the case of the hairy coat it is only necessary to produce a larger number of structures such as had existed previously; but how can it have been possible for the petals of flowers, with their peculiar and complex forms, to have been developed from stamens as a direct result of the insects which visit them in order to obtain pollen and nectar? How could the creeping of these insects and the small punctures made by them constitute stimuli for the production of an increased rate of growth? And how is it possible in any way to explain, by mere increase in growth, the origin of a structure in which each part has its own distinct meaning and plays a peculiar part in attracting insects and in the process of cross-fertilization effected by them? Even if the manifold peculiarities of form could be explained in this way, how can such an explanation possibly hold for the colours of flowers? How could the white colour of flowers which open at night be explained as the direct result of the creeping of insects? How can the suggestion of such a cause offer any interpretation of the fact that flowers which open by day are tinted with various colours, or of the fact that there is often a bright or highly coloured spot which shows the way to the hidden nectary?

There are, moreover, a large number of very striking adaptations in form and colour, for which no stimulus acting directly upon the organism can be found. Can we imagine that the green caterpillar[209], plant-bug, or grasshopper, sitting among green surroundings, is thus exposed to a stimulus which directly produces the green colour in the skin? Can the walking-stick insect, which resembles a brown twig, be subject to a transforming stimulus by sitting on such branches or by looking at them? Or again, if we consider the phenomena of mimicry, how can one species of butterfly, by flying about with another species, exercise upon the latter such an influence as to render it similar to the first in appearance? In many cases of mimicry, the mimicked and the mimicking species do not even live in the same place, as we see in the moths, flies, and beetles which resemble in appearance the much-dreaded wasps.

The interpretation of adaptation is the weak part of Nägeli's theory, and it is somewhat remarkable that so acute a thinker should not have perceived this himself. One very nearly gains the impression that Nägeli does not wish to understand the theory of natural selection. He says, for instance, in speaking of the mutual adaptation observable between the proboscis, the so-called 'tongue' of butterflies, and flowers with tubular corolla[210]:—'Among the most remarkable and commonest adaptations observable in the forms of flowers, are the corollas with long tubes considered in relation to the long "tongues" of insects, which suck the nectar from the bottom of the long narrow tubes, and at the same time effect the cross-fertilization of the plant. Both these arrangements have been gradually developed to their present degree of complexity— the long-tubed corollas from those without tubes, and from those with short ones, the long "tongues" from short ones. Undoubtedly both have been developed at the same rate so that the length of both sets of structures has always remained the same.'

No objection can be raised against these statements, but Nägeli goes on to say:—'But how can such a process of development be explained by the theory of natural selection, for at each stage in the process the adaptation was invariably complete. The tube of the corolla and the "tongue" must have reached, for instance, at a certain time, a length of 5 or 10 mm. If now the tube of the corolla became longer in some plants, such an alteration would have been disadvantageous because the insects would be no longer able to obtain food from them, and would therefore visit flowers with shorter tubes. Hence, according to the theory of natural selection, the longer tubes ought to have disappeared. If on the other hand the "tongue" became longer in some insects, such a change would be superfluous and should have been given up, according to the same theory, as unnecessary structural waste. The simultaneous change in the two structures must, according to the theory of natural selection, be due to the same principle as that by which Münchhausen pulled himself out of a bog by means of his own pig-tail.'

But, according to the theory of natural selection, the case appears in a very different light from that in which it is put by Nägeli. The flower and the insect do not compete for the greater length of their respective organs: all through the gradual process, the flower is the first to lengthen its corolla and the butterfly follows. Their relation is not like that between a certain species of animal and another which serves as its prey, where each strives to be the quicker, so that the speed of both is increased to the greatest possible extent in the course of generations. Nor do they stand in the same relation as that obtaining between an insectivorous bird and a certain species of butterfly which forms its principal food; in such a case two totally different characters may be continually increased up to their highest point, e.g. in the butterfly similarity to the dead and fallen leaves among which it seeks protection when pursued, in the bird keenness of sight. As long as the latter quality is still capable of increase, so long will it still be advantageous to any individual butterfly to resemble the leaf a little more completely than other individuals of the same species; for it will thus be capable of escaping those birds which possess a rather keener sight than others. On the other hand, a bird with rather keener sight will have the greatest chance of catching the better protected butterflies. It is only in this way that we can explain the constant production of such extraordinary similarities between insects and leaves or other parts of plants. At every stage of growth both the insect and its pursuer are completely adapted to each other; i.e. they are so far protected and so far successful respectively, as is necessary to prevent that gradual decrease in the average number of individuals which would lead to the extermination of the species[211]. But the fact that there is complete adaptation at each stage does not prevent the two species from increasing those qualities of protection and of pursuit upon which they respectively depend. So far from this being the case, they would be necessarily compelled to gradually increase these qualities so long as the physical possibility of improvement remained on both sides. As long as some birds possessed a rather keener sight than those which previously existed, so long would those butterflies possess an advantage in which the resemblance to leaf-veining was more distinct than in others. But from the moment at which the maximum keenness of eyesight attainable had been reached, at which therefore all butterflies resembled leaves so completely that even the birds with the keenest eyesight might fail to detect them when at rest,— from this very point any further improvement in the similarity to leaves would cease, because the advantage to be gained from any such improvement would cease at the same time.

Such reciprocal intensification of adaptive characters appears to me to have been one of the most important factors in the transformation of species: it must have persisted through long series of species during phylogeny: it must have affected the most diverse parts and characters in the most diverse groups of organisms.

In certain large butterflies of the Indian and African forests—*Kallima paralecta*, *K. inachis*, and *K. albofasciata*—it has been frequently pointed out that the deceptive resemblance to a leaf is so striking that an observer who has received no hint upon the subject believes that he sees a leaf, even when he is looking at the butterfly very closely. The similarity is nevertheless incomplete; for out of sixteen specimens in the collections at Amsterdam and Leyden, I could not find a single one which had more than two lateral veins on one side of the mid-rib of the supposed leaf, or more than three upon the other side; while about six or seven veins should have been present on each side. But from two to three lateral veins are amply sufficient to produce a high degree of resemblance; in fact so much so that it is a matter for wonder as to how it has been possible for such a relatively perfect copy to have been produced; or how the sight of birds can have become so highly developed that while flying rapidly they could perceive the vein-like markings; or to state the case more accurately, that they could detect those individuals with a less number of veins than others. It is possible that the process of increase in resemblance is still proceeding in the species of the genus *Kallima*; at all events, I was struck by the rather strong individual differences in the markings of the supposed leaf.

On the other hand, the cause of the increase in length of the tubular corolla and of the butterfly's 'tongue,' lies neither in the flower nor in the butterfly, but it is to be found in those other insects which visit the flower and steal its honey without being of any assistance in cross-fertilization. It may be stated shortly, that non-tubular corollas, with the honey freely exposed—for it must be assumed the ancestral form was of this kind—gradually developed into corollas with the honey deeply concealed. The whole process was presumably first started by the flower, for the gradual withdrawal of the honey to greater depths conferred the advantage of protection from rain

(Hermann Müller), while larger quantities of honey could be stored up, and this would also increase the number of insects visiting the flower and render their visits more certain. As soon as this withdrawal occurred, the mouth-parts of insects began to be subjected to a selective process whereby these organs in some of them were lengthened at the same rate as that at which the honey was withdrawn. When once the process had begun, its continuance was ensured, for as soon as flower-frequenting insects were divided into two groups with short and with long mouth-parts respectively, a further increase in the length of the corolla-tube necessarily took place in all those flowers which were especially benefited by the assured visits of a relatively small number of species of insects, viz., those flowers in which cross-fertilization was more certainly performed in this way than by the uncertain visits of a great variety of species. This would imply that a still further increase in length would take place, for it is obvious that the cross-fertilization of any flower would be more certainly performed by an insect when the number of species of plants visited by it became less; and hence the cross-fertilization would be rendered most certain when the insect became completely adapted—in size, form, character of its surface, and the manner in which it obtained the honey—to the peculiarities of the flower. Those insects which obtain honey from a great variety of flowers are sure to waste a great part of the pollen by carrying it to the flowers of many different species, while insects which can only obtain honey from a few species of plants must necessarily visit many flowers of the same species one after the other, and they would therefore more generally distribute the pollen in an effective manner.

Hence the tube of the corolla, and the 'tongue' of the butterfly which brings about fertilization, would have continued to increase in length as long as it remained advantageous for the flower to exclude other less useful visitors, and as long as it was advantageous for the butterfly to secure the sole possession of the flower. Hence there is no competition between the flower and the butterfly which fertilizes it, but between these two on the one side, and the other would-be visitors of the flower on the other. Further details as to the advantages which the flower gains by excluding all other visitors, and the butterfly by being the only visitor of the flower, and also as to the manifold and elaborate mutual adaptations between insects and flowers, and as to the advantages and disadvantages which follow from the concealment of the honey—will be found in Hermann Müller's[212] work on the fertilization of flowers, in which all these subjects are minutely discussed, and are clearly explained in a most admirable manner.

Appendix III. Adaptations in Plants[213].

It is well known that Christian Conrad Sprengel was the first to recognise that the forms and colours of flowers are not due to chance, that they are not the mere sport of nature, and that they are not made for the enjoyment of man, but that their purpose is to attract insects for the performance of cross-fertilization. It is also well known that this discovery—which was made at the end of the last century, and which caused much excitement at that time—was completely forgotten, and was brought to light again by Charles Darwin when attacking the same problem.

In his work entitled 'The Solution of Nature's Secret in the Structure and Fertilization of Flowers' ('Das entdeckte Geheimniss der Natur im Bau und der Befruchtung der Blumen'), published at Berlin, in 1793, Sprengel showed, in several hundred cases, that the peculiarities in the structure and colours of flowers were calculated to attract insects, and to ensure the fertilization of the flowers by their instrumentality. But it was due to his successor in this line of investigation that the whole significance of the cross-fertilization effected by insects was made clear. Darwin[214] showed that in many cases, although not in all, the intention of nature was to avoid self-fertilization, and he showed that stronger and more numerous descendants are produced after cross-fertilization.

After Darwin, several investigators, such as Kerner, Delpino and Hildebrand, have paid further attention to the subject, but it has been especially studied in a most thorough manner by Hermann Müller[215]. He looked at the subject from more than one point of view, and showed by direct observation the species of insects which effect cross-fertilization in various species of our native flowers: he also studied the structure of insects in relation to that of flowers, and attempted to establish the mutual adaptations which exist between them. In this way he succeeded in throwing much light upon the process of transformation in many species of flowers, and in proving that certain insects, although unconsciously, are, as it were, breeders of certain forms of flowers. He not only distinguished the disagreeably smelling, generally inconspicuous flowers ('Ekelblumen') produced by Diptera which live on putrid substances, and the flowers which are produced by butterflies; but he also distinguished the flowers bred by saw-flies, by Fossoria, and by bees. He even believes that in certain cases (*Viola calcarata*) he can prove that a flower which owed its original form to being bred by bees, was afterwards adapted to cross-fertilization by butterflies, when it had migrated into an Alpine region where the latter insects are far more abundant than the former.

Although there must of course be much that is hypothetical in the interpretations of the different parts of flowers offered by Hermann Müller, the majority of these explanations are certainly correct, and it is of the greatest interest to be able to recognise the adaptive character of details, even when apparently unimportant, in the structure and colours of flowers.

Sachs has offered a very convincing explanation as to the meaning of leaf-veining, and of its significance in relation to the functions of leaves[216]. He shows that the venation of a leaf is in every case exactly adapted for the fulfilment of its purpose. It has, in the first place, to conduct the nutrient fluid in both directions, and in the second place to support the thin layers of assimilating chlorophyll cells, and to stretch them out so as to expose as large a surface as possible to the light; lastly, it has to toughen the leaf as a protection against being torn. He shows in a very convincing manner that the whole diversity of leaf venation can be understood from these three principles. Here, again, we meet with purposeful arrangements in a class of structures in which it was formerly thought that there was only a chaos of accidental forms, or, as it were, the mere sport of nature with form.

Appendix IV. On the supposed transmission of acquired characters[217].

When I previously maintained that the proofs of the transmission of artificially produced diseases are inconclusive, I had in mind the only experiments which, as far as I am aware, can be adduced in favour of the transmission of acquired characters; viz. the experiments of Brown-Séquard[218] on guinea-pigs. It is well known that he produced artificial epilepsy in these animals by dividing certain parts of the central and also the peripheral nervous system. The descendants of the animals which acquired epilepsy sometimes inherited the disease of their parents.

These experiments have been since repeated by Obersteiner[219], who has described them in a very exact and entirely unprejudiced manner. The fact itself cannot be doubted: it is certain that some of the descendants of animals in which epilepsy has been artificially produced, have also themselves suffered from epilepsy in consequence of the disease of their parents. This fact may be accepted as proved, but in my opinion we have no right to conclude from it that acquired characters can be transmitted. Epilepsy is not a morphological character; it is a disease. We could only speak of the transmission of a morphological character, if a certain morphological change which was the cause of epilepsy had been produced by the nervous lesion, and if a similar change had re-appeared in the offspring, and had produced in them also the symptoms of epilepsy. But that this really occurs is utterly unproved; and is even highly improbable. It has only been proved that many descendants of artificially epileptic parents are small, weakly, and very soon die; and that others are paralysed in various parts of the body, i. e. in one or both of the posterior or anterior extremities; while others again exhibit trophic paralysis of the cornea leading to inflammation and the formation of pus. In addition to these symptoms, the descendants in very rare cases exhibit upon the application of certain stimuli to the skin, a tendency towards those tonic and clonic convulsions together with loss of consciousness which constitute the features of an epileptic attack. Out of thirty-two descendants of epileptic parents only two exhibited such symptoms, both of them being very weakly, and dying at an early age.

These experiments, although very interesting, do not enable us to assert that a distinct morphological change is transmitted to the offspring after having been artificially induced in the parents. The injury caused by the division of a nerve is not transmitted, and the part of the brain corresponding to that which was removed from the parent is not absent from the offspring. The symptoms of a disease are undoubtedly transmitted, but the cause of the disease in the offspring is the real question which requires solution. The symptoms of epilepsy are by no means invariably transmitted; they are in fact absent from the great majority of cases, and the very small proportion in which they do occur, exhibit the symptoms of other diseases in addition to those of epilepsy. The offspring are either quite healthy (thirteen out of thirty cases) or they suffer from disturbances of the nervous system, such as the above-mentioned motor and trophic paralysis,—symptoms which are not characteristic of epilepsy: however in some of the latter epilepsy is also present.

If therefore we wish to express the matter correctly we must not state that epilepsy is transmitted to the offspring, but we must express the facts in the following manner:—animals which have been rendered epileptic by artificial means, transmit to some of their offspring a tendency to suffer from various nervous diseases, viz. from motor paralysis, to a less degree from sensory, and to a high degree from trophic paralysis; in rare cases, when the symptoms of paralysis are very marked, epilepsy is also transmitted.

If we now remember that a considerable number of diseases are already known to be caused by the presence of living organisms in the body, and that these diseases may be transmitted from one organism to another in the form of germs, ought we not to conclude from the above-mentioned facts, that the symptoms are due to an unknown microbe which finds its nutritive medium in the nervous tissues, rather than to suppose that they are due to morphological changes, such as a modification of the histological or molecular structure of certain parts of the nervous system? At all events, it would be more difficult to understand the transmission of such a structural change, than the passage of a bacillus into the sperm- or germ-cell of the parent. There is no ascertained fact which supports the former assumption, but it is very probable that the transmission of syphilis, small-pox and tuberculosis[220] is to be explained by the latter method, although the bacilli have not yet been detected in the reproductive cells. Furthermore, this method of transmission has been rigidly proved in the case of the muscardine disease of the silkworm. At all events we can understand in this way how it happened that the offspring of artificially epileptic guinea-pigs were affected with various forms of nervous disease, a fact which would be quite unintelligible if we assume the occurrence of a true hereditary transmission of a morphological character, such as a pathological change in the structure of some nervous centre.

The manner in which artificial epilepsy becomes manifest after the operation, is also in favour of the explanation offered above. In the first place epilepsy does not result from any one single injury to the nervous system, but it may follow from a variety of different injuries. Brown-Séquard produced it by removing a portion of the grey matter of the brain, and by dividing the spinal cord, although the disease also resulted from a transverse section through half of the latter organ, or from the section of its anterior or posterior columns alone, or from simply puncturing its substance. The most striking effects appeared to follow when the spinal cord was injured in the region between the eighth dorsal and the second lumbar vertebrae, although the results were sometimes also produced by the injury of other parts. Epilepsy also followed the division of the sciatic nerve, the internal popliteal, and the posterior roots of all nerves which pass to the legs. The disease never appears at once, but only after the lapse of some days or weeks, and, according to Brown-Séquard, it is impossible to conclude that the disease will not follow the operation until after six or eight weeks have passed without an epileptic attack. Obersteiner did not witness in any case the first symptoms of the disease for several days after the division of the sciatic nerve. After the operation, sensibility decreases over a certain area on the head and neck, on the same side as the injury. If the animal be pinched in this region (which is called the epileptic area, 'zone epileptogène') it curves itself round towards the injured side, and violent scratching movements are made with the hind leg of the same side. After the lapse of several days or even weeks, these scratching movements which result from pinching in the above-mentioned area, form the beginning of a complete epileptic attack. Hence the changes immediately produced by the division of a nerve are obviously not the direct cause of epilepsy, but they only form the beginning of a pathological process which is conducted in a centripetal direction from the nerve to some centre which is apparently situated in the pons and medulla oblongata, although, according to others[221], it is placed in the cortex of the cerebrum. Nothnagel[222] considers that certain changes, the nature of which is still entirely unknown, but which may be histological or perhaps solely molecular in character, must be produced, leading to an increased irritability of the grey matter of the centres concerned.

Nothnagel thinks it possible or even probable that in those cases in which the division of nerves is followed by epilepsy, a neuritis ascendens—an inflammation passing along the nerves in a central direction—is the cause of the changes suggested by him in the epileptic centre. All our knowledge of bacteria and of the pathological processes induced by them, seems to indicate that such a neuritis ascendens, as is assumed by Nothnagel, would render important support to the hypothesis that the artificial epilepsy is due to infection. But when we further consider that the offspring of artificially epileptic animals may themselves become epileptic, although in most cases they suffer from a variety of other nervous diseases (in consequence of trophic paralysis), I hardly see how the facts can be rendered intelligible except by supposing that in these cases of what I may call traumatic epilepsy, we are dealing with an infectious disease caused by microbes which find their nutritive medium in the nervous tissues, and which bring about the transmission of the disease to the offspring by penetrating the ovum or the spermatozoon.

Obersteiner found that the offspring were more frequently diseased when the mother was epileptic, rather than the father. This is readily intelligible when we remember that the ovum contains an immensely larger amount of substance than the spermatozoon, and can therefore be more frequently infected by microbes and can contain a greater number of them.

Of course, I do not mean to assert that epilepsy always depends upon infection, or upon the presence of microbes in the nervous tissues. Westphal produced epilepsy in guinea-pigs by striking them once or twice sharply upon the head: the epileptic attack took place immediately and was afterwards repeated. It is obvious that the presence of microbes can have nothing to do with such an attack, but the shock alone must have caused morphological and functional changes in the centres of the pons and medulla oblongata, identical with those produced by microbes in the other cases. Nothnagel also distinctly expresses the opinion that epilepsy 'does not depend upon one uniform and invariable histological change, but that the symptoms which constitute the disease may in all probability be caused by various anatomical alterations, provided that they take place in parts of the pons and medulla which are morphologically and physiologically equivalent[223].' Just as a sensory nerve produces the sensation of pain under various stimuli, such as pressure, inflammation, infection with the poison of malaria, etc., so various stimuli might cause the nervous centres concerned to develope the convulsive attack which, together with its after-effects, we call epilepsy. In Westphal's case, such a stimulus would be given by a powerful mechanical shock, in Brown-Séquard's experiments, by the penetration of microbes.

However, quite apart from the question of the validity of this suggestion, we can form no conception as to the means by which an acquired morphological change in certain nerve-cells—a change which is not anatomical, and probably not even microscopical, but purely molecular in nature—can be possibly transferred to the germ-cells: for this ought to take place in such a manner as to produce in their minute molecular structure a change which, after fertilization and development into a new individual, would lead to the reproduction of the same epileptogenic molecular structure of the nervous elements in the grey centres of the pons and medulla oblongata as was acquired by the parent. How is it possible for all this to happen? What substance could cause such a change in the resulting offspring after having been transferred to the egg or sperm-cell? Perhaps Darwin's gemmules may be suggested; but each gemmule represents a cell, while here we have to do with molecules or groups of molecules. We must therefore assume the existence of a special gemmule for each group of molecules, and thus the innumerable gemmules of Darwin's theory must be imagined as increased by many millions. But if we suppose that the theory of pangenesis is right, and that the gemmules really circulate in the body, accompanied by other gemmules from the diseased parts of the brain, and that some of these latter pass into the germ-cells of the individual,—to what strange results would the further pursuit of this idea lead? What an incomprehensible number of gemmules must meet in a single sperm- or germ-cell, if each of them is to contain a representative of every molecule or group of molecules which has formed part of the body at each period of ontogeny. And yet such is the unavoidable consequence of the supposition that acquired molecular states of certain groups of cells can be transmitted to the offspring. This supposition could only be rendered intelligible by some theory of *preformation*[224], such as Darwin's pangenesis; for the latter theory certainly belongs to this category. We must assume that each single part of the body at each developmental stage is, from the first, represented in the germ-cell as distinct particles of matter, which will reproduce each part of the body at its appropriate stage as their turn for development arrives.

I will only briefly indicate some of the inevitable contradictions in which we are involved by such a theory. One and the same part of the body must be represented in the germ- or sperm-cell by many groups of gemmules, each group corresponding to a different stage of development; for if each part gives off gemmules, which ultimately reproduce the part in the offspring, it is clear that special gemmules must be given off for each stage in the development of the part, in order to reproduce that identical stage. And Darwin quite logically accepts this conclusion in his provisional hypothesis of pangenesis. But the ontogeny of each part is in reality continuous, and is not composed of distinct and separate stages. We imagine these stages as existing in the continuous course of ontogeny; for here, as in all departments of nature, we make artificial divisions in order to render possible a general conception, and to gain fixed points in the continuous changes of form which have in reality occurred. Just as we distinguish a sequence of species in the course of phylogeny, although only a gradual transition, not traversed by sharp lines of demarcation, has taken place, so also we speak of the stages of ontogeny, although we can never point out where any stage ends and another begins. To imagine that each single stage of a part is present in the germ, as a distinct group of gemmules, seems to me to be a childish idea, comparable to the belief that the skull of the young St. Laurence exists at Madrid, while the adult skull is to be found in Rome.

We are necessarily driven to such conceptions if we assume that the transmission of acquired characters takes place. A theory of preformation alone affords the possibility of an explanation: an epigenetic theory is utterly unable to render any assistance in reaching an interpretation. According to the latter theory, the germ does not contain any preformed gemmules, but it possesses, as a whole, such a chemical and molecular constitution that under certain circumstances, a second stage is produced from it. For example, the two first segmentation spheres may be regarded as such a second stage; these again possess such a constitution that a certain third stage, and no other, can arise from them, forming the four first segmentation spheres. At each of these stages the spheres produced are peculiar to a distinct species and a distinct individual. From the third stage a fourth arises, and so on, until the embryo is developed, and still later the mature animal which can reproduce itself. No one of the parts of such an animal was originally present as distinct parts in the egg from which it was developed, however minute we may imagine these parts to be. If now an inherited peculiarity shows itself in any organ of the mature animal, this will be the consequence of the preceding developmental stages, and if we were able to investigate the molecular structure of all these stages as far back as the egg-cell, we should trace back to the latter some minute difference of molecular constitution which would distinguish it from any other egg-cell of the same species, and was destined to be the cause of the subsequent appearance of the peculiarity in the mature animal. It is only by the aid of some such hypothesis that we can conceive the cause of hereditary individual differences and the tendencies towards hereditary diseases. Hereditary epilepsy would be intelligible in this way, that is, when the disease is congenital and not due to the presence of microbes, as is presumably the case with artificially induced epilepsy.

The question now arises as to whether we can conceive the communication of such traumatic and therefore acquired epilepsy to the germ-cells. This is obviously impossible under the epigenetic theory of development described above. In what way can the germ-cells be affected by molecular or histological changes in the pons varolii and medulla oblongata? Even if we assume, for the sake of argument, that the central nervous system exercises trophic influences upon the germ-cells, and that such influences may consist of something more than variations in nutritive conditions, and may even include the power of altering the molecular constitution of the germ-plasm in spite of its usual stability; even if we concede these suppositions, how is it conceivable that the changes produced would be of the exact nature and in the exact direction necessary in order to confer upon the germ-plasm the molecular structure of the first ontogenetic stage of an epileptic individual? How can the last ontogenetic stage of the ganglion cells in the pons and medulla of such an individual, stamp upon the germ-plasm in the germ-cells of the same animal—not indeed the peculiar structure of the stage itself—but such a molecular constitution as will ensure the ultimate appearance of epilepsy in the offspring? The theory of epigenesis does not admit that the parts of the full-grown individual are contained in the germ as preformed material particles, and therefore this theory cannot allow that anything is added to the germ-plasm; but in accepting the above-made supposition, we are compelled to assume that the molecular structure of the whole of the germ-plasm is changed to a slight extent.

Nägeli is quite right in maintaining that the solid protoplasm alone, as opposed to the fluid part, i.e. that part of the protoplasm which has passed into solution, can act as the bearer of hereditary tendencies. This appears to be undoubtedly proved by the fact that the amount of material provided by the male parent for the development of an embryo is in almost all animals far smaller than the amount provided by the female parent.

In Mammalia the share contributed by the father probably only forms about one hundred-billionth part of that contributed by the mother, and yet nevertheless the influence of the former in heredity is on an average equal to that exerted by the latter[225]. Now, from the point of view of epigenesis, no molecule of the brain of an epileptic animal can reach the germ-cell except in a state of solution, and therefore no direct increase in the germ-plasm can be referred to such molecules, quite apart from the fact that such addition, even if possible, could not be of any value, because the last stage of the epileptic tendency must be represented in the nerve-cells and nerve-fibres of the diseased brain, while the first stage ought to be represented in the germ-cell.

It may be safely asserted that according to the theory of epigenesis the germ-cells cannot be influenced except as regards their nutrition. Nutritive changes may be imagined to occur through the varying trophic influence of the nervous system upon the sexual organs, but the structure of the germ-plasm cannot be altered by mere nutritive changes, or at all events it cannot be altered in that distinct and definite direction which is required by the supposed transmission of acquired epilepsy.

Thus the transmission of artificially produced epilepsy can neither be explained upon the epigenetic theory, nor upon the theory of preformation; it can only be rendered intelligible if we suppose that the appearance of the disease in the offspring depends upon the introduction and presence of living germs, viz. of microbes. The supposed transmission of this artificially produced disease is the only definite instance which has been hitherto brought forward in support of the transmission of acquired characters. I believe that I have shown that such support is deceptive, not because there is any uncertainty about the fact of the transmission itself, but because it is a transmission which cannot depend upon heredity, and is in all probability due to infection.

Ever since I began to doubt the transmission of acquired characters, I have been unable to meet with a single instance which could shake my conviction. There were many instances in which hereditary transmission was clearly established, but in none of them was there any reason to suppose that the characters transmitted were really acquired. For example, Fritz Müller has recently informed me of an instance in which he believes that there can be no doubt of the transmission of acquired characters. His observations are so interesting in several respects that I will quote them here. He says in his letter, 'Among the bastards of two species of *Abutilon*, in which I had never observed hexamerous flowers, there was a single plant with a few such blossoms. As these flowers are sterile with the pollen of the same plant, I was obliged to fertilize it with pollen from another plant bearing only pentamerous flowers, in order to obtain seeds from the former. For three weeks I examined all the flowers from a plant grown from such seed, finding 145 pentamerous, 103 hexamerous, and 13 heptamerous flowers. I examined similarly the flowers of another plant produced from seed obtained from pentamerous flowers from the same parent plants. There were 454 pentamerous and 6 hexamerous flowers, and hence only 1·3 per cent. of the latter kind.'

It must certainly be admitted that the large proportion of abnormal hexamerous flowers depends upon heredity in the instance first quoted; but the hexamerous condition is not an acquired character; it is merely the first appearance of a new innate character. It is not due to the reaction of the vegetable organism under some external stimulus, for it appeared in a plant exposed to conditions similar to those which acted upon the other plant which only produced the normal pentamerous flowers. It must therefore have resulted from the tendencies which were present in the germ from which the plant itself developed, either as a spontaneous change in the germ-plasm or through the combination of two parental germ-plasms—a combination which may lead to the appearance or the reality of a new character. We know that the germ-plasm of each individual is not a simple substance, but possesses a very complex composition, for it consists of a number of ancestral germ-plasms represented in very different proportions. Now, although we cannot learn anything directly about the processes of growth of the germ-plasm, and its resulting ontogenetic stages, yet we do know, chiefly from observations upon man, that the characters of ancestors appear in the offspring in very different combinations and in very different degrees of strength. This may, perhaps, be explained by assuming that in the union of parental germ-plasms which takes place at fertilization, the contained ancestral germ-plasms unite in different ways, and thus come to grow with different strengths. Certain ancestral germ-plasms will meet and together produce a double effect: other opposed germ-plasms will neutralize each other; and between these two extremes all intermediate conditions will occur. And these combinations will not only take place at fertilization, but also at every stage of the whole ontogenetic history, for each stage is represented by its idioplasm, which is itself composed of ancestral idioplasms.

We do not yet know enough to be able to prove in detail the manner in which new characters may arise from such a combination of different kinds of germ-plasm. And yet it appears to me that such a view, e.g. in the case of the variation of buds, is by far the most natural. There is indeed a single example in which we can, to some extent, understand how it is that a new character may arise by these means. Certain canary-birds have a tuft of feathers on the head, but if two such birds are paired, their descendants are generally bare-headed, instead of having larger tufts[226]. The formation of a tuft depends upon the fact that the feathers are scanty and in fact absent from part of the skin of the head. Now when the scanty plumage of both parents is combined in the offspring the latter is bare-headed. Hence by the combination of ancestral characters a new character (bare-headedness) is produced, and one which is hardly likely to have ever occurred in the ancestors of existing canaries.

We do not know the causes which have been in operation when a flower possesses one petal more than the usual number, any more than we can explain why it is that one star-fish has five and another six rays. We cannot unravel the details of the mysterious relationship between two parent germ-plasms, each of which is composed of a countless number of ancestral germ-plasms from the first and second back to the nth degree. But we can nevertheless maintain in a general way that such irregularities are the result of this complex struggle between the germ-plasms in the ovum and the idioplasms in the subsequent stages of the developing organism, and that they are not the result of external influences.

If, however, acquired characters are brought forward in connexion with the question of the transformation of species, the term 'acquired' must only be applied to those characters which do not arise from within the organism, but which arise as the reaction of the organism under some external stimulus, most commonly as the consequence of the increased or diminished use of an organ or part. We have then to learn whether the altered conditions of life, by forcing an organism to adopt new habits, can by such means lead directly, and not indirectly through natural selection, to the transformation of the species; or whether the effects of increased or diminished use of certain parts, implied by the new habits, are restricted to the individual itself, and therefore powerless to effect any direct modification of the species.

Fritz Müller's observation is also interesting in another respect: it appears to controvert my views upon heredity as expressed in the theory of the continuity of the germ-plasm. If a single flower can transmit to its descendants special peculiarities which were not possessed by its ancestors, we seem to be driven to the conclusion that the ancestral germ-plasm has not passed into the flower in question, but that new germ-plasm has been formed, inasmuch as the new characters are derived from the flower itself, and not from any of its ancestors. I think, however, that the observation admits of another interpretation: a specimen of *Abutilon* with many hundred flowers is not a single individual, but a colony consisting of numerous individuals which have arisen by budding from the first individual developed from the seed.

I have not hitherto considered budding in relation to my theories, but it is obvious that it is to be explained from my point of view, by supposing that the germ-plasm which passes on into a budding individual consists not only of the unchanged idioplasm of the first ontogenetic stage (germ-plasm), but of this substance altered, so far as to correspond with the altered structure of the individual which arises from it—viz. the rootless shoot which springs from the stem or branches. The alteration must be very slight, and perhaps quite insignificant, for it is possible that the differences between the secondary shoots and the primary plant may chiefly depend upon the changed conditions of development, which takes place beneath the earth in the latter case, and in the tissues of the plant in the former. Thus we may imagine that the idioplasm, when it developes into a flowering shoot, produces at the same time the germ-cells which are found in the latter. We thus approach an understanding of Fritz Müller's observation; for if the whole shoot which produces the flower arises from the same idioplasm which also forms its germ-cells, we can readily understand why the latter should contain the same hereditary tendencies which were previously expressed in the flower which produced them. The fact that variations may occur in a single shoot depends upon the changes explained above, which occur in the idioplasm during the course of its growth, as a result of the varying proportions in which the ancestral idioplasms may be contained in it.

Fritz Müller's observation affords a beautiful confirmation of this view, for if the flower itself transmitted the hexamerous condition to its germ-cells, we could not understand why some of the extremely rare hexamerous flowers were produced by the crossing of two pentamerous flowers, in the control experiment. An explanation of this fact can only be found in the assumption that the germ-plasm contained in the mother plant, during its growth and consequent distribution through all the branches of the colony, became arranged into a combination of idioplasms, which, whenever it predominated (as it did at certain places), necessarily led to the formation of hexamerous flowers. I will not consider here the question as to whether this combination is to be looked upon as an instance of reversion, or whether it represents something new. Such a question is of no importance for our present purpose; but the hexamerous flowers of the control experiment prove, in my opinion, that germ-plasm containing the requisite combination was distributed in the mother plant and also existed, but in insufficient amount, in shoots which did not produce any hexamerous flowers.

Appendix V. On the Origin of Parthenogenesis[227].

The transformation of heterogeny into pure parthenogenesis has obviously been produced by other causes as well as by those mentioned in the main part of this paper. Other and quite different circumstances have also had a share in its production. Pure parthenogenesis may be produced without the intermediate condition of heterogeny. Thus, for example, the pure and exclusive parthenogenesis with which the large Phyllopod crustacean, *Apus*, is reproduced at most of its habitats, has not arisen from the loss of previously existent sexual generations, but simply from the non-appearance of males, accompanied by the simultaneous acquisition of the power, on the part of the females, of producing eggs which do not require fertilization. This is proved by the fact that males occur in certain scattered colonies of this species, and sometimes they are even present in considerable numbers. But even if we were not aware of these facts, the same conclusions might nevertheless have been drawn from the fact that *Apus* produces eggs of only one form—viz. resting eggs with hard shells. In every case in which parthenogenesis has been first introduced in alternation with sexual reproduction, the resting eggs are produced by the latter generations, while the parthogenetic generations produce eggs with thin shells, in which the embryo developes and hatches very rapidly. In this way parthenogenesis leads to a rapid increase of the colony. In *Apus* such increase in the number of individuals is gained in an entirely different manner, viz. by the fact that all the animals become females, which produce eggs at a very early age, and continue producing them in increasing fertility for the whole of their life. In this manner an enormous number of eggs collects at the bottom of the pool inhabited by the colony, so that after it has dried up, in spite of loss from various destructive agencies, there will still remain a sufficiency of eggs to reproduce a numerous colony, as soon as the pool has filled again.

This form of parthenogenetic reproduction is especially well suited to the needs of species inhabiting small pools which entirely depend upon rain-fall, and which may disappear at any time. In these cases the time during which the colony can live is often too short to permit the production of several generations even from rapidly developing summer-eggs. Under these circumstances the pool would often suddenly dry up before the series of parthenogenetic generations had been run through, and hence before the appearance of the sexual generation and resting eggs. In all such cases the colony would be exterminated.

This consideration might lead us to think that Crustacea, such as the *Daphnidae*, which develope by means of heterogeny, would hardly be able to exist in small pools filled by the rain; but here also nature has met the difficulty by another adaptation. As I have shown in a previous paper[228], the heterogeny of the species of *Daphnidae* which inhabit such pools is modified in such a manner, that only the first generation produced from the resting eggs consists of purely parthenogenetic females, while the second includes many sexual animals, so that resting eggs are produced and laid, and the continuance of the colony is secured a few days after it has been first founded; viz. after the appearance of the first generation.

But it is also certain that in the *Daphnidae*, heterogeny may pass into pure parthenogenesis by the non-appearance of the sexual generations. This seems to have taken place in certain species of *Bosmina* and *Chydorus*, although perhaps only in those colonies of which the continuance is secured for the whole year; viz. those which inhabit lakes, water-pipes, or wells in which the water cannot freeze. In certain insects also (e. g. *Rhodites rosae*) pure parthenogenesis seems to be produced in a similar manner, by the non-appearance of males.

But the utility which we may look upon as the cause of parthenogenesis is by no means so clear in all cases. Sometimes, especially in certain species of Ostracoda, its appearance seems almost like a mere caprice of nature. In this group of the Crustacea, one species may be purely parthenogenetic, while a second reproduces itself by the sexual method, and a third by an alternation of the two methods: and yet all these species may be very closely allied and may frequently live in the same locality and apparently with the same habit of life. But it must not be forgotten that it is only with the greatest difficulty that we can acquire knowledge about the details of the life of these minute forms, and that where we can only recognize the appearance of identical conditions, there may be highly important differences in nutrition, habits, enemies and the means by which they are resisted, and in the mode by which the prey is captured—circumstances which may place two species living in the same locality upon an entirely different basis of existence. It is not merely probable that this is the case; for the fact that certain species have modified their modes of reproduction is in itself a sufficient proof of the validity of the conclusions which have just been advanced.

The fact that different methods of reproduction may obtain in different colonies of the same species, although with thoroughly identical habits, may depend upon differences in the external conditions (as in *Bosmina* and *Chydorus* mentioned above), or upon the fact that the transition from sexual to parthenogenetic reproduction is not effected with the same ease and rapidity in all the colonies of the same species. As long as males continue to make their appearance in a colony of *Apus*, sexual reproduction cannot wholly disappear. Although we are unable to appreciate, with any degree of certainty, the causes by which sex is determined, we may nevertheless confidently maintain that such determining influences may be different in two widely separated colonies. As soon, however, as parthenogenesis becomes advantageous to the species, securing its existence more efficiently than sexual reproduction, it will not only be the case that the colonies which produce the fewest males will gain advantage, but within the limits of the colony itself, those females will gain an advantage which produce eggs that can develope without fertilization. When the males are only present in small numbers, it must be very uncertain whether any given female will be fertilized: if therefore the eggs of such a female required fertilization in order to develope, it is clear that there would be great danger of entire failure in this necessary condition. In other words:—as soon as any females begin to produce eggs which are capable of development without fertilization, from that very time a tendency towards the loss of sexual reproduction springs into existence. It seems, however, that the power of producing eggs which can develope without fertilization is very widely distributed among the Arthropoda.

Appendix VI. W. K. Brooks' Theory of Heredity[229].

The only theory of heredity which, at any rate in one point, agrees with my own, was brought forward two years ago by W. K. Brooks of Baltimore[230]. The point of agreement lies in the fact that Brooks also looks upon sexual reproduction as the means employed by nature in order to produce variation. The manner in which he supposes that the variability arises is, however, very different from that suggested in my theory, and our fundamental conceptions are also widely divergent. While I look upon the continuity of the germ-plasm as the foundation of my theory of heredity, and therefore believe that permanent hereditary variability can only have arisen through some direct change in the germ-plasm effected by external influences, or following from the varied combinations which are due to the mixture of two individually distinct germ-plasms at each act of fertilization, Brooks, on the other hand, bases his theory upon the transmission of acquired characters, and upon the idea which I have previously called 'the cyclical development of the germ-plasm.'

Brooks' theory of heredity is a modification of Darwin's pangenesis, for Brooks also assumes that minute gemmules are thrown off by each cell in the body of the higher organisms; but such gemmules are not emitted always, and under all circumstances, but only when the cell is subjected to unaccustomed conditions. During the persistence of the ordinary conditions to which it is adapted, the cell continues to perform its special functions as part of the body, but as soon as the conditions of life become unfavourable and its functions are disturbed, the cell 'throws off minute particles which are its germs or gemmules.'

These gemmules may then pass into any part of the organism; they may penetrate the ova in the ovary, or may enter into a bud, but the male germ-cells possess a special power of attracting them and of storing them up within themselves.

According to Brooks, variability arises as a consequence of the fact that each gemmule of the sperm-cell unites, during fertilization, with that part of the ovum which, in the course of development, is destined to become a cell corresponding to that from which the gemmule has been derived.

Now, when this cell developes in the offspring, it must, as a hybrid, have a tendency to vary. The ova themselves, as cells, are subject to the same laws; and the cells of the organism will continue to vary until one of the variations is made use of by natural selection. As soon as this is the case, the organism becomes, *ipso facto*, adapted to its conditions; and the production of gemmules ceases, and with it the manifestation of variability itself, for the cells of the organism then derive the whole of their qualities from the egg, and being no longer hybrid, have no tendency to vary. For the same reason the ova themselves will also cease to vary, and the favourable variation will be transmitted from generation to generation in a stereotyped succession, until unfavourable conditions arise, and again lead to a fresh disposition to vary.

In this way Brooks[231] attempts to mediate between Darwin and Lamarck, for he assumes, on the one hand, that external influences render the body or one of its parts variable, while, on the other hand, the nature of the successful variations is determined by natural selection. There is, however, a difference between the views of Brooks and Darwin, although not a fundamental difference. Darwin also holds that the organism becomes variable by the operation of external influences, and he further assumes that changes acquired in this way can be communicated to the germ and transmitted to the offspring. But according to his hypothesis, every part of the organism is continually throwing off gemmules which may be collected in the germ-cells of the animal, while, according to Brooks, this only takes place in those parts which are placed under unfavourable conditions or the function of which is in some way disturbed. In this manner the ingenious author attempts to diminish the incredible number of gemmules which, according to Darwin's theory, must collect in the germ-cells. At the same time he endeavours to show that those parts must always vary which are no longer well adapted to the conditions of life.

I am afraid, however, that Brooks is confounding two things which are in reality very different, and which ought necessarily to be treated separately if we wish to arrive at correct conclusions: viz., the adaptation of a part of the body to the body itself, and its adaptation to external conditions. The first of these adaptations may exist without the second. How can those parts become variable which are badly adapted to the external conditions, but are nevertheless in complete harmony with the other parts of the body? If the conditions of life of the cells which constitute the part in question must become unfavourable, in order that the gemmules which produce variation may be thrown off, it is obvious that such a result would not occur in the case mentioned above. Suppose, for example, that the spines of a hedgehog are not sufficiently long or sharply pointed to afford protection to the animal, how could such an unfavourable development afford the occasion for the throwing off of gemmules, and a resulting variability of the spines, inasmuch as the epidermic tissue in which these structures arise, remains under completely normal and favourable conditions, whatever length or sharpness the spines may attain? The conditions of the epidermis are not unfavourably affected because, as the result of short and blunt spines, the number of hedgehogs is reduced to far below the average. Or consider the case of a brown caterpillar which would gain great advantage by becoming green; what reason is there for believing that the cells of the skin are placed in unfavourable conditions, because, in consequence of the brown colour, far more caterpillars are detected by their enemies, than would have been the case if the colour were green? And the case is the same with all adaptations. Harmony between the parts of the organism is an essential condition for the existence of the individual. If it is wanting, the individual is doomed; but such harmony between any one part and all others, i. e. proper nutrition for each part, and adequate performance of its proper function, can never be disturbed by the fact that the part in question is insufficiently adapted to the outer conditions of life. According to Darwin, all the cells of the body are continually throwing off gemmules, and against such an assumption no similar objection can be raised. It can only be objected that the assumption has never been proved, and that it is extremely improbable.

A further essential difference between Darwin's theory of pangenesis and Brooks' hypothesis lies in the fact that Brooks holds that the male and female germ-cells play a different part, and that they tend to become charged with gemmules in different degrees, the egg-cell containing a far smaller number than the sperm-cell. According to Brooks the egg-cell is the conservative principle which brings about the permanent transmission of the true characters of the race or species, while he believes that the sperm-cell is the progressive principle which causes variation.

The transformation of species is therefore believed to take place, for the most part, as follows:— those parts which are placed in unfavourable conditions by the operation of external influences, and which have varied, throw off gemmules which reach the sperm-cells, and the latter by fertilization further propagate the variation. An increase of variation is produced because the gemmules which reach the egg through the sperm-cell may unite or conjugate with parts of the former which are not the exact equivalents of the cells from which the gemmules arose, but only very nearly related to them. Brooks calls this 'hybridization,' and he concludes that, just as hybrids are more variable than pure species, so such hybridized cells are also more variable than other cells.

The author has attempted to work out the details of his theory with great ingenuity, and as far as possible to support his assumptions by facts. Moreover, it cannot be denied that there are certain facts which seem to indicate that the male germ-cell plays a different part from that taken by the female germ-cell in the formation of a new organism.

For example, it is well known that the offspring of a horse and an ass is different when the male parent is a horse from what it is when the male parent is an ass. A stallion and a female ass produce a hinny which is more like a horse, while a male ass and a mare produce a mule which is said to be more like an ass[232]. I will refrain from considering here the opinion of several authors (Darwin, Flourens, and Bechstein) that the influence of the ass is stronger in both cases, only predominating to a less extent when the ass is the female parent; and I will for the sake of brevity accept Brooks' opinion that in these cases the influence of the father is greater than that of the mother. Were this so in all cross-breeding between different species and in all cases of normal fertilization, we should be compelled to conclude that the influences of the male and female germ-plasms upon the offspring differ at any rate in strength. But this is by no means always the case, for even in horses the reverse may occur. Thus it is stated that certain female race-horses have always transmitted their own peculiarities, while others allowed those of the stallion to preponderate.

In the human species the influence of the mother preponderates quite as often as that of the father, although in many families most of the children may take after either parent. There is nevertheless hardly any large family in which all the children take after the same parent. If we now try to explain the preponderating influence of one parent by the supposition of a greater strength in hereditary power, without first inquiring after some deeper cause, I think the only conclusion warranted by the facts before us is that this power is rarely or never equal in both of the conjugating germ-cells, but that even within the same species, sometimes the male and sometimes the female is the stronger, and that the strength may even vary in the different offspring of the same individuals, as we so frequently see in human families. The egg-cells of the same mother which ripen one after the other, and also the sperm-cells of the same father, must therefore present variations in the strength of their hereditary power. It is then hardly to be wondered at that the relative hereditary power of the germ-cells in different species should vary, although we cannot as yet understand why this should be the case.

It would not be very difficult to render these facts intelligible in a general way by an appeal to physiological principles. The quantity of germ-plasm contained in a germ-cell is very minute, and together with the idioplasms of the various ontogenetic stages to which it gives rise, it must be continually increased by assimilation during the development of the organism. If now this power of assimilation varied in intensity, a relatively rapid growth of the idioplasm derived from one of the parents would ensue, and with it the preponderance of the hereditary tendencies of the parent in question. Now, it is obvious that no two cells of the same kind are entirely identical, and hence there must be differences in their powers of assimilation. Thus the varying hereditary powers of the egg-cells produced from the same ovary become explicable, and still more easily the varying powers of the germ-cells produced in the ovaries or testes of different individuals of the same species; most easily of all the differences observable in this respect between the germ-cells of different species.

Of course, this hereditary power is always relative, as may be easily proved by cross-breeding between different species and races. Thus when a fantail pigeon is crossed with a laugher, the characters of the former preponderate, but when crossed with a pouter the characters of the latter preponderate[233]. The facts afforded by cross-breeding between hybrids and one of the pure parent species, together with a consideration of the resulting degree of variability, seem to me to be even more unfavourable to Brooks' view. They appear to me to admit of an interpretation different from that brought forward by him; and when he proceeds to make use of secondary sexual characters for the purpose of his theory, I believe that his interpretation of the facts can be easily controverted. It is hardly possible to conclude that variability is due to the male parent, because the males in many species of animals are more variable, or deviate further from the original type, than the females. It is certainly true that in many species the male sex has taken the lead in processes of transformation, while the female sex has followed, but there is no difficulty in finding a better explanation of the fact than that afforded by the assumption 'that something within the animal compels the male to lead and the female to follow in the evolution of new breeds.' Brooks has with great ingenuity brought forward certain instances which cannot be explained with perfect confidence by Darwin's theory of sexual selection, but this hardly justifies us in considering the theory to be generally insufficient, and in having recourse to a theory of heredity which is as complicated as it is improbable. The whole idea of the passage of gemmules from the modified parts of the body into the germ-cells is based upon the unproved assumption that acquired characters can be transmitted. The idea that the male germ-cell plays a different part from that of the female, in the construction of the embryo, seems to me to be untenable, especially because it conflicts with the simple observation that upon the whole human children inherit quite as much from the father as from the mother.

Footnotes for Appendices for Essay V.

199. Appendix to page 257.

200. l. c., p. 118.

201. l. c., p. 118.

202. Appendix to page 258.

203. l. c., p. 137.

204. Compare Brücke, 'Farbenwechsel des Chamäleon.' Wien. Sitzber. 1851. Also Leydig, 'Die in Deutschland lebenden Saurier,' 1872.

205. 'Philosophical Transactions,' vol. cxlviii. 1858, pp. 627-644.

206. Adler, 'Beiträge zur Naturgeschichte der Cynipiden,' Deutsche entom. Zeitschr. XXI., 1877, p. 209; and by the same author, 'Ueber den Generationswechsel der Eichen-Gallwespen,' Zeitschr. f. wiss. Zool., Bd. XXXV. 1880, p. 151.

207. Beyerinck, 'Beobachtungen über die ersten Entwicklungsphasen einiger Cynipidengallen,' Verhandl. d. Amsterd. Akad. d. Wiss. Bd. XXII. 1883.

208. l. c., p. 144.

209. [It is now known that many such caterpillars are actually modified in colour by their surroundings, but the process appears to be indirect and secondarily acquired by the operation of natural selection, like that of the change of colour in the chamaeleon, frogs, fish, etc.; although the stimulus of light acts upon the eyes of the latter animals and upon the skin of the caterpillar. See the seventh Essay (pp. 394-397) for a more detailed account.—E. B. P.]

210. l. c., p. 150.

211. In order to make the case as simple as possible, I assume that the insectivorous bird feeds upon a single species of insect, and that the insect is only attacked by a single species of bird.

212. English Edition, translated by D'Arcy W. Thompson, B.A. London, 1883, p. 509 et seqq.

213. Appendix to page 260.

214. Ch. Darwin, 'On the fertilization of Orchids by Insects.' London, 1877.

215. Compare Hermann Müller, 'Die Befruchtung der Blumen durch Insekten und die gegenseitigen Anpassungen beider.' Leipzig, 1873. See also many articles by the same author in 'Kosmos,' and other periodicals. These later articles are included in the English translation by D'Arcy W. Thompson.

216. 'Lectures on the Physiology of Plants,' translated by H. Marshall Ward, Oxford, 1887, p. 47.

217. Appendix to page 267.

218. Brown-Séquard, 'Researches on epilepsy; its artificial production in animals and its etiology, nature, and treatment.' Boston, 1857. Also various papers by the same author in 'Journal de physiologie de l'homme,' Tome I and III, 1858, 1860, and in 'Archives de physiologie normale et pathologique,' Tome I-IV, 1868-1872.

219. 'Oesterreichische medicinische Jahrbücher.' Jahrgang, 1875, p. 179.

220. A direct transmission of the germs of disease through the reproductive cells has lately been rendered probable in the case of tuberculosis, for the bacilli have been found in tubercles in the lungs of an eight-months' fœtal calf, the mother being affected at the time with acute tuberculosis. However it is not impossible that infection may have arisen through the placenta. See 'Fortschritte der Medicin,' Bd. III, 1885, p. 198.

221. Compare Unvericht, 'Experimentelle und klinische Untersuchungen über die Epilepsie.' Berlin, 1883. With regard to the question of hereditary transmission, the part of the brain in which the epileptic centre is placed is of no importance.

222. Compare Ziemssen's Handbuch der spec. Pathologie und Therapie.' Bd. XII. 2. Hälfte; Artikel 'Epilepsie und Eklampsie.' Leipzig, 1877.

223. l. c., p. 269.

224. It is generally known that the earlier physiologists believed in what was called the 'evolutionary theory,' or the 'theory of preformation.' This assumes that the germ contains, in a minute form, the whole of the fully-developed animal. All the parts of the adult are preformed in the germ, and development only consists in the growth of these parts and their more perfect arrangement. This theory was generally accepted until the middle of the last century, when Kaspar Friedrich Wolff brought forward the theory of 'epigenesis,' which since that time has been the dominant one. This assumes that no special parts of the germ are preformations of certain parts of the fully-developed animal, and that these latter arise by a series of changes in the germ, which gradually gives rise to them. In modern times the theory of preformation has been

revived in a less crude form, as is shown by the ideas of Nägeli, and by Darwin's 'pangenesis.'—A. W., 1888.

225. Nägeli, l. c. p. 110.

226. See Darwin, 'The Variation of Animals and Plants under Domestication.' 1875. Vol. I. p. 311.

227. Appendix to page 290.

228. Weismann, 'Naturgeschichte der Daphnoiden,' Zeitschrift f. wiss. Zool. XXIII. 1879.

229. Appendix to page 277.

230. Compare W. K. Brooks, 'The Law of Heredity, a Study of the Cause of Variation, and the Origin of living Organisms.' Baltimore, 1883.

231. l. c., p. 82.

232. This seems to be the general opinion (see the quotation from Huxley in Brooks' 'Heredity,' p. 127); but I rather doubt whether there is such a constant difference between mules and hinnies. Furthermore, I cannot accept the opinion that mules always resemble the ass more than the horse. I have seen many mules which bore a much stronger likeness to the latter. I believe that it is at present impossible to decide whether there is a constant difference between mules and hinnies, because the latter are very rarely seen, and because mules are extremely variable. I attempted to decide the question last winter by a careful study of the Italian mules, but I could not come across a single hinny. These hybrids are very rarely produced, because it is believed that they are extremely obstinate and bad-tempered. I afterwards saw two true hinnies at Professor Kühn's Agricultural Institute at Halle. These hinnies by no means answered to the popular opinion, for they were quite tractable and good-tempered. They looked rather more like horses than asses, although they resembled the latter in size. In this case it was quite certain that one parent was a stallion and the other a female ass.—A. W. 1889.

233. Darwin, 'Variation of Animals and Plants under Domestication,' 1875, Vol. II. p. 41.

VI.

ON THE NUMBER OF POLAR BODIES AND

THEIR SIGNIFICANCE IN HEREDITY.

1887.

ON THE NUMBER OF POLAR BODIES AND THEIR SIGNIFICANCE IN HEREDITY.

PREFACE.

The following paper stands in close relation to a series of short essays which I have published from time to time since the year 1881. The first of these treated of 'The Duration of Life,' and the last of 'The Significance of Sexual Reproduction.' The present essay is most intimately connected with that upon 'The Continuity of the Germ-plasm,' and has, in fact, grown out of the explanation of the meaning of polar bodies in the animal egg, brought forward in that essay. The explanation rested upon a trustworthy and solid foundation, as I am now able to maintain with even greater confidence than at that time. It rested upon the idea that in the egg-cell, a cell with a high degree of histological differentiation, two different kinds of nuclear substance exert their influence, one after the other. But continued investigation has shown me that the explanation built upon this idea is only correct in part, and that it does not exhaust the full meaning of the formation of polar bodies. In the present essay I hope to complete the explanation by the addition of essential elements, and I trust that, at the same time, I shall succeed in throwing new light upon the mysterious problems of sexual reproduction and parthenogenesis.

It is obvious that this essay can only contain an attempt at an explanation, an hypothesis, and not a solution which is above criticism, like the results of mathematical calculation. But no biological theory of the present day can escape a similar fate, for the mathematical key which opens the door leading to the secrets of life has not yet been found, and a considerable period of time must elapse before its discovery. But although I can only offer an hypothesis, I hope to be able to show that it has not been rashly adopted, but that it has grown in a natural manner from the secure foundation of ascertained facts.

Nothing impresses the stamp of truth upon an hypothesis more than the fact that its light renders intelligible not only those facts for the explanation of which it has been framed, but also other and more distantly related groups of phenomena. This seems to me to be the case with my hypothesis, since the interpretation of polar bodies and the ideas derived from it unite from very different points of view, the facts of reproduction, heredity and even the transformation of species, into a comprehensive system, which although by no means complete, is nevertheless harmonious, and therefore satisfactory.

Only the most essential elements of the new facts which form the foundation of the views developed in this essay will be briefly mentioned. My object is to show all the theoretical bearings of these new facts, not to describe them in technical detail. Such a description accompanied by the necessary figures will shortly be given in another place[234].

A. W.

Freiburg I. Br., *May 30, 1887.*

ON THE NUMBER OF POLAR BODIES, &c.

CONTENTS.

VI.

ON THE NUMBER OF POLAR BODIES AND THEIR SIGNIFICANCE IN HEREDITY.

I. Parthenogenetic and Sexual Egg.

Hitherto no value has been attached to the question whether an animal egg produces one or two polar bodies. Several observers have found two such bodies in many different groups of animals, both high and low in the scale of organization. In certain species only one has been observed, in others again three, four, or five (e. g. Bischoff, in the rabbit). Many observers did not even record the number of polar bodies found by them, and simply spoke of 'polar bodies.' As long as their formation was looked upon as a process of secondary physiological importance—as an 'excretion,' or a 'process of purification,' or even as the 'excreta' (!) of the egg, as a 'rejuvenescence of the nucleus,' or of mere historical interest as a reminiscence of ancestral processes, without any present physiological meaning—so long was it unnecessary to attach any importance to the number of these bodies, or to pay special attention to them. Of all the above-mentioned views, the one which explained polar bodies as a mere reminiscence of ancestral processes seemed to be especially well founded. Ten years ago we were far from being able to prove that polar bodies occurred in all animal eggs, and even in 1880, Balfour said in his excellent 'Comparative Embryology,' 'It is very possible, not to say probable, that such changes [the formation of polar bodies] are universal in the animal kingdom, but the present state of our knowledge does not justify us in saying so[235].'

Even at the present day we are not, strictly speaking, justified in making this assertion, for polar bodies have not yet been proved to occur in certain groups of animals, such as reptiles and birds; but they have been detected in the great majority of the large groups of the animal kingdom, and wherever they have been looked for with the aid of our modern highly efficient appliances, they have been found[236].

A deeper insight into the process of fertilization has above all led to a closer study of antecedent phenomena.

O. Hertwig[237] and Fol[238] showed that the formation of polar bodies was connected with a division of the nuclear substance of the egg. Hertwig and Bütschli[239] then proved that the body expelled from the egg possessed the nature of a cell, and thus led the way to the view that the formation of polar bodies is a process of cell-division, although a very unequal one. Even then there was no reason for attaching any special importance to the number of these bodies; nor should we have such a reason if we agreed with Minot[240], Balfour[241], and van Beneden in ascribing a high physiological significance to this process, and assumed that the expelled polar body is the male part of the previously hermaphrodite egg-cell. We should not know in what proportion the quantities of the 'male' and 'female' parts were present, and it would therefore be impossible to decide, *a priori*, whether the 'male' part had to be removed from the body of the egg-cell in one, two, or more portions.

Even after the view that the nuclear substance is the essential element in fertilization had gained ground—a view chiefly due to Strasburger's investigations on the process of fertilization in Phanerogams—and after Hertwig's opinion had been confirmed, that the process of fertilization is essentially the conjugation of nuclei, even then there appeared to be no reason why the number of divisions undergone by the nucleus of the mature egg should be looked upon as an essential feature.

This was the state of the subject at the time when I first made an attempt to ascertain the meaning of the formation of polar bodies. I based my views upon the idea, which was just then gaining ground, that Nägeli's idioplasm was to be sought for in the nucleus, and that the nucleoplasm must therefore contain the substance which determines the form and functions of the cell. Hence it followed that the germ-plasm—the substance which determines the course of embryonic development—must be identified with the nucleoplasm of the egg-cell. The conception of germ-plasm was brought forward by me before the appearance of Nägeli's work[242] which is so rich in fertile ideas; and germ-plasm does not exactly coincide with Nägeli's idioplasm[243]. Germ-plasm is only a certain kind of idioplasm—viz. that contained in the germ-cell—and it is the most important of all idioplasms, because all the other kinds are merely the results of the various ontogenetic stages into which it developes. I attempted to show that the molecular structure in these ontogenetic stages into which the germ-plasm developes would become more and more unlike that of the original structure of this substance, until it finally attains a highly specialized character at the end of embryonic development, corresponding to the production of specialized histological elements. It did not seem to me to be conceivable that the specialized idioplasm contained in the nuclei of the tissue cells could re-transform itself into the initial stage of the whole developmental series—that it could give up its specialized character and re-assume the generalized character of germ-substance. I will not repeat the reasons which induced me to adopt this opinion; they still seem to me to be conclusive. But let the above-mentioned theory be once accepted, and there follows from it another interesting conclusion concerning the germ-cell, or at least concerning those germ-cells which, like most animal eggs, possess a specific histological character. For obviously, such a character presupposes the existence of an idioplasm with a considerable degree of histological specialization, which must be contained in the nucleus of the egg-cell. We know, on the other hand, that when its growth is complete, after the formation of yolk and membranes, the egg contains germ-plasm, for it is capable of developing into an embryo. We have therefore, as it were, two natures in a single cell, which become manifest one after the other, and which, according to our fundamental conception, can only be explained by the presence of two different idioplasms, which control the egg-cell one after the other, and determine its processes of development. At first a nucleoplasm leading to histological specialization directs the development of the egg and stamps upon it a specific histological character; and then germ-plasm takes its place, and compels the egg to undergo development into an embryo. If then the histogenetic or ovogenetic nucleoplasm of the egg-cell can be derived from the germ-plasm, but cannot be re-transformed into it (for the specialized can be derived from the generalized, but not the generalized from the specialized), we are driven to the conclusion that the germ-plasm, which is already present in the youngest egg-cell, first of all originates a specific histogenetic or ovogenetic nucleoplasm which controls the egg-cell up to the point at which it becomes mature; that its place is then taken by the rest of the unchanged nucleoplasm (germ-plasm), which has in the meantime increased by growth; and that the former is removed from the egg in the form of polar bodies—a removal which has been rendered possible by the occurrence of nuclear division. Hence the formation of polar bodies signified, in my opinion, the removal of the ovogenetic part of the nucleus from the mature egg-cell. Such removal was absolutely necessary, if it is impossible that the ovogenetic nucleoplasm can be re-transformed into germ-plasm. Hence the former substance cannot be made use of after the maturation of the egg, and it must even be opposed to the commencement of embryonic development, for it is impossible that the egg can be controlled by two forces of different kinds in the same manner as it would have been by one of them alone. I therefore concluded that the influence of the ovogenetic idioplasm must be removed before embryonic development can take place. In this way it seemed to me that not only the ordinary cases of ovogenetic and embryonic development became more easily intelligible, but also the rarer cases in which one and the same species produces two kinds of eggs—'summer and winter eggs.' Such eggs not only differ in size but also in the structure of yolk and membranes, although identical animals are developed from each of them. This result presupposes that the nucleus in both eggs contains identical germ-plasm, while the formation of different yolks and membranes requires the supposition that their nucleoplasm is different, inasmuch as the two eggs differ greatly in histological character.

The fact that equal quantities are separated during nuclear division, led me to conclude further that the expulsion of ovogenetic nucleoplasm can only take place when the germ-plasm in the nucleus of the egg-cell has increased by growth up to a point at which it can successfully oppose the ovogenetic nuclear substance. But we do not know the proportion which must obtain between the relative quantities of two different nuclear substances in order that nuclear division may be induced; and thus, by this hypothesis at least, we could not conclude with certainty as to the necessity for a single or a double division of the egg. It did not seem to be altogether inconceivable that the ovogenetic nucleoplasm might be larger in amount than the germ-plasm, and that it could only be completely removed by means of two successive nuclear divisions. I admit that this supposition caused me some uneasiness; but since nothing was known which could have enabled us to penetrate more deeply into the problem, I was satisfied, for the time being, in having found any explanation of the physiological value of polar bodies; leaving the future to decide not only whether such explanation were valid, but also whether it were exhaustive. The explanation seems to have found but little favour with some of our highest authorities. Hensen[244] does not consider that my reasons for the distinction between germ-plasm and histogenetic nucleoplasm are conclusive, and it may be conceded that this objection was perhaps, at that time, well founded. O. Hertwig does not mention my hypothesis at all in his work on embryology[245], although he states in the preface: 'Among current problems I have chiefly taken into consideration the views which seem to me to be most completely justified, but I have not left unmentioned the views which I cannot accept.' Minot's hypothesis is discussed by Hertwig, but Bütschli's[246] is preferred by him, although these two hypotheses are not strictly opposed to each other; for the former is a purely physiological, the latter a purely morphological explanation. I desire to lay especial stress upon the fact that my hypothesis is simply a logical consequence from the conclusion that the nuclear substance determines the nature of a cell. How this takes place is quite another question, which need not be discussed here. If it is only certain that the nature of a cell is thus determined, it follows that a cell with a certain degree of histological specialization must contain a nucleoplasm corresponding to the specialization. But the mature egg also contains germ-plasm, and there are only two possibilities by which these facts can be explained: either the ovogenetic nucleoplasm is capable of re-transformation into germ-plasm, or it is incapable of such re-transformation. Now, quite apart from the arguments which might be advanced in favour of one of these two possibilities, the fact that a body is undoubtedly expelled from the mature egg seems to me of importance, while it is of even greater importance that this body contains nucleoplasm from the germ-cell.

It may be thought that the process, as supposed by me, is without analogy, but such a conclusion is wrong, for during every embryonic development there are numerous cell-divisions in which unequal nucleoplasms are separated from one another, and in all these cases we cannot imagine any way in which the process can take place, except by supposing that the two kinds of nucleoplasm were previously united in the mother-cell, although their differentiation probably took place only a short time before cell-division. Perhaps the new facts which will be mentioned presently, and the views derived from them, will make my hypothesis upon the histogenetic nucleoplasm of the germ-cells appear in a more favourable light to the authorities above-named.

My hypothesis has at all events the one merit that it has led me to fruitful investigations.

If the formation of polar bodies really means the removal of ovogenetic nucleoplasm from the mature egg, they must also be found in parthenogenetic eggs; inasmuch as the latter possess a specific histological structure equal to that found in eggs requiring fertilization. If, therefore, it were possible to observe the formation of polar bodies in eggs which develope parthenogenetically, such an observation would not form a proof of the validity of my interpretation; but it would be a fact which harmonized with it, and negatived a suggestion which, if confirmed, would have been fatal to the hypothesis. Minot, Balfour, and van Beneden, from the point of view afforded by their theories, were compelled to suppose that polar bodies are wanting in parthenogenetic eggs; and the facts which were known at that time favoured such an opinion, for in spite of many attempts, no one had ever succeeded in proving the formation of these bodies by parthenogenetic eggs.

During the summer of 1885 I first succeeded in ascertaining that a single polar body is expelled from the parthenogenetic summer-egg of one of the *Daphnidae,—Polyphemus oculus*[247]. Thus my interpretation of the process in question received support, while it seemed to me that Minot's interpretation of polar bodies had been refuted; for if these bodies are formed in the parthenogenetic eggs of a single species, just as in eggs which require fertilization, it follows that the expulsion of polar bodies cannot signify the removal of the male element from the egg.

The desire to throw light upon the significance of polar bodies has been the only cause of my investigation. At the same time I hoped by this means to gain further knowledge as to the nature of parthenogenesis.

In the third part of the essay on 'The Continuity of the Germ-plasm' (see p. 225) I attempted to make clear the nature of parthenogenesis, and I arrived at the conclusion that the difference between an egg which is capable of developing without fertilization, and another which requires fertilization, must lie in the quantity of nucleoplasm present in the egg. I supposed that the nucleus of the mature parthenogenetic egg contained nearly twice as much germ-plasm as that contained in the sexual egg, just before the occurrence of fertilization; or, more correctly, I believed that the quantity of nucleoplasm which remains in the egg, after the expulsion of the polar bodies, is the same in both eggs, but that the parthenogenetic egg possesses the power of doubling this quantity by growth, and thus produces from within itself the same quantity of germ-plasm as that contained in the sexual egg after the addition of the sperm-nucleus in fertilization.

This was only an hypothesis, and the considerations which had led to it depended, as far as they went into details, upon assumptions; but the fundamental view that the *quantity* of the nucleus decides whether embryonic development takes place with or without fertilization seemed to me, even at that time, to be correct, and to be a conclusion required by the facts of the case. Indeed, I thought it not unlikely that its validity might be proved by direct means: I pointed out that a comparison of the quantities of the nuclei in parthenogenetic and sexual eggs, if possible in the same species, would enable us to decide the question (*l. c.*, p. 234).

I had thus set myself the task of making this comparison. The result of this investigation was to show that, as already mentioned, polar bodies are formed in parthenogenetic eggs. But even the first species successfully investigated revealed a further fact, which, if proved to be wide-spread and characteristic of all parthenogenetic eggs, was certain to be of extreme importance:—the maturation of the parthenogenetic egg is accompanied by the expulsion of *one* polar body, or, as we might express it in another way, the substance of the female pronucleus is only *once* divided, and not *twice*, as in the sexual eggs of so many other animals. If this difference between parthenogenetic and sexual eggs was shown to be general, then the foundations of my hypothesis would indeed have been proved to be sound. The quantity of nuclear substance decides whether the egg is capable of undergoing embryonic development. This quantity is twice as large in the parthenogenetic as in the sexual egg. I had, however, been mistaken in a matter of detail; for the difference in the quantities of nuclear substance is not produced by the expulsion of two polar bodies, and the reduction of the nuclear substance to a quarter of its original amount, in both eggs, while the parthenogenetic egg then doubles its nuclear substance by growth; but the difference is produced because the reduction of nuclear substance originally present is less in one case than it is in the other. In the parthenogenetic egg the nuclear substance is only reduced to one-half by a single division; in the sexual egg it is reduced to a quarter by two successive divisions. It is an obvious conclusion from this fact, if proved to be wide-spread, that the significance of the first polar body must be different from that of the second. Only one polar body can signify the removal of ovogenetic nucleoplasm from the mature egg, and the second is obviously a reduction of the germ-plasm itself to half of its original amount. This very point seemed to me to be of great importance, because, as I had foreseen long ago, and as will be shown later on, the theory of heredity forces us to suppose that every fertilization must be preceded by a reduction of the ancestral idioplasms present in the nucleus of the parent germ-cell, to one-half of their former number.

But before the full bearing of the phenomena could be considered, it was necessary to ascertain how far they were of general occurrence. There were two ways in which this might be achieved, and in which it was possible to prove that parthenogenetic eggs expel only one polar body, while sexual eggs expel two. We might attempt to observe the phenomena of maturation in both kinds of eggs in a species which reproduces itself by the parthenogenetic as well as the sexual method. This would be the simplest way in which the question could be decided, if it were possible to make such observations on a sufficient number of species. But the other method was also open, a method which would have been the only one, if we did not know of any animals with two kinds of reproduction. We might attempt to investigate the phenomena of maturation in a large number of parthenogenetic eggs, if possible from different groups of animals, and we might compare the results with the facts which are already certain concerning the expulsion of polar bodies from the sexual eggs of so many species.

I have followed both methods, and by means of the second I have arrived already, indeed some time ago, at the certain conclusion that the above-mentioned difference is really general and without exception. The first polar body only is formed in all the parthenogenetic eggs which I investigated, with the valuable assistance of my pupil, Mr. Ischikawa of Tokio. On the other hand, an extensive examination of the literature of the subject convinced me that there is not a single undoubted instance of the expulsion of only one polar body from eggs which require fertilization, and that there are very numerous cases known from almost all groups of the animal kingdom in which it is perfectly certain that two polar bodies are formed, one after the other. A number of the older observations cannot be relied upon, for the presence of two polar bodies is mentioned without any explanation as to whether they are expelled from the egg one after the other, or whether they have merely resulted from the division of a single body after its expulsion. In parthenogenetic eggs two polar bodies are also formed in most cases, but they arise from the subsequent division of the single body which separates from the egg. But such subsequent division is only of secondary importance as far as the egg itself is concerned, and is also unimportant in the interpretation of the process. The essential nature of the process is to be found in the fact that the nucleus of the egg-cell only divides once when parthenogenesis occurs, but twice when fertilization is necessary, and it is of no importance whether the expelled part of the nucleus of the cell-body atrophies at once, or after it has undergone division. We have, therefore, to distinguish between primary and secondary polar bodies. If this distinction is recognized, and if we leave out of consideration all doubtful cases mentioned in literature, such a large number of well-established observations remain, that the existence of two primary polar bodies in sexual eggs, and neither a smaller nor a larger number, may be considered as proved.

Hence follows a conclusion which I believe to be very significant,—the difference between parthenogenetic and sexual eggs lies in the fact that in the former only one primary polar body is expelled, while two are expelled from the latter. When, in July, 1886, I published a short note[248] on part of the observations made upon parthenogenetic eggs, I confined myself to facts, and did not mention this conclusion. I took this course simply because I did not wish to bring it forward until I had made sufficient observations in the first of the two ways described above. I had hoped to be able to offer all the proofs that can be obtained before undertaking to publish the far-reaching consequences which would result from the above-mentioned conclusion. Unfortunately the material with which I had hoped to quickly settle the matter, proved less favourable than I had expected. Many hundred sections through freshly laid winter-eggs of *Bythotrephes longimanus* were made in vain; they did not yield the wished for evidence, and although continued investigation of other material has led to better results, the proofs are not yet entirely complete.

I should not therefore even now have brought forward the above-mentioned conclusion, if another observer had not alluded to this idea, referring to my observations and also to a new discovery of his own. In a recent number of the 'Biologische Centralblatt,' Blochmann[249] gives an account of his continued observations upon the formation of polar bodies. It is well known that this careful observer had previously shown that polar bodies do occur in the eggs of insects, although they had not been found before. Blochmann proved that they are found in the representatives of three different orders, so that we may indeed 'confidently hope to find corresponding phenomena in other insects.' This discovery is most important, and it was naturally very welcome to me, as I had for a long time ascribed a high physiological importance to the process of the formation of polar bodies, and it would not be in accordance with such a view if the process was entirely wanting from whole classes of animals. To fill up this gap in our knowledge, and to give the required support to my theoretical views, I had proposed to one of my pupils, Dr. Stuhlmann[250], that he should work out the maturation of the eggs of insects; and it is a curious ill-luck that he, like many other observers, did not succeed in observing the expected expulsion of polar bodies, in spite of the great trouble he had taken. It may be that the species selected for investigation were unfavourable: at all events, we cannot now doubt that a division of the egg-nucleus is quite universal among insects, for Blochmann, in his latest contribution to the subject, proves that the *Aphidae* also form polar bodies. He examined the winter-eggs of *Aphis aceris*, and ascertained that they form two polar bodies, one after the other. Even in the viviparous *Aphidae*, thin sections revealed the presence of a polar body, though Blochmann could not trace all the stages of its development. It appears that the polar body is here preserved for an exceptional period, and its presence can still be proved when the blastoderm has been formed, and sometimes when development is even further advanced. Skilled observers of recent times, such as Will and Witlaczil, have not been able to find a polar body in the parthenogenetic eggs of the *Aphidae*, and Blochmann's proof of its existence seems to me to be of especial value, because the eggs of *Aphidae* are in many respects so unusually reduced; for instance, the primary yolk is absent and the egg-membrane is completely deficient, so that we might have expected that if polar bodies are ever absent, they would be wanting in these animals—that is, if they were of no importance, or at any rate of only secondary importance.

Hence the presence of polar bodies in *Aphidae* is a fresh confirmation of their great physiological importance. As bearing upon the main question dealt with in this essay, Blochmann's observations have an especial interest, because only one polar body was found in the parthenogenetic eggs of *Aphis*, while the sexual eggs normally produce two. The author rightly states that this result is in striking accordance with my results obtained from the summer-eggs of different *Daphnidae*, and he adds the remark,—'It would be of great interest to know whether these facts are due to the operation of some general law.' To this remark I can now reply that there is indeed such a law: not only in the parthenogenetic eggs of *Daphnidae*, but also, as I have since found, in those of the Ostracoda and Rotifera[251], only one primary polar body is formed, while two are formed in all eggs destined for fertilization.

Before proceeding to the conclusions which follow from this fact, I will at once remove a difficulty which is apparently presented by the eggs which may develope with or without fertilization. I refer to the well-known case of the eggs of bees. It might be objected to my theory that the same egg cannot be prepared for development in more than one out of the two possible ways; it might be argued that the egg either possesses the power of entering upon two successive nuclear divisions during maturation, and in this case requires fertilization; or the egg may be of such a nature that it can only enter upon one such division and can therefore form only one polar body, and in that case it is capable of parthenogenetic development. Now there is no doubt, as I pointed out in my paper on the nature of parthenogenesis[252], that in the bee the very same egg may develope parthenogenetically, which under other circumstances would have been fertilized. Bessel's[253] experiments, in which young queens were rendered incapable of flight, and were thus prevented from fertilization, have shown that all the eggs laid by such females develope into drones (males) which are well known to result from parthenogenetic development. On the other hand, bee-keepers have long known that young queens which are fertilized in a normal manner continue for a long time to lay eggs which develope into females, that is to say, which have been fertilized. Hence the same eggs, viz. those which are lowest in the oviducts and are therefore laid first, develope parthenogenetically in the mutilated female, but are fertilized in the normal female. The question therefore arises as to the way in which the eggs become capable of adapting themselves to the expulsion of two polar bodies when they are to be fertilized, and of one only when fertilization does not take place.

But perhaps the solution of this problem is not so difficult as it appears to be. If we may assume that in eggs which are capable of two kinds of development the second polar body is not expelled until the entrance of a spermatozoon has taken place, the explanation of the possibility of parthenogenetic development when fertilization does not occur would be forthcoming. Now we know, from the investigations of O. Hertwig and Fol, that in the eggs of *Echinus* the two polar bodies are even formed in the ovary, and are therefore quite independent of fertilization, but in this and other similar cases a parthenogenetic development of the egg never takes place. There are, however, observations upon other animals which point to the fact that the first only and not the second polar body may be formed before the spermatozoon penetrates into the egg. It can be easily understood why it is that entirely conclusive observations are wanting, for hitherto there has been no reason for any accurate distinction between the first and the second polar body. But in many eggs it appears certain that the second polar body is not expelled until the spermatozoon has penetrated. O. Schultze, the latest observer of the egg of the frog, in fact saw the first polar body alone extruded from the unfertilized egg: a second nuclear spindle was indeed formed, but the second polar body was not expelled until after fertilization had taken place. A very obvious theory therefore suggests itself:—that while the formation of the second polar body is purely a phenomenon of maturation in most animal eggs, and is independent of fertilization,—in the eggs of a number of other animals, on the other hand, and especially among Arthropods, the formation of the second nuclear spindle is the result of a stimulus due to the entrance of a spermatozoon. If this suggestion be confirmed, we should be able to understand why parthenogenesis occurs in certain classes of animals wherever the external conditions of life render its appearance advantageous, and further, why in so many species of insects a sporadic parthenogenesis is observed, viz. the parthenogenetic development of single eggs (Lepidoptera). Slight individual differences in the facility with which the second nuclear spindle is formed independently of fertilization would in such cases decide whether an egg is or is not capable of parthenogenetic development. As soon, however, as the second nuclear spindle is formed, parthenogenesis becomes impossible. The nuclear spindle which gives rise to the second polar body, and that which initiates segmentation, are two entirely different things, and although they contain the same quantity, and the same kind of germ-plasm, a transformation of the one into the other is scarcely conceivable. This conclusion will be demonstrated in the following part of the essay.

II. The Significance of the Second Polar Body.

I have already discussed the physiological importance of the first polar body, or rather of the first division undergone by the nucleus of the egg, and I have explained it as the removal of ovogenetic nuclear substance which has become superfluous and indeed injurious after the maturation of the egg. I do not indeed know of any other meaning which can be ascribed to this process, now that we know of the occurrence of a first division of the nucleus in parthenogenetic as well as in sexual eggs. A part of the nucleus must thus be removed from both kinds of eggs, a part which was necessary to complete their growth, and which then became superfluous and at the same time injurious. In this respect the observations of Blochmann[254] upon the eggs of *Musca vomitoria* seem to me to be very interesting. Here the two successive divisions of the nuclear spindle arising from the egg-nucleus take place, but true polar bodies are not expelled, and the two nuclei corresponding to them (one of which divides once more) are placed on the surface of the egg, surrounded by an area free from yolk granules; and they break up at a later period. The essential point is obviously to eliminate from the egg-cell the influence of nucleoplasm which has been separated from the egg-nucleus as the first polar body; and this condition is satisfied whether the elimination is brought about by a process of true cell-division, as is the rule in the eggs of most animals, or by the division and removal of part of the egg-nucleus alone. The occurrence of the latter method of elimination certainly constitutes a still further proof of the physiological importance of the process, and this, taken together with the universal occurrence of polar bodies in all eggs—parthenogenetic and sexual—forces us to conclude that the process must possess a definite significance. No one of the various attempts which have been made to explain the significance of polar bodies generally is applicable to the *first* polar body except that which I have attempted.

But the case is different with the significance of the *second* nuclear division, or the *second* polar body. Here it might perhaps be possible to return to the view brought forward by Minot, Balfour, and van Beneden, and to consider the removal of this part of the nucleus as the expulsion of the male part of the previously hermaphrodite egg-cell. The second polar body is only expelled when the egg is to be fertilized, and at first sight it appears to be quite obvious that such a preparation of the egg for fertilization must depend upon its reduction to the female state. I believe however that this is not the case, and am of opinion that the process has an entirely different and much deeper meaning.

How can we gain any conception of this supposed hermaphroditism of the egg-cell, and its subsequent attainment of the female state? What are the essential characteristics of the male and female states? We know of female and male individuals, among both animals and plants: their differences consist essentially in the fact that they produce different kinds of reproductive cells; in part they are of a secondary nature, being adaptations of the organism to the functions of reproduction; they are intended to attract the other sex, or to ensure the meeting of the two kinds of reproductive cells, or to enable the fertilized egg to develope and sometimes to guide the development of the offspring until it has reached a certain period of growth. But all these differences, however great they may sometimes be, do not alter the essential nature of the organism. The blood corpuscles of man and woman are the same, and so are the cells of their nerves and muscles; and even the sexual cells, so different in size, appearance, and generally also in motile power, must contain the same fundamental substance, the same idioplasm. Otherwise the female germ-cell could not transmit the male characters of the ancestors of the female quite as readily as the female characters, nor could the male germ-cell transmit the female quite as readily as the male characters of the ancestors of the male. It is therefore clear that the nuclear substance itself is not sexually differentiated.

I have already previously pointed out that the above-mentioned facts of heredity contain the disproof of Minot's theory, inasmuch as the egg-cell transmits male as well as female characters. Strasburger[255] has also raised a similar objection. I consider this objection to be quite conclusive, for there does not seem to be any way in which the difficulty can be met by the supporters of the theory. The difficulty could indeed be evaded until we came to know that the essential part of the polar body is nuclear substance, and that the latter must be regarded as idioplasm,—as the substance which is the bearer of heredity. It might have been maintained that the male part, removed from the egg, consists only in a condition, perhaps comparable to positive or negative electricity; and that this condition is present in the substance of the polar body, so that the removal of the latter would merely signify a removal of the unknown condition. I do not mean to imply that any of those who have adopted Minot's theory have had any such vague ideas concerning this process, but even if any one were ready to adopt it, he would be unable to make any use of the idea. He would not be able to support the theory in this way, for we now know that nuclear substance is removed with the polar body, and this fact requires an explanation which cannot be afforded by the theory, if we are right in believing that the expelled nuclear substance is not merely the indifferent bearer of the unknown principle of the male condition, but hereditary substance. I therefore believe that Minot's, Balfour's, and van Beneden's hypothesis, although an ingenious attempt which was quite justified at the time when it originated, must be finally abandoned.

My opinion of the significance of the second polar body is shortly this,—a reduction of the germ-plasm is brought about by its formation, a reduction not only in quantity, but above all in the complexity of its constitution. By means of the second nuclear division the excessive accumulation of different kinds of hereditary tendencies or germ-plasms is prevented, which without it would be necessarily produced by fertilization. With the nucleus of the second polar body as many different kinds of idioplasm are removed from the egg as will be afterwards introduced by the sperm-nucleus; thus the second division of the egg-nucleus serves to keep constant the number of different kinds of idioplasm, of which the germ-plasm is composed during the course of generations.

In order to make this intelligible a short explanation is necessary.

From the splendid series of investigations on the process of fertilization, commenced by Auerbach and Bütschli, and continued by Hertwig, Fol, Strasburger, van Beneden, and many others, and from the theoretical considerations brought forward by Pflüger, Nägeli, and myself, at least one certain result follows, viz. that there is an hereditary substance, a material bearer of hereditary tendencies, and that this substance is contained in the nucleus of the germ-cell, and in that part of it which forms the nuclear thread, which at certain periods appears in the form of loops or rods. We may further maintain that fertilization consists in the fact that an equal number of loops from either parent are placed side by side, and that the segmentation nucleus is composed in this way. It is of no importance, as far as this question is concerned, whether the loops of the two parents coalesce sooner or later, or whether they remain separate. The only essential conclusion demanded by our hypothesis is that there should be complete or approximate equality between the quantities of hereditary substance derived from either parent. If then the germ-cells of the offspring contain the united germ-plasms of both parents, it follows that such cells can only contain half as much paternal germ-plasm as was contained in the germ-cells of the father, and half as much maternal germ-plasm as was contained in the germ-cells of the mother. This principle is affirmed in a well-known calculation made by breeders of animals, who only differ from us in their use of the term 'blood' instead of the term germ-plasm. Breeders say that half of the 'blood' of the offspring has been derived from the father and the other half from the mother. The grandchild similarly derives a quarter of its 'blood' from each of the four grandparents, and so on.

Let us imagine, for the sake of argument, that sexual reproduction had not been introduced into the animal kingdom, and that asexual reproduction had hitherto existed alone. In such a case, the germ-plasm of the first generation of a species which enters upon sexual reproduction must still be entirely homogeneous; the hereditary substance must, in each individual, consist of many minute units, each of which is exactly like the other, and each of which contains within itself the tendency to transmit, under certain circumstances, the whole of the characters of the parent to a new organism—the offspring. In each of the offspring of such a first generation, the germ-plasms of two parents will be united, and every germ-cell contained in the individuals of this second sexually produced generation will now contain two kinds of germ-plasm—one kind from the father, and the other from the mother. But if the total quantity of germ-plasm present in each cell is to be kept within the pre-determined limits, each of the two ancestral germ-plasms, as I may now call them, must be represented by only half as many units as were contained in the parent germ-cells.

In the third sexually produced generation, two new ancestral germ-plasms would be added by fertilization to the two already present, and the germ-cells of this generation would therefore contain four different ancestral germ-plasms, each of which would constitute a quarter of the total quantity. In each succeeding generation the number of the ancestral germ-plasms is doubled, while their quantities are reduced by one half. Thus in the fifth sexually produced generation, each of the sixteen ancestral germ-plasms will only constitute 1/16 of the total quantity; in the sixth, each of the thirty-two ancestral germ-plasms, only 1/32, and so on. The germ-plasm of the tenth generation would be composed of 1024 different ancestral germ-plasms, and that of the nth of 2^n. By the tenth generation each single ancestral germ-plasm would only form 1/1024 of the total quantity of germ-plasm contained in a single germ-cell. We know nothing whatever of the length of time over which this process of division of the ancestral germ-plasms may have endured, but even if it had continued to the utmost possible limit—so far indeed that each ancestral germ-plasm was only represented by a single unit—a time would at last come when any further division into halves would cease to be possible; for the very conception of a unit implies that it cannot be divided without the loss of its essential nature, which in this case constitutes it as the hereditary substance.

In the diagram represented in Fig. I. I have tried to render these conclusions intelligible. In generation I. each paternal and maternal germ-plasm is still entirely homogeneous, and does not contain any combination of different hereditary qualities, but the germ-plasm of the offspring is made up of equal parts of two kinds of germ-plasm. In the second generation this latter germ-plasm unites with another derived from other parents, which is similarly composed of two ancestral germ-plasms, and the resulting third generation now contains four different ancestral germ-plasms in its germ-cells, and so on. The diagram only indicates the fusion of ancestral germ-plasms as far as the offspring of the fourth generation, the germ-cells of which contain sixteen different ancestral germ-plasms. If we imagine the germ-plasm units to be so large that there is only room for sixteen of them in the nuclear thread, the limits of division would-be reached in the fifth generation, and any further division into halves of the ancestral germ-plasms would be impossible.

Now however minute the units may be, there is not the least doubt that the limits of possible division have been long since reached by all existing species, for we may safely assume that no one of them has acquired the sexual method of reproduction within a small number of recent generations. All existing species must therefore now contain as many different kinds of ancestral germ-plasms as they are capable of containing; and the question arises,—How can sexual reproduction now proceed without a doubling of the quantity of germ-plasm in each germ-cell, with every new generation?

There is only one possible answer to such a question:—sexual reproduction can proceed by a reduction in the *number* of ancestral germ-plasms, a reduction which is repeated in every generation.

Fig. I.

This *must* be so: the only question is, how and when does the supposed reduction take place.

Inasmuch as the germ-plasm is seated, according to our theory, in the nucleus, the necessary reduction can only be produced by nuclear division; and quite apart from any observation which has been already made, we may safely assert that there *must* be a form of nuclear division in which the ancestral germ-plasms contained in the nucleus are distributed to the daughter-nuclei in such a way that each of them receives only half the number contained in the original nucleus. After Roux's[256] elaborate review of the whole subject, we need no longer doubt that the complex method of nuclear division, hitherto known as karyokinesis, must be considered not merely as a means for the division of the total quantity of nuclear substance, but also for producing a division of the quantity and quality of each of its single elements. In by far the greater number of instances the object of this division is obviously to effect an equal distribution of nuclear substance in the two daughter-nuclei, so that each of the different qualities contained in the mother-nucleus is transferred to the two daughter-nuclei. This interpretation of ordinary karyokinesis is less uncertain than perhaps at first sight it may appear to be. We cannot, it is true, directly see the ancestral germ-plasms, nor do we even know the parts of the nucleus which are to be looked upon as constituting ancestral germ-plasm; but if Flemming's original discovery of the longitudinal division of the loops lying in the equatorial plane of the nuclear spindle is to have any meaning at all, its object must be to divide and distribute the different kinds of the minutest elements of the nuclear thread as equally as possible. It has been ascertained that the two halves produced by the longitudinal splitting of each loop never pass into the same daughter-nucleus, but always in opposite directions. The essential point cannot therefore be the division of the nucleus into absolutely equal quantities, but it must be the distribution of the different qualities of the nuclear thread, without exception, in both daughter-nuclei. But these different qualities are what I have called the ancestral germ-plasms, i.e. the germ-plasms of the different ancestors, which must be contained in vast numbers, but in very minute quantities, in the nuclear thread. The supposition of a vast number is not only required by the phenomena of heredity but also results from the comparatively great length of the nuclear thread: furthermore it implies that each of them is present in very small quantity. The vast number together with the minute quantity of the ancestral germ-plasms permit us to conclude that they are, upon the whole, arranged in a linear manner in the thin thread-like loops: in fact the longitudinal splitting of these loops appears to me to be almost a proof of the existence of such an arrangement, for without this supposition the process would cease to have any meaning.

This is the only kind of karyokinesis which has been observed until recently; but if the supposed nuclear division leading to a reduction in the number of ancestral germ-plasms has any real existence, there must be yet another kind of karyokinesis, in which the primary equatorial loops are not split longitudinally, but are separated without division into two groups, each of which forms one of the two daughter-nuclei. In such a case the required reduction in the number of ancestral germ-plasms would take place, for each daughter-nucleus would receive only half the number which was contained in the mother-nucleus.

Now there is more evidence for the existence of this second kind of karyokinesis than the fact that it is demanded by my theory; for I believe that it has been already observed, although it has not been interpreted in this sense.

It is very probable that this is true of van Beneden's[257] observation on the egg of *Ascaris megalocephala*: he found that the nuclear division which led to the formation of the polar body differs from the ordinary course of karyokinesis, in that the plane of division is at right angles to that usually assumed. Carnoy[258] has confirmed this observation in its main features, and he has made the further observation that out of the eight nuclear loops which are found at the equator of the spindle, four are removed with the first polar body, and that half of the remaining four are removed with the second polar body. The first of these two divisions would have to be looked upon as a reduction, if it is certain that each of the eight nuclear loops consists of different ancestral germ-plasms; but this assumption is impossible, although on the other hand it cannot be *directly* disproved: for we are not able to see the ancestral germ-plasms. But it must nevertheless be maintained that the removal of the first four loops does not imply a reduction in the number of ancestral germ-plasms in the nucleus; because, as I have already argued, two successive divisions of the number of ancestral germ-plasms into halves is inconceivable; and because the first polar body is also present in parthenogenetic eggs in which such division into halves cannot take place. But the karyokinetic process can readily be looked upon as a removal of ovogenetic nucleoplasm, for we know from the observations of Flemming and Carnoy, that, under certain circumstances, subsequent divisions may occur, involving an increase in the number of nuclear loops to double their number. These subsequent divisions of course take place in the daughter-nuclei. This fact proves, as I think, that there are nuclei in which the same ancestral germ-plasm occurs in two different loops: but such loops, identical as regards the composition of their ancestral germ-plasms, may very well contain different ontogenetic stages of this substance. This will be the case in the instance alluded to, if four loops of the first nuclear spindle are to be looked upon as ovogenetic nucleoplasm, and the four others as germ-plasm. It is therefore unnecessary to regard the first division of the egg-nucleus as a 'reducing division': it may be looked upon as an 'equal division'[259] entirely analogous to the kind of division which, in my opinion, directs the development of the embryo. This conclusion would receive direct proof if it were possible to show that the eight loops of the first division have arisen by the longitudinal splitting of four *primary* loops: for a longitudinal splitting of the nuclear thread would be the means by which the different ontogenetic stages of the germ-plasm could be separated from one another, without leading to any reduction in the number of ancestral germ-plasms in the daughter-nuclei. Thus I have previously attempted to prove that the ontogenetic development of the egg must be connected with a progressive transformation of the nucleoplasm during successive nuclear divisions, and this transformation will very frequently (but not always) occur in such a way that the different qualities of the nucleoplasm are separated from one another by the nuclear division. The nucleoplasm of the daughter-nuclei will be identical if the two daughter-cells are to potentially contain corresponding parts of the embryo; as for instance the first two segmentation spheres of the egg of the frog, which according to Roux[260] correspond to the right and left halves of the future animal. But the nucleoplasm must be unequal if the products of division are to develope into different parts of the embryo. In both cases, however, karyokinesis is connected with a longitudinal splitting of the nuclear threads, and we may conclude from this fact (which is also confirmed by the phenomena of heredity) that all such nuclei, whether they have entered upon the same or different ontogenetic transformations of their nucleoplasm, are identical as regards the ancestral germ-plasm which they contain. During the whole process of segmentation and the entire development of the embryo, the total number of ancestral germ-plasms which were at first contained in the germ-plasm of the fertilized egg-cell must still be contained in each of the succeeding cells.

Thus no objection can be raised against the view that the four loops of the first polar body contain the ovogenetic nucleoplasm, that is to say, an idioplasm which contains the total number of ancestral germ-plasms, but at an advanced and highly specialized ontogenetic stage.

The formation of the second polar body may be rightly considered as a 'reducing division,' as a division leading to the expulsion of half the number of the different ancestral germ-plasms, in the form of two nuclear loops, for no reason can be alleged in support of the assumption that the four loops of the second nuclear spindle are made up of identical pairs. Furthermore the facts of heredity require the assumption that the greatest possible number of ancestral germ-plasms is accumulated in the germ-plasm of each germ-cell, and thus that the small number of loops not only means an increase in quantity but a multiplication in the number of different ancestral germ-plasms present in each of them. If this conclusion be correct, there can be no doubt that the second division of the egg-nucleus means a reduction in the above-mentioned sense.

But there are yet other observations which, if correct, must also be considered as 'reducing divisions.' I refer to all those cases in which the longitudinal splitting of the loops is either entirely wanting, or does not occur until after the loops have left the equator of the spindle and have moved towards the poles. In both instances the bearing upon the question would be the same, for only half the number of primary loops would reach each pole in either case. If therefore the primary loops are not made up of identical pairs, it follows that the two daughter-nuclei can only contain half the number of ancestral germ-plasms which were contained in the mother-nucleus. Whether the loops divide on their way to the poles or at the poles themselves, no difference will be brought about in the number of ancestral germ-plasms which they contain, for this number can neither increase nor diminish. The *quantity* of the different ancestral germ-plasms can alone be increased in this way. I am here referring to observations made by Carnoy[261] on the cells which form the spermatozoa in various Arthropods. It must be admitted, however, that these divisions cannot be regarded as 'reducing divisions,' if Flemming's[262] suggestion be confirmed, that in all these observations the fact has been overlooked that the equatorial loops are not primary but secondary, and that they have arisen from the longitudinal splitting of the nuclear thread during previous stages of nuclear division. But this point can only be decided by renewed investigation. Although many excellent results have been obtained in the subject of karyokinesis, there is still very much to be learnt before our knowledge is complete; and this is not to be wondered at when we remember the great difficulties in the way of observation which are chiefly raised by the minute size of the objects to be investigated. Flemming's most recent publications prove that we are still in the midst of investigation, and that highly interesting and important processes have hitherto escaped attention. A secure basis of facts is only very gradually obtained, and there are still many conflicting opinions upon the details of this process. I should therefore consider it to be entirely useless, from my point of view, to enter into a critical examination of everything known about all the details of karyokinesis. I am quite content to have shown how it may be imagined that the reduction required by my theory takes place during nuclear division; and at the same time to have pointed out that there are already observations which may be interpreted in this sense. But even if I am mistaken in this interpretation, the theoretical necessity for a reduction in the number of ancestral germ-plasms, a reduction repeated in every generation, seems to me to be so securely founded that the processes by which it is effected *must* take place, even if they are not supplied by the facts already ascertained. There must be two kinds of karyokinesis according to the different physiological effect of the process. First, a karyokinesis by means of which all the ancestral germ-plasms are equally distributed in each of the two daughter-nuclei after having been divided into halves: secondly, a karyokinesis by means of which each daughter-nucleus receives only half the number of ancestral germ-plasms possessed by the mother-nucleus. The former may be called 'equal division,' the latter 'reducing division.' Of course these two processes, which differ so greatly in their effects, must also be characterized by morphological differences, but we cannot assume that the latter are necessarily visible. Just as, during the division of the first and second nuclear spindle in the egg of *Ascaris megalocephala*, karyokinesis takes, upon the whole, the same morphological course, although we must ascribe different physiological meanings to the two processes of division,—so it may be in other cases. The 'reducing division' must be always accompanied by a reduction of the loops to half their original number, or by a transverse division of the loops (if such division ever occurs); although reduction can only occur when the loops are not made up of identical pairs. And it will not always be easy to decide whether this is the case. On the other hand, the form of karyokinesis in which a longitudinal splitting of the loops takes place *before* they separate to form the daughter-nuclei must always, as far as I can see, be considered as an 'equal division.' In the accompanying figures II and III, diagrams are given illustrating these two forms of karyokinesis, but I do not mean to imply that it is impossible to imagine any other form in which they may occur.

In Figure II a nuclear spindle is seen at A, and at its equatorial zone there are twelve primary loops. The transverse cross-lines and other markings on the loops indicate that they are composed of different ancestral germ-plasms. The loops are shaded differently in order to render the diagram clear. At B six of the loops are seen to have moved to either pole, so that the figure is a representation of the 'reducing division.' Figure III is a diagrammatic representation of 'equal division.' The six loops at the equatorial zone of A are shown by different cross-lining and shading to be composed of different ancestral germ-plasms. The loops split longitudinally in a direction indicated by the longitudinal line upon each of them. In B the halves of the loops are seen to have moved to the opposite poles of the spindle, so that there are not only six loops at each pole, but also all the six combinations of ancestral germ-plasms.

Figs. II, III.

Perhaps some may be inclined to look upon direct nuclear division as a 'reducing division,' but I believe that such a view would be incorrect. It is only approximately true that the nuclear thread is divided into two halves of equal quantity by direct division, and exact equality would only happen as it were accidentally; so that we cannot speak of a perfectly equal distribution of the ancestral germ-plasm in the two daughter-nuclei. But the 'reducing division' must obviously effect an exactly regular and uniform distribution of the ancestral germ-plasms, although this does not imply that every ancestral germ-plasm of the mother-nucleus would be represented in each of the two daughter-nuclei. But if out of e.g. eight nuclear loops at the equatorial plane, four pass into one, and the other four into the other daughter-nucleus, each of the latter will contain an equal number of ancestral germ-plasms, although different ones. This is indeed part of the foundation of the theory, for the 'reducing division' must remove exactly half of the original number of ancestral germ-plasms, and precisely the same number must be replaced at a later period by the sperm-nucleus. This could hardly be achieved with sufficient precision by direct nuclear division.

I now come to inquire whether the expulsion of the second polar body is in reality, as I have already maintained, a reduction in the number of ancestral germ-plasms present in the nucleus of the egg. The view itself is sufficiently obvious, and it would supply an explanation of the meaning of the process which is still greatly wanted; but it will nevertheless be not entirely useless to consider other possible theories.

It would be quite conceivable to suppose that the youngest egg-cells, which multiply by division, may undergo one 'reducing division' in addition to the ordinary process. Of course this should occur once only, for if repeated, the number of ancestral idioplasms in the nucleus of the germ-cell would undergo a decrease greater than could be afterwards compensated by the increase due to fertilization. Thus the number of ancestral germ-plasms would continually decrease in the course of generations,—a process which would necessarily end with their complete reduction to a single kind, viz. to the paternal or the maternal germ-plasm. But the occurrence of such a result is disproved by the facts of heredity. Although such an early occurrence of the 'reducing division' would offer advantages in that nothing would be lost, for both daughter-nuclei would become eggs, instead of one of them being lost as a polar body, nevertheless I do not believe that it really occurs: weighty reasons can be alleged against it.

Above all, the facts of parthenogenesis are against it. If the number of ancestral germ-plasms received from the parents were reduced to half in the ovary of the young animal, how then could parthenogenetic development ever take place? It is true that we cannot at once assert the impossibility of an early 'reducing division' on this account, for as I have shown above, the power to develope parthenogenetically depends upon the quantity of germ-plasm contained in the mature egg; the necessary amount might be produced by growth, quite independently of the number of different kinds of ancestral germ-plasms which form its constituents. The size of a heap of grains may depend upon the number of grains, and not upon the number of different kinds of grains. But in another respect such a supposition would lead to an unthinkable conclusion. In the first place, the number of ancestral germ-plasms in the germ-cells would be diminished by one half in each new generation arising by the parthenogenetic method; thus after ten generations only 1/1024 of the original number of ancestral germ-plasms would be present.

Now, it might be supposed that the 'reducing division' of the young egg-cells was lost at the time when the parthenogenetic mode of reproduction was assumed by a species; but this suggestion cannot hold, because there are certain species in which the same eggs can develope either sexually or parthenogenetically (e.g. the bee). It seems to me that such cases distinctly point to the fact that the reduction in the number of ancestral germ-plasms must take place immediately before the commencement of embryonic development, or, in other words, at the time of maturation of the egg. It is only decided at this time whether the egg of the bee is to develope into an embryo by the parthenogenetic or the sexual method; such decision being brought about, as was shown above, by the fact that only one polar body is expelled in the first case, while two are expelled in the second. But if we are obliged to assume that reproduction by means of fertilization, necessarily implies a reduction to one half of the number of ancestral germ-plasms inherited from the parents,—the further conclusion is obvious, that the second division of the egg-nucleus and the expulsion of the second polar body represent such a reduction, and that this second division of the egg-nucleus is unequal in the sense mentioned above, viz. one half of the ancestral germ-plasms remains in the egg-nucleus, the original number being subsequently restored by conjugation with a sperm-nucleus; while the other half is expelled in the polar body and perishes.

I may add that observations, so far as they have extended to such minute processes, do indeed prove that the number of loops is reduced to one half. It has been already mentioned that, according to Carnoy, such reduction occurs in *Ascaris megalocephala*, but the same author also describes the process of the formation of polar bodies in a large number of other *Nematodes*[263], and his descriptions show that the process occurs in such a way that the number of ancestral germ-plasms must be reduced by half. Sometimes half the number of primary loops pass into the nucleus of the polar body, while the other half remains in the egg. In other cases, as in *Ophiostomum mucronatum*, the primary nuclear rods divide transversely,—a process which must produce the same effect. It is true that these observations require confirmation, and since, with unfavourable objects, the difficulties of observation are extremely great, there may have been errors of detail; but I do not think that there is any reason for doubting the accuracy of the essential point. And this essential point is the fact that the number of primary loops is divided into half by the formation of the polar body.

But even if we could not admit that such a conclusion is securely founded, it cannot be doubted that the formation of the second polar body reduces to one half the quantity of the nucleus which would have become the segmentation-nucleus in the parthenogenetic development of the egg. This is a simple logical conclusion from the two following facts: first, parthenogenetic eggs expel only *one* polar body; secondly, there are eggs (such as those of the bee) in which it is absolutely certain that the same half of the nucleus—which is expelled as the second polar body in the egg requiring fertilization—remains in the egg when it is to develope parthenogenetically, and acts as half of the segmentation-nucleus. But this proves that the expelled half of the nucleus must consist of true germ-plasm, and thus a secure foundation is laid for the assumption that the formation of the nucleus of the second polar body must be considered as a 'reducing division.'

I was long ago convinced that sexual reproduction must be connected with a reduction in the number of ancestral germ-plasms to one half, and that such reduction was repeated in each generation. When, in 1885, I brought forward my theory of the continuity of the germ-plasm, I had long before that time considered whether the formation and expulsion of polar bodies must not be interpreted in this sense. But the two divisions of the egg-nucleus caused me to hesitate. The two divisions did not seem to admit of such an interpretation, for by it the quantity of the nucleus is not divided into halves, but into quarters. But a division of the number of ancestral germ-plasms into quarters would have caused, as was shown above, a continuous decrease, leading to their complete disappearance; and such a conclusion is contradicted by the facts of heredity. For this reason I was led at that time to oppose Strasburger's view that the expulsion of the polar bodies means a reduction of the quantity of nuclear substance by only half. My objection to such a view was valid when I said that the quantity of idioplasm contained in the egg-nucleus is not, as a matter of fact, reduced to one half, but to one quarter, inasmuch as two successive divisions take place. I may add that I had also considered whether the two successive divisions might not possess an entirely different meaning,—whether one of them led to the removal of ovogenetic nucleoplasm, while the other resulted in a reduction in the number of ancestral germ-plasms. But at that time there were no ascertained facts which supported the supposition of such a difference, and I did not wish to bring forward the idea, even as a suggestion, when there was no secure foundation for it. The morphological aspects of the formation of the first and second polar bodies are so extremely similar that such a supposition might have been considered as a mere effort of the imagination.

Hensen[264] also rejected the second part of the supposition that reduction must take place in the number of the hereditary elements of the egg, and that such reduction is caused by the expulsion of polar bodies, because he believed it to be incompatible with the fact, which had just been discovered, that polar bodies are formed by parthenogenetic eggs. He concludes with these words: 'If this striking fact be confirmed, the hypothesis which assumes that the egg must be divided into half before maturation, is refuted, and there only remains the rather vague explanation that a process of purification must precede the development of the embryo.' Nevertheless Hensen is the only writer who has hitherto taken into consideration the idea that sexual reproduction causes a regularly occurring 'diminution in the hereditary elements of the egg.'

If the result of the previous considerations be correct, and if the number of ancestral germ-plasms contained in the nucleus of the egg-cell destined for fertilization must be reduced by one half, there can be no doubt that a similar reduction must also take place, at some time and by some means, in the germ-plasms of the male germ-cells. This must be so if we are correct in maintaining that the young germ-cells of a new individual contain the same nuclear substance, the same germ-plasm, which was contained in the fertilized egg-cell from which the individual has been developed. The young germ-cells of the offspring must contain this substance if my theory of the continuity of the germ-plasm be well founded, for this theory supposes that, during the development of a fertilized egg, the whole quantity of germ-plasm does not pass through the various stages of ontogenetic development, but that a small part remains unchanged, and at a later period forms the germ-cells of the young organism, after having undergone an increase in quantity. According to this supposition therefore the germ-plasm of the parents must be found unchanged in the germ-cells of the offspring. If this theory were false, if the germ-plasm of the germ-cells were formed anew by the organism, perhaps from Darwin's 'gemmules' which pour into the germ-cells from all sides, it would be impossible to understand why it has not been long ago arranged that each germ-cell should receive only half the number of the ancestral gemmules present in the body of the parent. Hence the expulsion of the second polar body—assuming the validity of my interpretation—is an indirect proof of the soundness of the theory of the continuity of the germ-plasm, when contrasted with the theory of pangenesis. If furthermore, a kind of cyclical development of the idioplasm took place, as supposed by Strasburger, and if its final ontogenetic stage resulted in the re-appearance of the initial condition of the germ-plasm, we should fail to understand how any of the ancestral germ-plasms could be lost during such a course of development.

Whichever view, the latter or the theory of the continuity of the germ-plasm, be correct, in either case the male germ-cells of the young animal must contain the same germ-plasm as that which existed in the fertilized maternal egg, that is to say, they must contain all the ancestral germ-plasms of the father and the mother. Here therefore a reduction must occur, for otherwise the number of ancestral germ-plasms would be increased by one half at every fertilization. The egg-cell would furnish 1/2, but the sperm-cell 2/2 of the total quantity of germ-plasm present in the germ-cells of the parents. But there is no reason for believing that the reduction of germ-plasm in the sperm-cell must proceed in precisely the same way as in the egg-cell, viz. by the expulsion of a polar body. On the contrary, the processes of spermatogenesis are so remarkably different from those of ovogenesis that we may expect to find that reduction is also brought about in a different manner.

The egg-cell does not expel the superfluous ancestral germ-plasms until the end of its development, and in a form which induces the destruction of the separated portion. This is certainly remarkable, for germ-plasm is a most important substance, and although it seems to be wasted in the production of enormous quantities of sperm- and egg-cells, such waste is only apparent, and is in reality the means which renders the species capable of existence. It may perhaps be possible to prove that in this case also the waste is only apparent. Such proof would be forthcoming if it could be shown that the means by which reduction is brought about in eggs is advantageous, and therefore also, *ceteris paribus*, necessary. We see that everywhere, as far as our observation extends, the useful is also the actual, unless indeed it is impossible of attainment or can only be attained by the aid of processes which are injurious to the species. And if it be asked why germ-plasm is wasted in the maturation of egg-cells, the following may perhaps be a satisfactory answer.

Let us suppose that the necessary reduction of the germ-plasm does not take place by the separation of the second polar body, but that it happens during the first division of the first primitive-germ-cell which is found in the embryo, so that the two first egg-cells resulting from this division would already contain only half the number of ancestral germ-plasms from the father and the mother, contained in the fertilized egg-cell. In this case the main object, the reduction of the ancestral germ-plasms, would be gained by a single division, and all the succeeding nuclear divisions, causing the multiplication of these two first germ-cells, might take place by the ordinary form of nuclear division, viz. 'equal division.' But perhaps nature not only cares for this one main object alone, but also secures certain secondary advantages at the same time. In the case which we have supposed the egg-cells of the mature ovary would only contain two different combinations of germ-plasm, which we may call combinations A and B. Even if millions of egg-cells were formed, every one of them would contain either A or B, and hence (at least as far as the female pronucleus is concerned) only two kinds of individuals could arise from such eggs—viz. offspring A' and B'. All the offspring A' would be as similar to one another as identical twins, and the same would be true of offspring B'.

But if the 100th instead of the 1st embryonic germ-cell entered upon the 'reducing division,' a hundred cells would undergo this division at the same time, and thus two hundred different combinations of ancestral germ-plasm would arise, and two hundred different kinds of germ-cells would be found in the mature ovary. A still greater number of different combinations of hereditary tendencies would arise if the 'reducing division' occurred still later; but undoubtedly the diversity in the composition of the germ-plasm must be greatest of all when the 'reducing division' does not take place during the period in which the germ-cells undergo multiplication, but at the end of the entire course of ovarian development, and separately in each full-grown mature egg ready for embryonic development. In such a case there will be as many different combinations of ancestral germ-plasms as there are eggs, for, as I have shown above, it is hardly conceivable that such a complex body as the nuclear substance of the egg-cell—composed of innumerable different units—would ever divide twice in precisely the same manner. Every egg will therefore contain a somewhat different combination of hereditary tendencies, and thus the offspring which arise from the different germ-cells of the same mother can never be identical. Hence by the late occurrence of the 'reducing division' the greatest possible variability in the offspring is secured.

If my interpretation of the second polar body be accepted, it is obvious that the late occurrence of the 'reducing division' is proved. At the same time we receive an explanation of the advantage gained by the postponement of the reduction of the germ-plasm until the end of the ovarian development of the egg; because the greatest possible number of individual variations in the offspring are produced in this way.

If I am not mistaken, this argument lends additional support to the idea which I have previously propounded,—that the most important duty of sexual reproduction is to preserve and continually call forth individual variability, the foundation upon which the transformation of species is built[265].

But if it be asked whether the postponement of the 'reducing division' to the end of the ovarian development of the egg is inconsistent with the preservation of the other half of the dividing nucleus, I should be inclined to reply that a 'reducing division' of the mature egg, resulting in the production of two eggs, was probably the phyletic precursor of the present condition. I imagine that the division of the mature egg-cell—although it is now so extremely unequal—was equal in very remote times; but that for reasons of utility, connected with the specialization of the eggs of animals, it gradually became more and more unequal. It is now hardly possible to give in detail the various reasons of utility which have brought about this condition, but it may be assumed that the enormous size attained by many animal egg-cells has been especially potent in producing the change.

A careful consideration of this last point seems to me to be demanded by a comparison of the egg-cells with the male germ-cells. Just as the female germ-cells of animals are distinguished by the attainment of a large size, the male germ-cells are generally remarkable for their minute proportions. In most cases it would be physiologically impossible for a large egg-cell, rich in yolk, to attain double its specific size in order to undergo division into two equal halves and yet to remain of the characteristic size. Even without the additional difficulties imposed by the necessity for such a division, all means—such as cells used as food, or the passage of food from follicular cells into the ovum, etc.—are employed in order to bring the egg-cell to the greatest attainable size. Furthermore, the 'reducing division' of the nucleus cannot take place before the egg has attained its full size, because the ovogenetic nucleoplasm still controls the egg-cell, and must be removed before the germ-plasm can regulate its development. By arguments such as these I should attempt to render the whole subject intelligible.

But the case is entirely different with the sperm-cells, which are generally minute: here it is quite conceivable that a 'reducing division' of the nuclei may take place by an equal division of the sperm-cells, occurring towards the end of the period of their formation; that is to say, in such a way that both products of division remain sperm-cells, and neither of them perishes like the polar bodies. But the other possibility also demands consideration, viz. that the reducing division may occur at an earlier stage in the development of sperm-cells. At all events, the arguments adduced above, which proved that the consequence would be a want of variability in the egg-cells, would not apply to an equal extent in the case of the male germ-cells. Among the egg-cells it may be very important that each one should have its special individual character, produced by a somewhat different composition of its germ-plasm, inasmuch as a considerable proportion of the eggs frequently developes, although this is never the case with all of them. But the production of sperm-cells is in most animals so enormous that only a very small percentage can be used for fertilization. If, therefore, e. g. ten or a hundred spermatozoa contained germ-plasm with exactly the same composition, so that, as far as the paternal influence is concerned, ten or a hundred identical individuals would result if they were all used in fertilization, such an arrangement would be practically harmless, for only one spermatozoon out of an immense number would be employed for this purpose. From this point of view we might expect that the 'reducing division' of the sperm-nucleus would not take place at the end of the development of the sperm-cell, but at some earlier period. There is no necessary reason for the assumption that this division must take place at the end of development, and without some cause natural selection cannot operate. It is, of course, conceivable that the causes of other events may also involve the occurrence of this division at the end of development; but we do not at present know of any such causes. I should not consider the influence of the specific histogenetic nucleoplasm, i.e. the spermatogenetic nucleoplasm, to be such a cause, because the quantitative proportions are very different from those which obtain in the formation of egg-cells, and because it is not inconceivable that the small quantity of true germ-plasm which must be present in the nuclei of the sperm-cells at every stage in their formation might enter upon a 'reducing division' with the spermatogenetic nucleoplasm, even when the latter preponderated.

As soon as we can recognize with certainty the forms of nuclear division which are 'reducing divisions,' the question will be settled as far as spermatogenesis is concerned. It has been already established that various forms of nuclear division occur at different periods of spermatogenesis. I make this assertion, not only from my own observations, but also from observations which have been made and insisted upon by others. Thus, van Beneden and Julin[266] stated in 1884 that direct and karyokinetic nuclear divisions alternate with each other in the spermatogenesis of *Ascaris megalocephala*. Again, Carnoy[267] distinctly states that the different cell-generations in the same testis may not uncommonly exhibit considerable differences as regards karyokinesis. 'This may go so far that direct and indirect division may proceed simultaneously.' Platner[268], in his excellent paper on karyokinesis in Lepidoptera, also points out that the karyokinesis of the spermatocytes is essentially different from that of the spermatogonia. According to his description, the latter form may be very well interpreted as a 'reducing division,' for no equatorial plate is formed, and the chromatin rods (or granules, as they are better called in this case) remain from the first on both sides of the equatorial plane, and finally unite at the opposite poles to form the two daughter-nuclei. Furthermore, if Carnoy has correctly observed, the form of karyokinesis which I have previously interpreted as a 'reducing division' occurs in the sperm-mother-cells—a karyokinesis in which the chromatin rods either do not divide longitudinally, or else divide in this way after they have left the equatorial plate and are proceeding towards the poles. Carnoy does not himself attach any special importance to these observations, for he only considers them as proofs that the longitudinal splitting of the loops may occur at various periods in different species—either at the equator, or on the way towards the poles, or even at the poles themselves. We cannot conclude from the author's statements whether this form of nuclear division only occurs in a single cell-generation during spermatogenesis, as it must do if it really represents a 'reducing division.' Until this point is settled, we cannot decide with certainty whether the described form of karyokinesis is to be considered as the 'reducing division' for which we are seeking. Fresh investigations, undertaken from these points of view, are necessary in order to settle the question. It would be useless to seek further support for the theory by going into further details, and by critically examining the numerous observations upon spermatogenesis which have now been recorded.

I will only mention that among the various nuclei and other bodies in different animals which have been considered by different observers as the polar bodies of the sperm-cells, or the cells which form the latter—in my opinion the paranucleus ('Nebenkern') of the 'spermatides' described by La Valette St. George[269] has the highest claim to be considered as the homologue of a polar body. But I am inclined to identify it with the first rather than the second polar body of the egg-cells, and to regard it as the histogenetic part of the nucleoplasm which has been expelled or rendered powerless by internal transformations. There are two reasons which lead me to this conclusion: first, as I have tried to show above, it is probable that the ancestral germ-plasms are not removed by expulsion, but by means of equal cell-division; secondly, my theory asserts that the histogenetic nucleoplasm cannot be rendered powerless until the close of histological differentiation.

The whole question of the details of the transformations undergone by the nucleus of the male germ-cells is not ready for the expression of a mature opinion. From the very numerous and mostly minute and careful observations which have been hitherto recorded, we cannot conclude with any degree of certainty when and how the 'reducing division' of the nucleus takes place, nor can we decide upon the processes which signify the purification of the germ-plasm from the merely histogenetic part of the nucleoplasm. But perhaps it has not been without value as regards future investigation that I have tried to apply to the male germ-cells the views gained from our more certain knowledge of the corresponding structures in the female, and thus to indicate the problems which now chiefly demand solution.

IV. The Foregoing Considerations applied To Plants.

It remains to briefly consider the case of plants. Obviously, the 'reducing division' of the germ-nuclei, if it takes place at all, cannot be restricted to the germ-cells of animals. There must be a corresponding process in plants, for sexual reproduction is essentially the same in both kingdoms; and if fertilization must be preceded by the expulsion of half the number of ancestral germ-plasms from the eggs of animals, the same necessity must hold in the case of plants.

But whether the process always takes place in the form of polar bodies, and not perhaps principally, or at any rate frequently, in the form of equal cell-division, is another question. It is true that polar bodies occur in numerous plants, as we chiefly know from Strasburger's researches[270]. Strasburger shows that cells are separated by division from the germ-cells, and perish. But it seems to me doubtful whether we must always regard their formation as the removal of half the number of ancestral germ-plasms rather than the histogenetic nucleoplasm of the germ-cell. It appears to me that histogenetic nucleoplasm must be present in the highly differentiated vegetable germ-cells, especially in the male cells, and also that it must be removed during the maturation of the cell, if my idea of the histogenetic nucleoplasm be accepted. It is very possible, as I have already mentioned, that there may be quite indifferent germ-cells, viz. cells which are entirely without specific histological structure, and in such cases histogenetic nucleoplasm would be absent; and during the maturation of such germ-cells no polar body would be formed for its removal. This view accords with the fact that polar bodies are absent in many plants. Furthermore, I am far from maintaining that in the cases where polar bodies occur, they must have the above-mentioned significance. I only wish to point out that the reduction assumed to be necessary for the nucleus of the vegetable germ-cells is not necessarily to be sought for at the close of their maturation, but perhaps even more frequently in an equal division of the germ-cells during some period of their development.

It also seems to me to be not impossible that a number of these vegetative 'polar bodies' may have an entirely different significance, viz. to perform some special function accessory to fertilization, as in the so-called 'ventral canal-cells' of the higher cryptogams and conifers. As we know that even the two polar bodies of the animal egg are not identical—although externally they are extremely similar, and although they arise in a precisely similar manner—I am even more inclined than before to consider that the very various 'polar bodies' of plants possess very different meanings.

But I do not feel justified in criticizing in detail the results of botanical investigation. I must leave the decision of such questions to botanists, and I only desire to state distinctly that a 'reducing division' of the nuclei of germ-cells must occur in plants as well as in animals.

V. Conclusions with regard to Heredity.

The ideas developed in the preceding paragraphs lead to remarkable conclusions with regard to the theory of heredity,—conclusions which do not harmonize with the ideas on this subject which have been hitherto received. For if every egg expels half the number of its ancestral germ-plasms during maturation, the germ-cells of the same mother cannot contain the same hereditary tendencies, unless of course we make the supposition that corresponding ancestral germ-plasms are retained by all eggs—a supposition which cannot be sustained. For when we consider how numerous are the ancestral germ-plasms which must be contained in each nucleus, and further how improbable it is that they are arranged in precisely the same manner in all germ-cells, and finally how incredible it is that the nuclear thread should always be divided in exactly the same place to form corresponding loops or rods,—we are driven to the conclusion that it is quite impossible for the 'reducing division' of the nucleus to take place in an identical manner in all the germ-cells of a single ovary, so that the same ancestral germ-plasms would always be removed in the polar bodies. But if one group of ancestral germ-plasms is expelled from one egg, and a different group from another egg, it follows that no two eggs can be exactly alike as regards their contained hereditary tendencies: they must all differ. In many cases the differences will only be slight, that is, when the eggs contain very similar combinations of ancestral germ-plasms. Under other circumstances the differences will be very great, viz. when the combinations of ancestral germ-plasms retained in the egg are very different. I might here mention various other considerations; but this would lead me too far from my subject, into new theories of heredity. I hope to be able at some later period to develope further the theoretical ideas which are merely indicated in the present essay. I only wish to show that the consequences which follow from my theory upon the second division of the egg-nucleus, and the formation of the second polar body, are by no means opposed to the facts of heredity, and even explain them better than has hitherto been possible.

The fact that the children of the same parents are never entirely identical could hitherto only be rendered intelligible by the vague suggestion that the hereditary tendencies of the grandfather predominate in one, and those of the grandmother in another, while the tendencies of the great-grandfather predominate in a third, and so on. Any further explanation as to why this should happen was entirely wanting. Others even looked for an explanation to the different influences of nutrition, to which it is perfectly true that the egg is subjected in the ovary during its later development, according to its position and immediate surroundings. I had myself referred to these influences as a partial explanation[271], before I recognized clearly how extremely feeble and powerless are the influences of nourishment, as compared with hereditary tendencies. According to my theory, the differences between the children of the same parents become intelligible in a simple manner from the fact that each maternal germ-cell (I shall speak of the paternal germ-cells later on) contains a peculiar combination of ancestral germ-plasms, and thus also a peculiar combination of hereditary tendencies. These latter by their co-operation also produce a different result in each case, viz. the offspring, which are characterized by more or less pronounced individual peculiarities.

But the theory which explains individual differences by referring to the inequality of germ-cells, may be proved with a high degree of probability by an appeal to facts of an opposite kind, viz. by showing that identity between offspring only occurs when they have arisen from the same egg-cell. It is well known that occasionally some of the children of the same parents appear to be almost exactly alike, but such children are without exception twins, and there is every reason to believe that they have been derived from the *same* egg. In other words, the two children are exactly alike because they have arisen from the same egg-cell, which could of course only contain a single combination of ancestral germ-plasms, and therefore of hereditary tendencies[272]. The factors which by their co-operation controlled the construction of the organism were the same, and consequently the results were also the same. Twins derived from a single egg are identical: this is a statement which, although not mathematically proved, may be looked upon as nearly certain. But there are also twins which do not possess this high degree of similarity, and these are even far commoner than the others. The explanation is to be found in the fact that the latter were derived from two egg-cells which were fertilized at the same time. In most cases, indeed, each twin is enclosed in its own embryonic membranes, while much less frequently both twins are enclosed in the same membranes. In one point only the proof is incomplete; for it has not yet been shown that identical twins are always derived from a single egg, since such an origin, together with a high degree of similarity, could only be established as occurring together in a small proportion of the cases. We therefore see that under conditions of nutriment which are as identical as possible, *two* egg-cells develope into unlike twins, *one* into identical twins; although we cannot yet affirm that the latter result invariably follows. It is conceivable that the stimulus for the production of two eggs from one may be afforded by the entrance of two spermatozoa, but these latter, as was shown above, could hardly contain identical hereditary tendencies, and thus two identical twins would not arise. It appears indeed that some cases have been observed in which differences have been exhibited by twins which were enclosed in the same embryonic membranes; but nevertheless I believe that two spermatozoa are not necessary to cause the formation of twins by a single egg. We know, it is true, from the investigations of Fol[273], that multiple impregnation produces the simultaneous beginning of several embryos in the eggs of star-fishes. But several embryos and young animals are not developed in this way, for embryonic development soon ceases, and the egg dies.

The recent observations of Born[274] upon the eggs of the frog also make it very probable that a double development is produced by the entrance of two spermatozoa into the egg, but here also only monstrosities, and not twins, were produced. On the other hand, it has been shown that in birds twins may be produced from the same egg, and there is no reason for the belief that their production is due to multiple impregnation. But if it may be assumed that human twins, when identical, have been derived from a single egg, it seems to me to be extremely probable that fertilization was also effected by a single sperm-cell. We cannot understand how such a high degree of similarity could have been produced if two sperm-cells had been made use of, for we are compelled to assume that two such cells would very rarely contain identical germ-plasms.

It is most probable that the egg-nucleus coalesces with the nucleus of a single spermatozoon, but the resulting segmentation-nucleus divides together with the cell-body itself, without the occurrence of those ontogenetic changes in the germ-plasm which normally take place. The nucleoplasm of the two daughter-cells still remains in the condition of germ-plasm, and its ontogenetic transformation begins afterwards—a transformation which must of course proceed in the same way in both cells, and must lead to the production of identical offspring. This is at least a possible explanation which we may retain until it has been either confirmed or disproved by fresh observations,—an explanation which is moreover supported by the well-known process of budding in the eggs of lower animals.

VI. Recapitulation.

To bring together shortly the results of this essay:—the fundamental fact upon which everything else is founded is the fact that *two* polar bodies are expelled, as a preparation for embryonic development, from all animal eggs which require fertilization, while only *one* such body is expelled from all parthenogenetic eggs.

This fact in the first place refutes every purely morphological explanation of the process. If it were physiologically valueless, such a phyletic reminiscence of the two successive divisions of the egg-nucleus must have been also retained by the parthenogenetic egg.

In my opinion the expulsion of the first polar body implies the removal of ovogenetic nucleoplasm when it has become superfluous after the maturation of the egg has been completed. The expulsion of the second polar body can only mean the removal of part of the germ-plasm itself, a removal by which the number of ancestral germ-plasms is reduced to one half. This reduction must also take place in the male germ-cells, although we are not able to associate it confidently with any of the histological processes of spermatogenesis which have been hitherto observed.

Parthenogenesis takes place when the whole of the ancestral germ-plasms, inherited from the parents, are retained in the nucleus of the egg-cell. Development by fertilization makes it necessary that half the number of these ancestral germ-plasms must be first expelled from the egg, the original quantity being again restored by the addition of the sperm-nucleus to the remaining half.

In both cases the beginning of embryogenesis depends upon the presence of a certain, and in both cases equal, quantity of germ-plasm. This certain quantity is produced by the addition of the sperm-nucleus to the egg requiring fertilization, and the beginning of embryogenesis immediately follows fertilization. The parthenogenetic egg contains within itself the necessary quantity of germ-plasm, and the latter enters upon active development as soon as the single polar body has removed the ovogenetic nucleoplasm. The question which I have raised on a previous occasion—'When is the parthenogenetic egg capable of development?'—now admits of the precise answer—'Immediately after the expulsion of the polar body.'

From the preceding facts and considerations the important conclusion results that the germ-cells of any individual do not contain the same hereditary tendencies, but are all different, in that no two of them contain exactly the same combinations of hereditary tendencies. On this fact the well-known differences between the children of the same parents depend.

But the deeper meaning of this arrangement must doubtless be sought for in the individual variability which is thus continuously kept up and is always being forced into new combinations. Thus sexual reproduction is to be explained as an arrangement which ensures an ever-varying supply of individual differences.

Footnotes for Essay VI.

234. See Berichten der Naturforschenden Gesellschaft zu Freiburg i. B., Band III. (1887) Heft I, 'Ueber die Bildung der Richtungskörper bei thierischen Eiern,' by August Weismann and C. Ischikawa.

235. Vol. I. p. 60.

236. The most recent example of this kind is afforded by the excellent work of O. Schultze, 'Ueber die Reifung und Befruchtung des Amphibieneies,' Zeitschr. f. wiss. Zool., Bd. XLV. 1887. Schultze has proved that two polar bodies are expelled from the egg of the Axolotl and of the frog, although all previous observers, including O. Hertwig, had been unable to find them. Thus the latter authority states as the result of an investigation specially directed towards this point, that the nucleus is transformed in a peculiar manner ('Befruchtung des thierischen Eies,' III. p. 81).

237. O. Hertwig, 'Beiträge zur Kenntniss der Bildung, Befruchtung, und Theilung des thierischen Eies,' Morpholog. Jahrbuch, I, II, and III. 1875-77.

238. H. Fol, 'Recherches sur la fécondation et le commencement de l'hénogénie chez divers animaux.' Genève, Bâle, Lyon, 1879.

239. Bütschli, 'Entwicklungsgeschichtliche Beiträge,' Zeitschr. f. wiss. Zool. Bd. XXIX. p. 237. 1877.

240. C. S. Minot, 'Account, etc.' Proceedings Boston Soc. Nat. Hist., vol. xix. p. 165. 1877.

241. F. M. Balfour, 'Comparative Embryology.'

242. Nägeli, 'Mechanisch-physiologische Theorie der Abstammungslehre,' München und Leipzig, 1884.

243. See the second and fourth Essays in the present volume.

244. Hensen, 'Die Grundlagen der Vererbung,' Zeitschr. f. wiss. Landwirthschaft. Berlin, 1885, p. 749.

245. O. Hertwig, 'Lehrbuch der Entwicklungsgeschichte des Menschen und der Wirbelthiere.' Jena, 1886.

246. Bütschli, 'Gedanken über die morphologische Bedeutung der sog. Richtungskörperchen,' Biol. Centralblatt, Bd. VI. p. 5. 1884.

247. This observation was first published as a note at the end of the fourth Essay in the present volume. See p. 249.

248. Weismann, 'Richtungskörper bei parthenogenetischen Eiern,' Zool. Anzeiger, 1886, p. 570.

249. Blochmann, 'Ueber die Richtungskörper bei den Insekteneiern,' Biolog. Centralblatt., April 15, 1887.

250. F. Stuhlmann, 'Die Reifung des Arthropodeneies nach Beobachtungen an Insekten, Spinnen, Myriapoden und Peripatus,' Berichte der naturforschenden Gesellschaft zu Freiburg i. Br., Bd. I. p. 101.

251. In the summer-eggs of Rotifera I have, together with Mr. Ischikawa, observed one polar body, and we were able to establish for certain that a second is not formed. The nuclear spindle had already been observed by Tessin, and Billet had noticed polar bodies in *Philodina*, but without attaching any importance to their number. These latter observations were not conclusive proofs of the formation of polar bodies in parthenogenetic eggs, so long as it was not known whether the summer-eggs of Rotifera may develope parthenogenetically, or whether they can only develope in this way. Knowing now that parthenogenetic eggs expel only one polar body, we may perhaps be permitted to draw the conclusion that the summer-egg of a Rotifer (*Lacinularia*) which expelled only one polar body must have been a parthenogenetic egg. But I may add that we have also succeeded in directly proving the occurrence of parthenogenesis in Rotifera, as will be described in detail in another paper.

252. See Essay IV, Part III. p. 225.

253. E. Bessels, 'Die Landois'sche Theorie, widerlegt durch das Experiment.' Zeitschr. f. wiss. Zool. Bd. XVIII. p. 124. 1868.

254. l. c., p. 110.

255. Strasburger, 'Neue Untersuchungen über den Befruchtungsvorgang bei den Phanerogamen als Grundlage einer Theorie der Zeugung.' Jena, 1884.

256. Wilhelm Roux, 'Ueber die Bedeutung der Kerntheilungsfiguren.' Leipzig, 1884.

257. E. van Beneden, 'Recherches sur la maturation de l'œuf, la fécondation et la division cellulaire.' Gand et Leipzig, Paris, 1883.

258. J. B. Carnoy, 'La Cytodiérèse de l'œuf, la vésicule germinative et les globules polaires de l'Ascaris megalocephala.' Louvain, Gand, Lierre, 1886.

259. See p. 364.

260. Wilhelm Roux, 'Beiträge zur Entwicklungsmechanik des Embryo,' No. 3, Breslauer ärztliche Zeitschrift, 1885, p. 45.

261. Carnoy, 'La Cytodiérèse chez les Arthropodes.' Louvain, Gand, Lierre, 1885.

262. Flemming, 'Neue Beiträge zur Kenntniss der Zelle.' Arch. f. mikr. Anat. Bd. XXIX, 1887.

263. Carnoy, 'La Cytodiérèse de l'œuf; la vésicule germinative et les globules polaires chez quelques Nématodes.' Louvain, Gand, Lierre. 1886.

264. Hensen, 'Die Grundlagen der Vererbung nach dem gegenwärtigen Wissenskreis,' Zeitschr. f. wissenschaftl. Landwirthschaft, Berlin, 1885, p. 731.

265. See the preceding Essay on 'The Significance of Sexual Reproduction in the theory of Natural Selection.'

266. E. van Beneden and Julin, 'La Spermatogénèse chez l'Ascaride mégalocéphale.' Brussels, 1884.

267. Carnoy, 'La Cytodiérèse chez les Arthropodes.'

268. Gustav Platner, 'Die Karyokinese bei den Lepidopteren als Grundlage für eine Theorie der Zelltheilung.' Internation. Monatsschrift f. Anatomie und Histologie, Bd. III. Heft 10. Leipzig, 1886.

269. La Valette St. George, 'Ueber die Genese der Samenkörper.' Fünfte Mittheilung. Die Spermatogenese bei den Säugethieren und dem Menschen,' Archiv f. mikrosk. Anat. Bd. XV. 1878.

270. Weismann, 'Studien zur Descendenztheorie,' ii. p. 306, Leipzig, 1876, translated by Meldola; see 'Studies in the Theory of Descent,' p. 680.

271. l. c., p. 92.

272. [The similar conclusion that identical ova lead to the appearance of identical individuals was drawn from the same data by Francis Galton in 1875. See 'The history of the Twins, as a criterion of the relative powers of Nature and Nurture,' by Francis Galton, F.R.S., Journal of the Anthropological Institute, 1875, p. 391; also by the same author, 'Short Notes on Heredity, etc. in Twins,' in the same Journal, 1875, p. 325.

The author investigated about eighty cases of close similarity between twins, and was able to obtain instructive details in thirty-five of these. Of the latter there were no less than seven cases 'in which both twins suffered from some special ailment or had some exceptional peculiarity;' in nine cases it appeared that 'both twins are apt to sicken at the same time;' in eleven cases there was evidence for a remarkable association of ideas; in sixteen cases the tastes and dispositions were described as closely similar. These points of identity are given in addition to the more superficial indications presented by the failure of strangers or even parents to distinguish between the twins. A very interesting part of the investigation was concerned with the after-lives of the thirty-five twins. 'In some cases the resemblance of body and mind had continued unaltered up to old age, notwithstanding very different conditions of life,' in the other cases 'the parents ascribed such dissimilarity as there was, wholly, or almost wholly, to some form of illness.'

The conclusions of the author are as follows: 'Twins who closely resembled each other in childhood and early youth, and were reared under not very dissimilar conditions, either grow unlike through the development of natural characteristics which had lain dormant at first, or else they continue their lives, keeping time like two watches, hardly to be thrown out of accord except by some physical jar. Nature is far stronger than nurture within the limited range that I have been careful to assign to the latter.' And again, 'where the maladies of twins are continually alike, the clocks of their two lives move regularly on, and at the same rate, governed by their internal mechanism. Necessitarians may derive new arguments from the life histories of twins.'

The above facts and conclusions held for twins of the same sex, of which at any rate the majority are shown by Kleinwächter's observations to have been enclosed in the same embryonic membranes, and therefore presumably to have been derived from a single ovum; but in rarer cases the twins, although also invariably of the same sex, were marked by remarkable differences, greater than those which usually distinguish children of the same family. Mr. Galton met with twenty of these cases. In such twins the conditions of training, etc. had been as similar as possible, so that the evidence of the power of nature over nurture is strongly confirmed. Mr. Galton writes, 'I have not a single case in which my correspondents speak of originally dissimilar characters having become assimilated through identity of nurture. The impression that all this evidence leaves on the mind is one of wonder whether nurture can do anything at all, beyond giving instruction and professional training.'

The fact that twins produced from a single ovum seem to be invariably of the same sex is in itself extremely interesting, for it proves that the sex of the individual is predetermined in the fertilized ovum.—E. B. P.]

273. Fol, Recherches sur la fécondation et le commencement de l'hénogénie: Genève, Bâle, Lyon. 1879.

274. Born, 'Ueber Doppelbildungen beim Frosch und deren Entstehung.' Breslauer ärztl. Zeitschrift, 1882.

VII.

ON THE SUPPOSED BOTANICAL PROOFS

OF THE

TRANSMISSION OF ACQUIRED CHARACTERS.

1888.

From 'Biologisches Centralblatt,' Bd. VIII. Nr. 3 and 4, pages 65

and 97: April 1888.

VII.

ON THE SUPPOSED BOTANICAL PROOFS

OF THE

TRANSMISSION OF ACQUIRED CHARACTERS.

In a lecture on heredity, delivered in 1883[275], I first brought forward the opinion that acquired characters cannot be transmitted; and I then stated that there are no proofs of such transmission, that its occurrence is theoretically improbable, and that we must attempt to explain the transformation of species without its aid. Since that time many biologists have expressed their opinions upon the subject, some of them agreeing with me, while others have taken the opposite side. It is unnecessary to allude to those who have attacked my opinions without first understanding the real point in dispute, which turns upon the true meaning of the phrase 'acquired character.' I think it is now generally admitted that a very important problem is involved in this question, the solution of which will contribute in a decisive manner towards the formation of ideas as to the causes which have produced the transformation of species. For if acquired characters cannot be transmitted, the Lamarckian theory completely collapses, and we must entirely abandon the principle by which alone Lamarck sought to explain the transformation of species,—a principle of which the application has been greatly restricted by Darwin in the discovery of natural selection, but which was still to a large extent retained by him. Even the apparently powerful factors in transformation—the use and disuse of organs, the results of practice or neglect—cannot now be regarded as possessing any direct transforming influence upon a species. And the same is true of all the other direct influences, such as nutrition, light, moisture, and that combination of different influences which we call climate. All these, with use and disuse, may perhaps produce great effects upon the body (*soma*) of the individual, but cannot produce any effect in the transformation of the species, simply because they can never reach the germ-cells from which the succeeding generation arises. But if—as it seems to me— the facts of the case compel us to reject the assumption of the transmission of acquired characters, there only remains one principle by which we can explain the transformation of species—the direct alteration of the germ-plasm, however we may imagine that such alterations have been produced and combined to form useful modifications of the body.

The difficulty of understanding these processes of transformation is by no means lightened by abandoning the Lamarckian theory. The difficulty in fact becomes much greater, for we are now compelled to seek a different explanation of many phenomena which were previously believed to be understood. But this can hardly be regarded as a reason for not accepting the view: for we are in want of a correct explanation rather than one which is easy and convenient. We seek truth, and when we recognize that our path is leading in a wrong direction, we must leave it and take another road even if it presents more difficulties.

My theory rests, on the one hand, upon certain theoretical considerations which will be mentioned below, and which I have attempted to develope in previous papers[276]. On the other hand, it rests upon the want of any actual proof of the transmission of acquired characters. My theory might be disproved in two ways,—either by actually proving that acquired characters are transmitted, or by showing that certain classes of phenomena admit of absolutely no explanation unless such characters can be transmitted. It will be admitted, however, that we must be very cautious in accepting proofs of this latter kind, for the impossibility of explaining a given phenomenon may be merely temporary, and may disappear with the progress of science. No one could have explained the useful adaptations so common in animals and plants, before the light of the theory of natural selection had fallen on these phenomena; at that time we should have been far from right if we had assumed that organisms possess a power which causes them to respond to external influences by useful modifications, a power unknown elsewhere, entirely unproved and only supported by the fact that at that time it did not seem possible to explain the phenomena in any other way.

Although my theory has not been disproved, I will nevertheless attempt to bring into further accordance with it certain phenomena which seem at first sight to oppose it. I first began to take this course in my paper 'On Heredity[277].' In that paper I attempted to show how the fact that disused organs become rudimentary may be readily explained without assuming the transmission of acquired characters; and also that the origin of instincts may in all cases be referred to the process of natural selection[278], although many observers had followed Darwin in explaining them as inherited habits,—a view which becomes untenable if the habits adopted and practised in a single life cannot be transmitted.

Other phenomena which appeared to present difficulties were also considered and brought into accordance with the theory, and I think that I have been successful in showing that adequate and simple explanations may be given.

There certainly remain many phenomena which seem to be opposed to my theory and for which a new explanation must be found. Thus Romanes[279], following Herbert Spencer[280], has recently pointed to the phenomena of correlation as a proof of the transmission of acquired characters; but, at no distant time, I hope to be able to consider this objection, and to show that the apparent support given to the old idea is in reality insecure and breaks down as soon as it is critically examined. I believe that I shall be able to prove that correlation cannot be used as the indirect proof of an hypothesis, of which all direct evidence is still completely wanting. It must not be forgotten that the *onus probandi* rests with my opponents: they defend the assertion that acquired characters can be transmitted, and they ought therefore to bring forward actual proofs; for the mere fact that the assertion has been hitherto accepted as a matter of course by almost everyone, and has only been doubted by a very few (such as His, du Bois-Reymond, and Pflüger), cannot be taken as any proof of its validity. Not a single fact hitherto brought forward can be accepted as a proof of the assumption. Such proofs ought to be found: facts ought to be discovered which can only be understood with the aid of this hypothesis. If, for instance, it could be shown that artificial mutilation spontaneously re-appears in the offspring with sufficient frequency to exclude all possibilities of chance, then such proof would be forthcoming. The transmission of mutilations has been frequently asserted, and has been even recently again brought forward, but all the supposed instances have broken down when carefully examined. I think I may here safely omit all further reference to the proofs dependent upon transmitted mutilations, especially as Döderlein[281] has already, in the most convincing manner, disposed of the argument derived from the tailless cats which were so triumphantly exhibited at the last meeting of the Association of German Naturalists[282].

I now come to the real subject of this paper—the supposed botanical proofs of the transmission of acquired changes. The botanist Detmer has recently brought forward certain phenomena in vegetable physiology[283], as a support for the transmission of such changes, and although I do not believe that they will bear this interpretation, the discussion of them may perhaps be useful. I am even inclined to think that these and a few other phenomena in vegetable physiology, upon which I shall also touch, are very likely to throw new light upon the whole question which has been so frequently misunderstood. I should have preferred to leave this discussion to a botanist, but I do not know whether my views will meet with any support from the followers of this subject, and I must therefore attempt the discussion myself. And perhaps it is of some assistance in clearing up the question, for one who is not accustomed to the usual botanical views, and is more conversant with other classes of biological knowledge, to consider the facts brought to light by modern botany, from a general point of view. Of course I shall not attempt to question the validity of the observations, nor even the accuracy with which the facts have been interpreted. I shall only deal with the conclusions which may be drawn from the facts, and I do not think that it is absolutely necessary that such criticism should be made by a botanist. Questions of general biological significance such as that of heredity cannot be entirely solved within the single domain of either zoological or botanical facts. Both botanists and zoologists must give due weight to the facts of the province which is not their own, and must see whether the views which they have chiefly gained in the one province can be applied to the other, or whether phenomena occur in the latter which are in opposition to their previously formed views and which cause them to be abandoned or modified.

Detmer begins by bringing forward certain facts which prove, as he believes, that rather important changes in the organism can be directly produced by external influences. He is of opinion that I under-estimate the weight of these influences, and that I make light of the changes which may thus arise in a single individual life. But obviously, it is of no importance for the question of the transmission of acquired characters, whether the changes directly produced by external influences upon the *soma* of an individual are greater or smaller: the only question is whether they can be transmitted. If they can be transmitted, the smallest changes might be increased by summation in the course of generations, into characters of the highest degree of importance. It is in this way that Lamarck and Darwin have supposed that an organism is transformed by external influences. It is therefore interesting to see what Detmer considers to be a change which has been directly effected. We can in this way gain a very distinct appreciation of the difference in views which is caused by the different spheres of experience which belong to botany and zoology. It will be useful to gain a clear idea of the differences which are thus caused.

Detmer first alludes to the dorso-ventral structure of the shoots of *Thuja occidentalis*, chiefly shown in the fact that the upper sides of these shoots contain the green palisade cells, while the under sides which are turned away from the light possess green spheroidal (isodiametric) cells. If the branches of *Thuja* are turned upside down and fixed in this position before the production of new shoots, it is found that the anatomical structure of the latter, when developed, is reversed. The side of the shoot which was destined to become the under side, but which was artificially compelled to become the upper side, assumes the structure of the upper side and developes the characteristic palisade parenchyma; and on the other hand, the under side which was intended to become the upper side developes the spongy parenchyma which is characteristic of the under side. From these facts Detmer concludes that the dorso-ventral structure of the shoots of *Thuja* has resulted from the continual operation of an external force, and that the light must be considered as the cause of the structural change.

But such a conclusion obviously depends upon a confusion of ideas. No one will doubt that the light was the stimulus which led to the reversal of the structures in the shoot, but this is a very different thing from maintaining that it was the cause which conferred upon the *Thuja*-shoot the power of producing palisade and spongy parenchyma. When a phenomenon only occurs under certain conditions, it does not follow that these conditions are the cause of the phenomenon. A certain temperature is necessary for the development of a bird in the egg, but surely no one will maintain that the temperature is the cause of the capacity for such development. It is obvious that the egg has acquired the power of producing a bird chiefly as the result of a long phyletic course of development which has led to such a chemical and physical structure in the egg and the fertilizing sperm-cell, that after their union and development, a bird, and only a bird of a particular species, must be produced. But of course certain conditions must be fulfilled in order that such development may take place; and a definite temperature is one of these conditions of development. Thus we may briefly say that the physical nature of the egg is the cause of its development into a bird, and we may similarly maintain that the physical nature of a *Thuja*-shoot, and not the influence of light, is the cause of the development of tissues which are characteristic of the species. In the development of such a shoot the light plays precisely the same part which is played by temperature in the development of a bird: it is one of the conditions of development.

There is nevertheless a difference between these two cases in that the *Thuja*-shoot possesses the possibility of development in two different ways instead of only one. The upper side of the shoot can assume the structure of the under side and *vice versa*, and this structural reversal depends upon the way in which the light is thrown upon the shoot. But even if the light causes the structural reversal, does this justify us in assuming that the structure itself is also the direct consequence of the influence of light? I see no reason for rejecting the supposition that the physical nature of part of a plant may be of such a kind that this or that structure may be produced according as this or that condition of development prevails. Thus with stronger light the structure of the upper side of the shoot developes; with weaker light, the structure of the under side. But this physical nature of the *Thuja*-bud depends, like that of a bird's egg, upon its phyletic history, as we must assume to be the case with the germs producing all individual developments. It is therefore quite impossible to interpret the reversal of the structure in the *Thuja*-shoot as the result of modification produced by the direct influence of external conditions. It is an instance of double adaptation—one of those cases in which the specific nature of a germ, an organism, or a part of an organism, possesses such a constitution that it reacts differently under the incidence of different stimuli. An entirely analogous example of reversal occurs in the climbing shoots of the Ivy, and is described in Sachs' lectures on the physiology of plants. Such shoots produce leaves only on the side directed towards the light, and roots (which are made use of in climbing) only upon the opposite side. If however the position of the plant be altered so that the root-bearing side is turned towards the light, while the leafy side is shaded, a reversal occurs, so that from that time the former only produces leaves, and the latter nothing but roots. In other words, the Ivy-shoot reacts under strong light with the production of leaves and under weak light with the production of roots, just as litmus-paper becomes red with an acid and blue with an alkali. The physical nature of the Ivy-shoot was present before the production of either structure, and was no more due to the action of light itself, than the physical nature of litmus-paper is due to an acid or an alkali. But this is quite consistent with the possession of a physical nature which reacts differently under the two different conditions afforded by light and shade.

No one would think of bringing forward the changes in the colour of the green frog (*Hyla*) as a proof of the power of direct influences in causing structural modifications in the animal body. The frog is light green when it is resting upon green leaves, but it becomes dark brown or nearly black when transferred to dark surroundings. This is an obvious instance of adaptation, for the changes in the colour of the frog depend upon a complex reflex mechanism. The changes in the shape of the chromatophores of the skin are not produced by the direct influence of the different rays of light upon the body-surface, but in consequence of the action of these rays upon the retina. Blind frogs do not react under the changes of light. Hence it is impossible that any one can maintain that the skin of the frog has gained its green colour as the direct result of the green light reflected from its usual surroundings. It must be admitted that in this and in all similar cases, there is only one possible explanation, viz. an appeal to the operation of natural selection. It may be objected that we are not here dealing, as in the *Thuja* and Ivy, with changes in the course of ontogenetic development following upon the occurrence of this or that external condition, but only with the different reactions of a mature organism. But nevertheless, cases of the former kind appear to be also present in the animal kingdom.

Thus the very careful and extensive investigations of Poulton[284] upon the colours of certain caterpillars have distinctly shown that some species possess the possibility of development in two directions, and that the actual direction taken by the individual is decided by the influence of external conditions. Poulton surrounded certain larvae of Geometrae with an abundance of dark branches, in addition to the leaves upon which they fed. When such conditions prevailed from the beginning of larval life, the caterpillars as they developed, gradually assumed the dark colour of the twigs and branches upon which they rested. When other larvae of the same species (and in many experiments hatched from the same batch of eggs) were similarly exposed to the green leaves of the same food-plant, they did not indeed become bright green like the leaves, but were invariably of a much lighter colour than the other larvae, while many of them gained a brownish-green tint. The larvae of *Smerinthus ocellatus*[285] also possess the power of assuming different shades of green and of thus approaching, to some extent, the green of the plant upon which they happen to live. It is quite impossible to explain the phyletic development of the green colour of these and other caterpillars as due to the direct action upon the skin of the green light reflected from the leaves upon which they sit. The impossibility of such an effect was pointed out long ago by Darwin, and also followed from my own investigations. Here, as in the other cases, the only possible solution is afforded by natural selection. The colour of the caterpillars has become gradually more and more perfectly adapted to the colour of the leaves,—and often to the particular side of the leaves upon which these animals rest,—not by the direct effect of reflected light, but by the selection of those individuals which were best protected. Poulton's experiments quoted above prove that certain species which occur upon different plants with different colours (or even in some cases upon the differently coloured parts of the same plant), present us with a further complication in the process of adaptation, inasmuch as each individual has acquired the power of assuming a lighter or darker colour[286]. The light which falls upon a single individual caterpillar during the course of its growth determines whether the lighter or darker colour shall be developed. Here therefore we have a case exactly parallel to that of the *Thuja*-shoot in which the palisade or spongy parenchyma is developed according to the position in which the shoot is fixed.

As far as it is possible in the present condition of our knowledge to offer any opinion upon the origin of sex in bisexual animals, it may be suggested that this problem is also capable of an essentially similar solution. Each germ-cell may possess the possibility of developing in either of two directions, the one resulting in a male individual, and the other resulting in a female, while the decision as to which of the two possible alternatives is actually taken may rest with the external conditions. We must, however, include among the external circumstances everything which is not germ-plasm. Moreover, this explanation is by no means certain, and I only mention it as an instance which, if we assume it to be correct, further illustrates my views upon the phenomena presented by the *Thuja*-shoot.

The two other facts brought forward by Detmer as proofs of the transforming power of external influences can be explained in precisely the same manner. These instances are—the fact that *Tropaeolum* when grown in moist air produces leaves with anatomical characters different from those produced when the plant is grown in dry air; and the differences in the structure of the leaves of many plants, according as they have been grown in the sun or shade respectively. Such differences do not by any means afford proof of the direct production of structural changes by means of external influences. How would such an explanation be consistent with the fact that the leaves are, in all these cases, changed in a highly purposeful manner? Or is it assumed that these organs were so constituted from the beginning, that they are compelled to respond to external conditions by the production of useful changes? Any one who made such an assertion nowadays, or who even thought of such a thing as a possibility, would prove that he is entirely ignorant of the facts of organic nature, and that he has no claim to be heard upon the question of the transformation of species. The very first necessity in any scientific question is to gain acquaintance with that which has been thought and said upon the subject. And it has been frequently shown that whole groups of useful characters cannot by any possibility have been produced by the direct action of external influences. If a caterpillar, which hides itself by day in the crevices of the bark, possesses the same colour as the latter, while other caterpillars which rest on leaves are of a green colour, these facts cannot be explained as the results of the direct influence of the bark and leaves. And it would be even less possible to explain upon the same principle all the details of marking and colour by which these animals gain still further protection. If the upper side of the upper wings of certain moths is grey like the stone on which they rest by day, while in butterflies the under side of both wings which are exposed during rest, exhibits analogous protective colours, these facts cannot be due to the direct influence of the surroundings which are resembled, but, if they have arisen in any natural manner, they must have been indirectly produced by the surroundings. One may reasonably complain when compelled to repeat again and again these elements of knowledge and of thought upon the causes of transformation!

Any one who remembers these things, and is aware of the countless number of purposeful characters which cannot possibly depend upon such direct influences, will be very cautious in yielding to any single instance which at first sight appears to be the direct consequence of external conditions. If Detmer had been thus cautious he would hardly have written the following sentence as a *résumé* of the physiological experiments on plants which have been already alluded to: 'In certain cases it is possible, as we have seen, to artificially modify the anatomical structure of certain parts of plants. In such cases the relation between the structure and the external influences is undoubtedly clear: the latter act as the cause; the anatomical structure of the members of the plant is the consequence of this cause.' A little more logic would have prevented the author from expressing such an opinion, for, as has been already shown, it is founded on a confusion between the true cause of a phenomenon and one of the conditions which are necessary for its production. We might as well consider the phenomena of *geotropism*, *hydrotropism*, and *heliotropism*—which have been established, and investigated in such a brilliant way by modern vegetable physiologists—as the direct results of the attraction of the earth, of water, and of light; and it is not improbable that some botanists are even inclined to make this assumption. And yet it is perfectly easy to show that this cannot be the case. By geotropism we mean the power possessed by the parts of a plant of growing along lines which make certain angles with the direction of the earth's attraction. For example, the chief root grows parallel with the earth's attraction, viz. towards the centre of the earth, and it is described as positively geotropic: conversely the main shoot grows along the same line but in an opposite direction, and it is negatively geotropic. But geotropism is not a primitive attribute of the plant, and it is even now absent from those plants which, like many Algae, have no definite position. Geotropism cannot have arisen before plants first became fixed in the earth. If any one were to assume that the direct influence of gravity, continuous through countless generations, had at length conferred upon the root the power of growing in a geotropic direction, how would it be possible to explain the fact that the shoot which has been under precisely the same influence has acquired the power of growing in an exactly opposite direction? The characteristic differences between root and shoot cannot have appeared until the plant became fixed in the ground, and how can we imagine that the same influence of gravity has since that time directly produced the two antagonistic results of positive and negative geotropism, in two structures, which were originally and essentially similar? It should also be remembered that it is only the main root which exhibits true positive geotropism. The lateral roots form angles with the main root, and do not therefore grow towards the earth's centre; and the same is true of the lateral shoots which grow obliquely, and not perpendicularly upwards, like the main shoot. Moreover the angles which the lateral roots make with the main root, and the lateral shoots with the main shoot, are quite different in different species. How is it possible that all these different modes of reaction witnessed in the different parts of plants can be the direct results of one and the same external force? It is quite obvious that these are all cases of adaptation. The main root has not acquired the power of growing perpendicularly downwards under the stimulus of gravity, because this force has acted upon it for numberless generations, but because such a direction for such a part was the most useful to the plant. Hence natural selection has conferred upon the root the power of reacting under the stimulus of gravity by growing in a direction parallel to this force. For the main shoot, the opposite reaction was the most useful and has been established by natural selection, while still another reaction has been similarly established for the lateral roots and another for the lateral shoots.

Each part of a plant has received its special mode of reacting under the stimulus of gravity because it was useful for the whole plant, inasmuch as the position of its different parts relatively to one another and to the soil became thus fixed and regulated. These modes of reaction have become different in different species, because the conditions of life peculiar to each require special arrangements.

The same argument also holds with regard to heliotropism. The power of growing towards the light possessed by green shoots cannot be a primitive character of the plant: it must have arisen secondarily. If it were an essential and original character it could not be reversed in certain parts of the plant; but, the roots are negatively heliotropic, for they grow away from the light. There are also shoots, such as the climbing shoots of Ivy, which are similarly negatively heliotropic. Whenever the heliotropic power is thus reversed in shoots, the change is of a useful kind. Thus the shoots of the Ivy gain the power of clinging closely to a perpendicular wall or to some horizontal plane[287]. In this case, however, it is only the shoot which is negatively heliotropic, its leaves turn towards the light; and the same is true of the flower-bearing shoots which do not climb. All these are clearly adaptations and not the results of direct influence. The light only provides the stimulus which calls forth the characteristic reaction from each part of the plant, but the cause of each peculiar reaction lies in the specific nature of the part itself which has not been produced by light, but as we believe by processes of natural selection. If this explanation does not account for the facts we may as well abandon all attempts at understanding the useful arrangements in organisms.

Sachs has used the term *anisotropism* to express the fact that the various organs of a plant assume the most diverse directions of growth under the influence of the same forces. He also states that anisotropism is one of the most general characteristics of vegetable organization, and that it is quite impossible to form any idea as to how plants would appear or how they could live if their different organs were not anisotropic. Since anisotropism is nothing more than the expression of different kinds of susceptibility to the action of gravity, light, &c., it is obvious that the configuration of the plant is to be traced to such specific susceptibilities.

Now these specific susceptibilities cannot have been produced by the direct effect of the various external influences (as was shown above), and the only other possible explanation is to recognise them as adaptations, and to admit that they have arisen by the operation of natural selection upon the general variability of plant organization.

Simple as these conclusions are, I have failed to meet with them in any of the writings of botanists, and they may perhaps be of use in helping to shake the vaguely-felt opinion that the characters of plants are to be chiefly referred to the direct action of external influences.

At all events it cannot be maintained that the phenomena of anisotropism support the opinion mentioned above; and the mere assertion that it is highly probable that hereditary characters arise as the result of external influences, is no more than the expression of an unfounded individual opinion. It is remarkable that Detmer should make such an assertion as the outcome of his discussion of the reversed *Thuja*-shoot, &c., for even if we admit that the dorso-ventral structure of the shoot is—as Detmer believes—the direct and primary effect of the action of light, the experiment with the reversed shoot would prove that no part of this effect has become hereditary. Although the upper side of the shoot has produced the palisade parenchyma under the influence of light for thousands of generations, there is nevertheless no tendency towards the establishment of any hereditary effect, for as soon as the upper side of the growing shoot is artificially transformed into the under side, its normal structure is at once abandoned. Hence so far from lending any support to the assumption that acquired characters can be transmitted, Detmer's experiment rather tends to disprove this opinion.

I think I have sufficiently shown that Detmer's reproach—that I have under-estimated the effects of external influences upon an organism—may be fairly directed against its author. If we can believe that every structural arrangement in plants, which depends upon certain external conditions, has been produced in a phyletic sense by these latter, it becomes very easy to explain the transformation of species; but in accepting such an explanation we are building without any foundation, for the proof that acquired characters can be transmitted has yet to be given.

As a further disproof of my views Detmer quotes the so-called phenomena of correlation in plants, and he believes that these instances help us to conceive how the acquired changes of the body (*soma*) of the plant may also influence the sexual cells. If the apical shoot of a young spruce fir be cut off, one of the lateral shoots of the whorl next below the section rises and becomes an apical shoot: it not only assumes the orthotropic growth of such a shoot, but also its mode of branching. The phenomenon itself is well known, and I have often observed it myself in my garden without making any botanical experiments; for this experiment is not uncommonly made by Nature herself, when the apical shoot is destroyed by insects (for example the gall-making *Chermes*). The change of the lateral into an apical shoot occurs here in consequence of the loss of the true apical shoot, and is therefore really dependent upon it. The only difficulty is to understand how these and many other kindred phenomena can be considered to prove the transmission of acquired characters. That correlation exists between the parts of an organism, that correlated changes are not only common but nearly always accompany some primary change, has been perfectly well known since Darwin's time, and I am not aware that it has been disputed by any one. I further believe that hardly any one would maintain that it is impossible for the reproductive organs to be influenced by correlation. But this is very far from the admission that such changes would occur in the germ-cells as would be necessary for the transmission of acquired characters. For such transmission to occur it would be necessary for the germ-plasm (the bearer of hereditary tendencies) to undergo a transformation corresponding to that produced by the external influences;—such a transformation as would cause the future organism to spontaneously develop changes similar to those which its parent had acquired. But since the germ-plasm is not an organism in the sense of being a microscopic facsimile which only has to increase in size in order to become a mature organism, it is obvious that the developmental tendencies must exist in the specific molecular structure, and perhaps also in the chemical constitution of the germ-plasm itself. It therefore follows that the changes in the germ-plasm which would be required for the transmission of an acquired character must be of an entirely different nature from the change itself acquired by the body of the parent plant: and yet it is supposed that the former is produced by the latter as a result of correlation. I will illustrate this by an example. Let us suppose that the influence of climate had caused a plant to change the form of its leaves from an ovate into a lobate shape: now such a change could not be transferred to the germ-plasm in the pollen and the ovules, as anything similar to leaves or the form of leaves; for such specialized morphological features have no existence in the germ-plasm. The only thing which could happen would be changes in its molecular structure which bear no resemblance to those changes which are implied by the direct alteration of the form of the leaf in the parent plant. Any one who clearly appreciates this difficulty will hesitate in admitting the possibility of the transmission of acquired characters, because it is possible that the sexual cells may be affected by correlated influences. If the change in the form of a leaf exercises any influence at all upon the germ-plasm, why should it produce a corresponding (in the above-mentioned sense) change in its molecular structure? Why should it not produce some other out of the immense number of possible changes? There must be as many possible changes in the structure of germ-plasm as there are possible variations in each part of a plant that arises from it. Why then should the corresponding change always occur,—a change which had never previously existed in the whole phyletic development of the organic world; for the plant with the latest modification can have never existed before? The occurrence of a particular change out of the countless possible changes would be about as likely as if one out of a hundred thousand pins thrown out of a window were to balance on its point when it reached the ground. The assumption scarcely deserves to be called a scientific hypothesis, and yet it must be made by all who accept the transmission of acquired characters,—that is unless they adopt the hypothesis of pangenesis, which is quite as improbable, and which even Darwin did not look upon as a real, but only as a formal explanation.

Detmer is also greatly mistaken when he says that I refuse to admit the transmission of acquired characters, because I am prejudiced in favour of my doctrine of the continuity of the germ-plasm. This doctrine is either right or wrong, and there is no middle course: to this extent I quite admit that I am prejudiced. But the question as to whether acquired characters can be impressed upon the germ and thus transmitted would not be by any means settled in this way; for even if we admit that the germ-plasm is not continuous from one generation to another, but that it must be produced afresh in each individual, this would by no means necessarily imply that it would potentially receive and retain every change produced in every part of the individual, and at any time in its life. It seems to me that the problem of the transmission or non-transmission of acquired characters remains, whether the theory of the continuity of the germ-plasm be accepted or rejected.

I will now proceed to examine the last group of phenomena which Detmer brings forward in favour of the transmission of acquired characters. He charges me with not having taken into account, in discussing the problem of heredity, the very important facts which are known about the strange phenomena of 'after-effect' in plants. Among these 'after-effects' are the following.

If vigorous plants of the sun-flower, grown in the open air, be cut off close to the ground and transferred to complete darkness, the examination of a tube fixed to the cut surface of the stem will show that the escape of sap does not take place uniformly, but undergoes periodical fluctuation, being strongest in the afternoon and weakest in the early morning. Now the cause of this daily periodicity in the flow of sap depends upon the periodical changes due to the light to which the plant was exposed when it was growing under normal conditions. When plants which have been grown in darkness from the first are similarly treated, the flow of sap does not exhibit any such periodicity.

Another instance is as follows:—it is well known that darkness accelerates, while light retards the growth of plants, and therefore plants usually grow more strongly by night than by day. If now plants are transferred from the open air into constant darkness, the periodicity in their growth does not immediately disappear, and often persists for a long time as a phenomenon of after-effect.

The opening and closing of the leaves of *Mimosa pudica* also takes place periodically under natural conditions, the leaves closing at dusk as a result of changes in the stimulus provided by the light. In this case also, when the plants are transferred to constant darkness, the periodicity in the movements of the leaves continues for several days.

All this is certainly very interesting, and it proves that periodical stimuli produce periodical processes in the plant, which are not immediately arrested when the stimulus is withdrawn, and only become uniform gradually and after the lapse of a considerable time. But I certainly claim the right to ask what connexion there is between these facts and the transmission of acquired characters. All these peculiarities produced by external influences remain restricted to the individual in which they arose; most of them disappear comparatively soon, and long before the death of the individual. No example of the transmission of such a peculiarity is known. Although successive generations of sunflowers have been exposed for thousands of years to the daily alternation of light and darkness, the periodicity in the flow of sap has not become hereditary, and does not take place at all in plants which have always been kept in darkness. Detmer specially tells us that we can even reverse the periods of opening and closing the leaves in *Mimosa pudica* by keeping them in darkness during the day, but exposed to light at night; an experiment which was performed by Pfeffer. Here again we see the proof that influences which have acted upon countless generations have left no impression whatever upon the germ-plasm.

Detmer himself admits this when he says that the after-effects are only witnessed during the life of the individual, but he nevertheless adds that he has been for many years convinced that the phenomena of heredity and after-effect differ in degree and not in kind. He even goes so far as to assert that, in spite of the obvious non-transmission of after-effect, the similarity between the natures of these two classes of phenomena cannot escape the intelligent observer.

It seems to me that this question does not demand the attention of the observer (for the observations have already been made) so much as that of the thinker. It is not a correct train of reasoning to conclude that after-effect and heredity are identical in nature, from the fact that certain periodical influences, acting upon a single individual, set up periodical physiological processes which continue for a time after the influences have ceased to act. We might almost as well argue that the oscillations of a pendulum, which continue as after-effects when the pendulum has been set going, are of an identical nature with the process of heredity. All these phenomena have indeed this much in common:—a cause which acted at some time in the past, but which is no longer visible at the time when the phenomenon appears. But the likeness ends here, and the supposed identity in nature merely depends upon wild speculation. One difference is very obvious, for the phenomena of after-effect gradually cease after the withdrawal of the stimulus, just like the oscillations of the pendulum, while the phenomena of heredity continue without any interruption. As far as heredity is concerned the physiological processes of after-effect are not distinguishable from any of the other well-known acquired characters which are recognizable as morphological changes. After-effects are not transmitted, and compared with this fact but little importance can be attached to the use of vague analogies by Detmer, who would wish to conclude that heredity is only the after-effect of processes which had been set going in the parent organism.

At the end of his paper Detmer applies the ideas which he has gained from the consideration of after-effect to certain phenomena in the normal life of plants. He suggests that the periodical change of leaf in trees and shrubs may have been produced by the direct effect of climate. If branches bearing winter buds are cut off in the autumn and are placed in a hot-house, with their cut ends in water, the buds do not at once develope, and months may often elapse before they begin to break. He argues that this experiment proves that the annual periodicity of the plant no longer depends directly upon external influences; these latter produced the periodicity at some earlier time, but it has been gradually fixed in the organism by after-effect and heredity(!), so that its disappearance does not now take place when the stimulus is withdrawn, and changes would only happen very gradually under the influence of changed climatic conditions. He considers that this is proved from the fact that our cherry has become an evergreen in Ceylon.

Such are Detmer's opinions, and every one will agree with him in believing that the periodical change of leaf in temperate climates has been produced in relation to the recurring alternation of summer and winter. This is certainly the case, and it cannot be doubted that the character has become fixed by heredity. Where, however, is the proof that this hereditary character has been produced by the direct influence of climate? What right have we to look upon the hereditary appearance of the character as an after-effect of the direct influence exerted by changes of temperature upon previous generations? Such an opinion derives but little support from the previously described experiments upon after-effect, which showed that these phenomena were never hereditary.

It appears to me that there are certain points in this change of leaf and its accompanying phenomena, which distinctly indicate that natural selection has been at work. Can Detmer imagine that the brown scales which form the characteristic protective covering of winter buds have been produced by the direct action of the cold? If, however, the peculiar structure of these buds is to be referred to the specific constitution of the individual rather than to the direct effects of climate, would it be so very improbable for their physiological peculiarity of lying dormant for several months to have been developed simultaneously with the structure, by the operation of natural selection? And if this explanation be correct, we can at once see why the character has become hereditary, for natural selection works upon variations of the germ-plasm, and these are transferred from one generation to another with the germ-plasm itself.

But Detmer attempts to establish the converse conclusion, and he argues that the hereditary change of leaf has been abandoned under the long-continued effect of changed climatic conditions; but this opinion is based upon the single instance of the alteration in the habit of the European cherry in Ceylon. If it were proved that our cherry, grown from seed in Ceylon and propagated by seed for several generations, became evergreen gradually and not suddenly in the first generation: if, under such circumstances, it came to retain its leaves in the autumn and ceased to produce the dormant winter buds:—then indeed the transmission of acquired characters could hardly be doubted. I am not a botanist, but I believe I am right in supposing that the wild cherry reproduces itself by seeds, while the edible domesticated cherry is propagated by grafting. Grafts are, however, parts of the *soma* of a previously existent tree, and we are not therefore concerned, in this method of propagation, with a succession of generations, but with the successive distribution of one and the same individual over many wild stocks. But no one will doubt that one and the same individual can be gradually changed during the course of its life, by the direct action of external influences. The really doubtful point is whether such changes can be transmitted by means of the germ-cells. If, as I presume, the English in Ceylon do not care to eat wild cherries but prefer the cultivated kinds, it follows that the branches which bear fruit in that island have not been developed from germ-cells, at any time since their introduction, and there is nothing to prevent them from gradually changing their anatomical and physiological characters in consequence of the direct influence of climate.

Hence the instance which Detmer looks upon as plainly conclusive, can hardly be accepted in support of such a far-reaching assumption as the transmission of acquired characters.

It is therefore clear that none of the facts brought forward by Detmer really afford the proofs which he believes that they offer. But another botanist, Professor Hoffman of Marburg, well known for his long-continued experiments on variation, has recently called attention to certain other botanical facts in support of the transmission of acquired characters. These facts are indeed conclusive, if we accept the author's use of the term 'acquired,' but it will be found that they lead to hardly any modification in the state of existing opinion upon the subject.

In a short note, dated Jan. 1, 1888, the author communicated to this journal ('Biologisches Centralblatt') the statement that changes in the structure of flowers caused by poor nutrition can be proved to be hereditary to a greater or less extent[288].

A more elaborate account of the experiments will be found in several numbers of the 'Botanische Zeitung,' and the author expresses his final results in the following words (see Bot. Zeit. 1887, p. 773):—'These experiments prove with certainty (1) that insufficient nutrition may cause considerable morphological changes (viz. qualitative variations) which are in the first place acquired by the sexual apparatus of the flower, (2) that the "transient" (Weismann) characters acquired by the individual can be transmitted[289].'

The data upon which Hoffman bases these opinions are certain experiments conducted upon various plants, in order to determine the conditions of life under which abnormal flowers or any other variations occur most frequently: to decide, in short, how far variations are caused by the change of conditions.

It is obvious that the attention of the author was not at first directed to the question of the transmission of acquired characters. His experiments are of a much older date than the present condition and significance of the question before us. Hoffmann has, in fact, re-examined his former results from the new point of view, and this explains why his proofs are not always sufficiently convincing when applied to the present issue. But this is of no great importance, inasmuch as there is no necessity for me to question the correctness of his assumptions.

The essential details of the experiments to which he directs attention are as follows.

Different plants with normal flowers were subjected to greatly changed conditions of life for a series of generations. They were, for example, crowded together in small pots. Under these circumstances the plants were of course poorly nourished, and in the course of generations, several species produced a variable proportion of abnormal—viz. double-flowers. This, however, was not always the case, for such flowers did not appear in *Matthiola annua* and *Helianthemum polifolium*. In other species, such as *Nigella damascena*, *Papaver alpinum* and *Tagetes patula*, they appeared and often increased in numbers in the course of generations, although this was not a constant result. For instance, four successive generations of *Nigella damascena*, when closely sown, produced the following results:—

1883. No double flowers.

1884. No double flowers.

1885. 23 typical flowers: 6 double flowers.

1886. 10 typical flowers: 1 double flower.

But it was not always the case that the double flowers continued to appear after they had been once produced. In *Papaver alpinum*, which Hoffman has cultivated in successive generations since 1862, other changes in addition to the doubling of the flowers first appeared in 1882, viz. a slight variability in the form of the leaf, and a greater variability in the colours of the flowers. The production of double flowers appeared to be favoured by poor nutrition caused by crowding the plants. The results as regards the number of double flowers produced in this species by close sowing, from 1882-1886, have been as follows:—

Experiment XI.   1881. 40 per cent. of double flowers.

1882. 4   per cent. of double flowers.

1883. 5·3 per cent. of double flowers.

Experiment XVII. 1884. 13· per cent. of double flowers.

1885. 0·0 per cent. of double flowers.

1886. 0·0 per cent. of double flowers.

Although in these and some other series of generations the double flowers again disappeared in the later generations, yet there can be hardly any doubt that their first appearance was due to the abnormal conditions of nutrition. This conclusion is also unaffected by the fact that double flowers appeared in nearly the same proportions in consequence of cultivation in ordinary garden soil. The plants which were crowded in pots produced 2879 normal flowers, and 256 (=8·8 per cent.) abnormal and mostly double ones, while 867 normal and 62 (=7·0 per cent.) abnormal ones were produced on garden beds. Hoffman will not indeed admit that such a comparison can be fairly made, for the plants in the garden beds were raised from seed which was in part taken from the double flowers, and was therefore, he believed, under a strong hereditary influence. But this latter assumption is not supported by the results of his own experiments.

Thus experiment XVIII., conducted upon *Papaver alpinum*, is described in these words,—'Seeds yielded by double flowers from experiment XI. (1883), were sown in pots, and the resulting plants produced from 1884-1886, fifty-three single flowers and no double ones.'

In the converse experiment XIX. 'The seeds of single flowers from different stocks were sown in pots, and the resulting plants produced in 1885 and 1886 forty-three flowers, of which all were typical except one;' while plants produced in the garden by seed from the same sources, yielded 166 single and five double flowers. Hoffman also describes other experiments in which the seeds from double flowers produced plants which also yielded many double flowers. Thus, for example, in experiment XXI. seeds yielded by the double flowers of *Papaver alpinum* were sown in the garden and produced numerous plants, which in 1885 and 1886 bore 284 single and twenty-one double flowers, that is 7 per cent. of the latter.

It will therefore be seen that the transmission of the abnormality is by no means proved beyond the possibility of doubt, for who can decide between the effects due to heredity and changed conditions in the last experiment? I have no doubt however that the results are at any rate in part due to the operation of heredity, for I do not see how the phenomena can be otherwise understood. Nevertheless I cannot admit the transmission of acquired characters on this evidence, for the changes which have appeared are not 'acquired' in the sense in which I use the term and in the sense required by the general theory of evolution. It is true that they may be described by the use of this word: inasmuch as they are characters which the plant has come to possess; we are not however engaged in a mere dispute about terms, but in the discussion of a weighty scientific question. Our object is to decide whether changes in the *soma* (the body, as opposed to the germ-cells) which have been produced by the direct action of external influences, including use and disuse, can be transmitted; whether they can influence the germ-cells in such a manner that the latter will cause the spontaneous appearance of corresponding changes in the next generation. This is the question which demands an answer; and, as has been shown above, such an answer would decide whether the Lamarckian principle of transformation must be retained or abandoned.

I have never doubted about the transmission of changes which depend upon an alteration in the germ-plasm of the reproductive cells, for I have always asserted that these changes, and these alone, must be transmitted. If any one makes the contrary assertion, he merely proves that he does not understand what I have said upon the subject. In what other way could the transformation of species be produced, if changes in the germ-plasm cannot be transmitted? And how could the germ-plasm be changed except by the operation of external influences, using the words in their widest sense; unless indeed we assume with Nägeli, that changes occur from internal causes, and imagine that the phyletic development of the organic world was planned in the molecular structure of the first and simplest organism, so that all forms of life were compelled to arise from it, in the course of time, and would have arisen under any conditions of life. This is the outcome of Nägeli's view, against which I have contended for years.

If we now use the term 'acquired characters' for changes in the soma which, like spontaneous abnormalities, depend upon previous changes in the germ-plasm—it is of course easy to prove that acquired characters are transmitted; but this is hardly the way to advance science, for nothing but confusion would be produced by such a use of terms[290]. I am not aware that any one has ever doubted that spontaneous characters, such as extra fingers or toes, patches of grey hair, moles, etc., can be transmitted. It is true that such characters are sometimes called 'acquired' in pathological works, but His has rightly insisted that such an obviously inaccurate use of the term ought to be avoided, in order to prevent misunderstanding. If every new character is said to be 'acquired' the term at once loses its scientific value, which lies in the restricted use. If generally used, it would mean no more than the word 'new'; but new characters may arise in various ways,—by artificial or natural selection, by the spontaneous variations of the germ, or by the direct effect of external influences upon the body, including the use and disuse of parts. If we assume that these latter characters are transmitted, the further 'assumption of complicated relations between the organs and the essential substance of the germ becomes necessary' (His), while the transmission of the other kinds of characters do not involve any theoretical difficulties. There is therefore obviously a wide difference between these two groups of characters as far as heredity is concerned, quite apart from the question as to whether acquired characters are really transmitted. It is at all events necessary to have distinct terms which cannot be misunderstood. His[291] has proposed to call those characters which are due to selection 'changes produced by breeding' ('erzüchtete Abänderungen'), those which appear spontaneously—'spontaneous changes' ('eingesprengte Abänderungen'), and these two groups of characters would then be opposed to those which he calls 'acquired changes' ('erworbene Abänderungen'), of course using the term in the restricted sense. Science has always claimed the right of taking certain expressions and applying them in a special sense, and I see no reason why it should not exercise this right in the case of the term 'acquired.' It appears moreover that this word has not always been used in this vague sense by pathological anatomists, such as Virchow and Orth; for Weigert and Ernst Ziegler have employed it in precisely the same sense as that in which it has been used by Darwin, du Bois-Reymond, Pflüger, His and many others, including myself.

It is certainly necessary to have two terms which distinguish sharply between the two chief groups of characters—the primary characters which first appear in the body itself, and the secondary ones which owe their appearance to variations in the germ, however such variations may have arisen. We have hitherto been accustomed to call the former 'acquired characters,' but we might also call them '*somatogenic*,' because they follow from the reaction of the *soma* under external influences; while all other characters might be contrasted as '*blastogenic*,' because they include all those characters in the body which have arisen from changes in the germ. In this way we might perhaps prevent the possibility of misunderstanding. We maintain that the '*somatogenic*' characters cannot be transmitted, or rather, that those who assert that they can be transmitted, must furnish the requisite proofs. The *somatogenic* characters not only include the effects of mutilation, but the changes which follow from increased or diminished performance of function, and those which are directly due to nutrition and any of the other external influences which act upon the body. Among the *blastogenic* characters, we include not only all the changes produced by natural selection operating upon variations in the germ, but all other characters which result from this latter cause.

If we now wish to place Hoffmann's results in their right position, we must regard all of them as '*blastogenic*' characters, for no one of them can be considered as belonging to the group which has been hitherto spoken of as 'acquired,' in the literature of evolution: they are not due to *somatogenic* but to *blastogenic* changes. The body of the plant—the *soma*—has not been directly affected by external influences, in Hoffman's experiments, but changes have been wrought in the germ-plasm of the germ-cells and, only after this, in the *soma* of succeeding generations.

There is no difficulty in finding facts in support of this statement, among Hoffmann's experiments. The proof chiefly lies in the fact that in no one of his numerous experiments did any change appear in the first generation. The seeds of different species of wild plants, with normal flowers, were cultivated in the garden and in pots (thickly sown in the latter case), but no one of the plants produced by these wild seeds possessed a single double flower. It was only after a greater or less number of generations had elapsed that a variable proportion of double flowers appeared, sometimes accompanied by changes in the leaves and in the colours of the flowers. This fact admits of only one interpretation;—the changed conditions at first produced slight and ineffectual changes in the idioplasm of the individual, which was transmitted to the following generation: in this again the same causes operated and increased the changes in the idioplasm which was again handed down. Thus the idioplasm was changed more and more, in the course of generations, until at last the change became great enough to produce a visible character in the *soma* developed from it, such as, for example, the appearance of a double flower. Now the idioplasm of the first ontogenetic stage (viz. germ-plasm) alone passes from one generation to another, and hence it is clear that the germ-plasm itself must have been gradually changed by the conditions of life until the alteration became sufficient to produce changes in the *soma*, which appeared as visible characters in either the flower or leaf[292].

In addition to the above-mentioned cases Hoffmann also quotes some facts of a somewhat different kind. He succeeded in inducing considerable changes in the structure of the root of the wild carrot (*Daucus carota*) by means of the changes in nutrition implied by garden cultivation. These changes also proved to be hereditary.

Unfortunately, I have not the literature of the subject at hand, and hence I am unable to read the accounts of these older experiments *in extenso*; but it is sufficiently obvious that in this case we are also concerned with a change which did not become visible until after some generations had elapsed, and which was therefore a change in the germ-plasm.

Many instances of a precisely similar kind have been long known, and one of them is to be found in the history of the garden pansy, which Hoffmann has succeeded in producing from the wild form, *Viola tricolor*, in the course of eighteen years. Darwin some time ago pointed out in his work upon 'The Variation of Animals and Plants under Domestication,' that, in the case of the pansy and all other 'improved' garden flowers, the wild form remained unchanged for many generations after its transference to the garden, apparently uninfluenced by the new conditions of life. At length single varieties began to appear, and these were further developed by artificial selection and appropriate crossing, into well-marked races distinguished by peculiar colours, forms, etc.

In these cases also, changes in the germ-plasm are the first results of the new conditions, and there is no evidence for the occurrence of acquired characters, using the term in its restricted sense.

I now come to the last botanical fact brought forward by Hoffmann in support of the transmission of acquired characters. He states that specimens of *Solidago virgaurea* brought from the Alps of the Valais, commenced flowering in the botanical garden at Giessen, at a time which differed by several weeks from that at which specimens from the surrounding country, planted beside them, began to flower. In other words, the time of flowering must have been fixed by heredity in the alpine *Solidago*, for the external conditions would have favoured a time which was simultaneous with that of the Giessen plants.

What conclusions can be drawn from these facts? Hoffmann of course sees in them the proof of the transmission of acquired characters, but this presupposes that the time of flowering was originally an acquired character. Hoffmann indeed appears to entertain this opinion when he somewhat vaguely states that the time at which flowering begins has been acquired by accommodation—that is by the influence of climate—during a long series of generations, and has become hereditary. But what does Hoffmann mean by 'accommodation'? He presumably means that which, since the appearance of Darwin's writings, has been generally called adaptation:—that is a purposeful arrangement, suited to certain conditions. The majority of biologists have followed Darwin in believing that such adaptations have been produced by processes of natural selection. Hoffmann seems to imagine that they have arisen in some other way: perhaps he believes, with Nägeli, that they have been directly produced by external influences.

The fixation of the time at which flowering begins, is an adaptation which formerly could have been very well explained as the direct result of external conditions. The question we have to decide is whether such an explanation is the true one. We might imagine that the plant would be forced into quicker development by an earlier appearance of the warm season. Hence when transferred into a warmer climate the plant would at first flower rather earlier, the habit would then be transmitted, and would increase in successive generations from the continued influence of climate, until it advanced as far as the organization of the plant permitted. But in this explanation, as in so many others of the same kind, it has unfortunately been forgotten that the transmission of acquired characters which is presupposed in the explanation is a totally unproved hypothesis. It is sufficiently obvious that by interpreting a phenomenon in a manner which presupposes the transmission of acquired characters, we cannot furnish a proof of the existence of such transmission.

It always seemed to me that the fixation of the commencement of flowering, together with similar physiological phenomena in the animal kingdom (for example, the hatching of insects from winter eggs), could be explained very satisfactorily by the operation of natural selection: and even now this explanation appears to me to be the simplest and most natural. In Freiburg, where the vine is largely grown, the harvest is often injured by frosts in spring, which kill the young shoots, buds and flowers. Accordingly, different kinds of vine, which do not push their buds so early, have now been planted. Any one, who has seen all the shoots of the former destroyed by the frosts at the end of April, while the latter, not having opened their buds, were spared, would not doubt that the former must have been long ago exterminated, if they had been compelled to struggle for existence with the others, under natural conditions. Now the time of flowering fluctuates slightly in the individuals of every species of plant, and can therefore be modified by natural selection. It is therefore difficult to see why the time at which each plant flowers should not have been fixed in the most favourable manner for each habitat, by natural selection alone.

Hoffmann is obviously unaware of the fundamental distinction between the characters primarily acquired by the *soma*, and the secondary characters which follow from changes in the germ-plasm.

If the author had appreciated this distinction he would not have attempted to strengthen his opinions by following up the botanical facts which exclusively belong to the second class of characters, with the enumeration of certain instances selected from the animal kingdom (viz., the supposed transmission of mutilations), all of which belong to the first class. I will not discuss these latter instances, for most of them are old friends, and they are all far too uncertain and inaccurate to have any claim on scientific consideration.

I believe that I have shown that no botanical facts have been hitherto brought forward which prove the transmission of acquired characters (in the restricted sense), and that there are not even any facts which render such transmission probable.

A. W.

Naples, Zoological Station,

Jan. 11, 1888.

Footnotes for Essay VII.

275. See the second Essay.

276. Consult 'Ueber die Vererbung,' Jena, 1883; 'Die Kontinuität des Keimplasmas,' Jena, 1885; 'Ueber die Zahl der Richtungskörper und über ihre Bedeutung für die Vererbung,' Jena, 1887. These papers are translated as the second, fourth and sixth Essays in the present volume.

277. See the second Essay.

278. [See R. Meldola in Ann. and Mag. Nat. Hist., 1878, vol. i. pp. 158-161. The author discusses many cases among insects in which instinct is related to protective structure or colouring: he also considers that instinct is to be explained by the principle of natural selection which accounts for the other protective features.—E. B. P.]

279. [See 'Nature,' vol. 36, pp. 491-507.—E. B. P.]

280. [See 'The Factors of organic Evolution' in 'The Nineteenth Century' for April and May 1886.—E. B. P.]

281. See 'Biol. Centralbl.' Bd. VII. No. 23.

282. See the next Essay (VIII).

283. Detmer, 'Zum Problem der Vererbung,' Pflüger's Archiv f. Physiologie, Bd. 41, (1887), p. 203.

284. [Dr. Weismann is here alluding to experiments upon the larvae of *Rumia Crataegata*. A short account of the results will be found in the Report of the British Association at Manchester (1887), and in 'Nature,' vol. 36, p. 594. I have now obtained similar results with many other species (see Trans. Ent. Soc., Lond. 1888, p. 553); but many of the results are as yet unpublished.—E. B. P.]

285. [See the editorial notes by Raphael Meldola, in his translation of Weismann's 'Studies in the Theory of Descent' (the Essay on 'The Origin of the Markings of Caterpillars,' pp. 241 and 306): also E. B. Poulton, in 'Proc. Roy. Soc.,' vol. xxxviii. pp. 296-314; and in 'Proc. Roy. Soc.,' vol. xl. p. 135.—E. B. P.]

286. [Professor Meldola first called attention to the scattered instances of the kind here alluded to by Professor Weismann, in 1873: see 'Proc. Zool. Soc.,' 1873, p. 153. The author explains the relation of this 'variable protective colouring' to other protective appearances, and he is strongly of the opinion that the former as well as the latter is to be explained by the action of the 'survival of the fittest.'

The validity of Dr. Weismann's interpretation of these effects as due to adaptation, through the operation of natural selection, is conclusively proved by the following facts. The light reflected from green leaves becomes the stimulus for *the production of dark brown pigment* in those cases in which the leaves constitute the surroundings for many months. Under these circumstances the leaves of course become brown at a relatively early date, and protection is thus afforded for the remainder of the period, although the dark pigment is produced before the change in the colour of the leaf. Instances of this kind are seen in the colours of cocoons spun among leaves by certain lepidopterous larvae (see 'Proc. Ent. Soc. Lond.,' 1887, pp. l, li, and 1888, p. xxviii), the cocoons of the same species being of a creamy white colour when spun upon white paper.

Conversely, the light reflected from the same surfaces serves as the stimulus for *withholding pigment* in the cases alluded to by Dr. Weismann (larvae of *R. Crataegata*, &c.), in all of which the organism only remains in contact with the leaves while they are green, viz. at a time when the dark colour would be disadvantageous.

Hence precisely opposite effects are produced by the operation of the same force; the nature of the effect which actually follows in any case being solely determined by the advantage afforded to the organism.—E. B. P.]

287. Compare Sachs, 'Lectures on the Physiology of Plants,' translated by H. Marshall Ward, p. 710.

288. Compare Biol. Centralbl. Bd. VII. No. 21.

289. I have used the expression 'transient' ('passant') in the same sense as 'acquired,' in order to enforce the conclusion that they are merely temporary, and disappear with the individual in which they arise. Since the characters of which Hoffmann speaks are hereditary, the term cannot be rightly applied to them, and I shall prove later on that they cannot be regarded as acquired characters in the sense required by the theory of descent.

290. Compare a paper by J. Orth, 'Ueber die Entstehung und Vererbung individueller Eigenschaften,' Leipzig, 1887. This author considers my theory of the non-transmission of acquired characters to be incorrect, because he will insist upon using the term 'acquired' for those characters which are due to spontaneous changes in the germ; although he considers that they are only indirectly acquired. He also reproaches me with not having discriminated with sufficient clearness between the two modes in which new characters are acquired by the body, and with having altogether failed to take into account the class of characters which are due to variations in the germ. On the very same page he quotes the following sentence from my writings:—'Every change of the germ-plasm itself, however it may have arisen, must be transmitted to the following generation by the continuity of the germ-plasm; and hence also any changes in the *soma* which arise from the germ-plasm must be transmitted to the following generation.' Not only does the transmission of Orth's 'indirectly acquired characters' necessarily follow from this sentence, but it is even distinctly asserted by it. I cannot understand how any one who is aware of what happened at the meeting of the Association of German naturalists at Strassburg in 1885, can charge me with the confusion of ideas which has prevailed since Virchow took part in the discussion of this question.

291. His, 'Unsere Körperform,' Leipzig, 1874, p. 58.

292. Compare on this point Nägeli in his 'Theorie der Abstammungslehre.' This writer also concludes from similar facts that external influences have wrought in the idioplasm, changes which were at first ineffectual, and which only increased during the course of generations up to a

point at which they could produce visible changes in the plant. He does not, however, draw the further conclusion that these changes only influence the germ-plasm, for he was not aware of the distinction between germ-plasm and somatoplasm.

VIII.

THE SUPPOSED TRANSMISSION OF

MUTILATIONS.

1888.

A lecture delivered at the Meeting of the Association of

German Naturalists at Cologne, September 1888.

VIII.

THE SUPPOSED TRANSMISSION OF MUTILATIONS.

We know well the manner in which Lamarck imagined that the gradual transformation of species occurred, when he first made the attempt to penetrate into the mechanism of the process of evolution, and to ascertain the causes by which it is produced. In his opinion, a change in the structure of any part of an organism was chiefly brought about when the species in question met with new conditions of life and was thus forced to assume new habits. Such habits caused an increased or diminished activity, and therefore a stronger or weaker development, of certain parts, and the modified parts were then transmitted to the offspring. Inasmuch as the offspring continued to live under the same changed conditions, and kept up the altered manner of using the part in question, the inherited changes would be increased in the same direction during the course of their life, and would be further increased in each successive generation, until the greatest possible change had been effected.

In this way Lamarck was able to give an apparently satisfactory explanation of at any rate those changes which consist in the mere enlargement or diminution of a part; such, for instance, as the great length of neck in the swan and other swimming birds, which he believed to have been produced by the habit of stretching after food at the bottom of the water; or the webbed feet of the same animals, supposed to be produced by the habit of striking the water with outspread toes, etc. In this way he was also able to explain the disappearance of a part after it had ceased to be of use; as, for instance, the degeneration of the eyes of animals inhabiting caves or the sunless depths of lakes or the sea.

But it is obvious that such an explanation tacitly assumes that changes produced by use or disuse can be transmitted to the offspring; *it assumes the transmission of acquired characters.*

Lamarck made this assumption as a matter of course, and when half a century later Charles Darwin, his more fortunate successor, refounded the theory of organic evolution, he also believed that we could not entirely dispense with the Lamarckian principle of explanation, although he added the new and extremely far-reaching principle of natural selection. But he certainly attempted to decide whether the Lamarckian principle of the effects of use and disuse is truly efficient, by asking himself the question whether such changes, as for example those produced by exercise during an individual life, can be transmitted to the offspring. Many observations appeared to him, if not to prove the transmission directly, yet to render it extremely probable; and he thus came to the conclusion that there is no sufficient reason for denying the transmission of acquired changes. Hence, in Darwin's works, use and disuse still play important parts as direct factors of transformation, in addition to natural selection.

Darwin was not only an original genius, but also an extraordinarily unbiassed and careful investigator. Whatever he expressed as his opinion had been carefully tested and considered. This impression is gained by every one who has studied Darwin's writings, and perhaps it in part explains the fact that doubts as to the correctness of the Lamarckian principle adopted by him have only arisen during the last few years. These doubts have, however, culminated in the decided denial of the assumption that changes acquired by the body can be transmitted. I for one frankly admit that I was in this respect under the influence of Darwin for a long time, and that only by approaching the subject from an entirely different direction was I led to doubt the transmission of acquired characters. In the course of further investigations I gradually gained a more decided conviction that such transmission has no existence in fact.

Doubts on this point have been expressed not only by me but also by others, such as du Bois-Reymond and Pflüger. Indeed, concerning a certain class of acquired characters, viz. mutilations, the great German philosopher, Kant, has distinctly denied that transmission can take place[293]; and in more recent times Wilhelm His has expressed the same opinion[294].

But if the transmission of acquired characters is truly impossible our theory of evolution must undergo material changes. We must completely abandon the Lamarckian principle, while the principle of Darwin and Wallace, viz. natural selection, will gain an immensely increased importance.

When I first expressed this opinion in my essay 'On Heredity[295],' I was well aware of the consequences of such an idea. I knew well that apparently insurmountable obstacles would be raised against any explanation of evolution, from which the principle of the direct transformation of the species by external influences had been excluded. I therefore endeavoured to show that these difficulties are not in reality insurmountable, and that it is quite possible to explain certain phenomena, such as the degeneration of useless parts, without the aid of the Lamarckian principle. Furthermore it can be shown that a not inconsiderable number of instincts, viz. all those which are exercised only once in a lifetime, cannot possibly have arisen by transmitted practice. This fact renders it unnecessary to make use of the Lamarckian principle for the explanation of other kinds of instinct. I do not mean to deny the existence of phenomena for which such an explanation has not yet been found, or at least has not been brought forward; but on the other hand it appears to me that it has never been proved that we cannot dispense with the Lamarckian principle in the explanation of these phenomena. At any rate, I do not know of any facts which could induce us to abandon from the first any hope of finding an explanation without the aid of this hypothesis.

If we are able to prove that we may dispense with the assumption of the transmission of acquired characters in explaining such phenomena, of course it by no means follows that we *must* dispense with it; or, in other words, it does not follow that the transmission of acquired changes cannot take place. It would be as unsafe to make this assertion as to state of a ship seen at a great distance, that it is only moving by sails and not by steam simply because the movement appears to be explicable by sails alone. We ought first to attempt to show that the ship does not possess a steam-engine, or at least that the existence of such an engine cannot be proved.

I believe that I am able to show that the actual existence of the transmission of acquired characters cannot be directly proved; that there are no direct proofs supporting the Lamarckian principle.

If we ask for the facts which can be brought forward by the supporters of the theory of the transmission of acquired characters, if we inquire for the observations which induced Darwin, for instance, to adopt such an hypothesis, or which at least prevented him from rejecting it,—a very brief answer can be given. There are a small number of observations made upon man and the higher animals which seem to prove that injuries or mutilations of the body can, under certain circumstances, be transmitted to the offspring.

A cow which had accidentally lost its horn, produced a calf with an abnormal horn; a bull which had accidentally lost its tail, from that time begat tailless calves: a woman whose thumb had been crushed and malformed in youth, afterwards had a daughter with a malformed thumb, and so on.

In a great number of such cases every guarantee for the trustworthiness of the statements is entirely wanting, and, as His and still earlier Kant have already said, they are of no greater value as evidence than the merest tales. But in other cases this assertion cannot be made without further examination; and a small number of such observations can indeed claim a scientific investigation and value. I shall presently discuss this point in greater detail, but I wish now to lay stress upon the fact that, as far as direct evidence goes, we cannot bring forward any proofs in favour of the transmission of acquired characters, except these cases of mutilations. There are no observations which prove the transmission of functional hypertrophy or atrophy, and it is hardly to be expected that we shall obtain such proofs in future, for the cases are not of a kind which lend themselves to an experimental investigation. The hypothesis that acquired characters can be transmitted is therefore only directly supported by the above-mentioned instances of the transmission of mutilations. For this reason, the defenders of the Lamarckian principle, who have come forward in rather large numbers recently[296], have endeavoured to show that these observations are conclusive, and therefore of the highest importance. For the same reason I believe that it is my duty, as I take the opposite view, to explain what I think of the value of these apparent proofs of transmitted mutilations.

It can hardly be doubted that mutilations are acquired characters: they do not arise from any tendency contained in the germ, but are merely the reaction of the body under external influences. They are, as I have recently expressed it, purely somatogenic characters[1297], viz. characters which emanate from the body (*soma*) only, as opposed to the germ-cells; they are therefore characters which do not arise from the germ itself.

If mutilations must necessarily be transmitted, or even if they might occasionally be transmitted, a powerful support would be given to the Lamarckian principle, and the transmission of functional hypertrophy or atrophy would thus become highly probable. For this reason it is absolutely necessary that we should try to come to a definite conclusion as to whether mutilations can or cannot be transmitted.

We will now consider in greater detail the facts which have hitherto been brought forward upon this point. It is not my purpose to discuss every single case which has been mentioned anywhere or by anybody; such a discussion would hardly lead to any result. I propose to select a small number of such instances, in order to show why they cannot be maintained as proofs. I shall chiefly deal with cases which have been brought forward as especially strong proofs by my opponents, and which have been carefully and completely examined. I shall attempt to show that these are not conclusive and that they must be explained in an entirely different manner. The insufficiency of the proof does not always depend upon the same circumstances, and, according to the latter, we may distinguish different classes of cases.

First of all we may briefly mention those instances in which the necessary precautions have not been taken before drawing conclusions.

To this class belong the tailless cats which were shown at last year's (1887) Meeting of the Association of German Naturalists, at Wiesbaden. These cats had inherited their taillessness, or rather their rudimentary tails, from the mother cat, which 'was said' to have lost her tail by the wheel of a cart having passed over it. Not only did the owner of the cats, Dr. Zacharias, consider them as a proof of the transmission of mutilations, but in a recently-published work, entitled 'On the Origin of Species, based upon the Transmission of acquired characters' ('Ueber die Entstehung der Arten auf Grundlage des Vererbens erworbener Eigenschaften'), the author, Prof. Eimer, speaks of these cats in the preface as a 'valuable' instance of the transmission of mutilations: these examples therefore form part of the foundation upon which the author builds up his theoretical views.

Certainly, the want of tails in young cats, of which the mother had lost its tail by an accident, would have been well worth consideration, but unfortunately there is no trustworthy record as to how the mother cat became tailless. Without absolute certainty upon this point the evidence becomes utterly worthless; and Dr. Zacharias has acted very wisely in afterwards admitting that this is the case, for inherent taillessness has been known in cats for a long time. The tailless race of the Isle of Man is mentioned in the first edition of 'The Origin of Species'; of course I am referring to Darwin's work, and not to the above-mentioned book of the same name, by Prof. Eimer. As to the first origin of the tailless Manx breed we know no more than about the origin of that remarkable race of cats with supernumerary toes, which E. B. Poulton has recently described from Oxford, and has traced through several generations[1298]. These are innate monstrosities which have arisen from unknown changes in the germ. Similar monstrosities have been known for a long time, and no one has ever doubted that they can be transmitted.

It would be equally justifiable to derive the cats with extra toes from an ancestor of which the toes had been trodden upon, as to derive the tailless cats of the Isle of Man from an ancestor of which the tail had been cut off by a cart passing over it, and thus to regard the existence of the race as a proof of the transmission of mutilations.

But even if it were certain that the tail of the mother cat had been mutilated, such a fact would not necessarily prove that the rudimentary tails of the offspring were due to transmission from the mother: they might have been transmitted from the unknown father. This is probably not the case with Dr. Zacharias' cat, for tailless kittens occurred in several families produced by the same mother; but in other cases the possibility of the possession of innate taillessness by the father must be taken into account. The following case is, in this respect, very instructive.

Last summer, my friend, Prof. Schottelius, of Freiburg, brought me a kitten with an innate rudimentary tail, which he had accidentally discovered as one of a family of kittens at Waldkirch, a small town in the southern part of the Black Forest. The mother of the kitten possessed a perfectly normal tail; the father could not be identified.

A closer investigation resulted in the following rather unexpected discovery. For some years past, tailless kittens have frequently appeared in the families of many different mother cats at Waldkirch, and this fact is explained in the following manner. A clergyman, who lived for some time at Waldkirch, had married an English lady who possessed a tailless male Manx cat. The probability that all the tailless cats in Waldkirch are more or less distant descendants of that male cat almost amounts to certainty. Since a male Manx cat has reached the Black Forest, it might equally well arrive at some other place.

But we will now leave observations such as these, which do not prove the transmission of a mutilation, because the mutilation itself has not been established; and we will turn to more serious 'proofs.'

Let us still consider the tails of domesticated animals. In these animals a spontaneous and considerable reduction of the tail occurs not uncommonly, and since the habit of cutting off part of the tail of young animals prevails in many countries, the coincidence has been explained as a causal relation, and the question has been raised whether the disposition towards the spontaneous appearance of rudimentary tails has not arisen in consequence of the artificial mutilation practised through many generations. This supposition appears very plausible at first sight, but the keen scientific criticism of Döderlein, Richter, and Bonnet, together with careful anatomical investigations, have shown that, at least in the cases which were carefully examined, such a causal connection did not exist. It has been shown that the spontaneous rudimentary tails which occasionally appear in cats and dogs have an entirely different origin from the transmission of artificial mutilation. They depend upon an innate peculiarity of the germ, a peculiarity which is easily and strongly transmitted. They are monstrosities, like the sixth finger or toe, or, rather, like the rudimentary fingers and toes, which also occasionally appear. Bonnet[299] has shown that the rudimentary tails of dogs depend upon the absence of several vertebrae, together with an abnormal ossification, and sometimes also with a premature coalescence, of the vertebrae of the tail.

Bonnet states that in the two first cases examined by him the reduction occurred at the distal end of the vertebral column in the tail, the more or less malformed vertebrae being anchylosed. A membranous appendage extended beyond the end of the reduced caudal vertebrae, as the so-called 'soft tail.' These characters were shown to have been inherited from the mother and to have undergone progressive development as regards the number of missing vertebrae and the proportion of individuals with rudimentary tails.

In a third instance Bonnet found that four to seven of the normal caudal vertebrae were absent, and that the column in the region of the tail was characterised by a tendency towards premature anchylosis along its whole length and not merely in its distal portion. Furthermore the last three to four vertebrae were distorted and were either placed transversely to the long axis of the tail, or were so greatly curved that the tip of the tail was directed forwards.

It is obvious that these changes are not such as we should expect as a result of the transmission of the mutilation of the tail which is so commonly practised. If the artificial injury were transmitted we should not expect that a variable number of the mesial vertebrae would be absent, but rather those of the tip. There would be no reason why the existing vertebrae should be degenerate as in the majority of the caudal vertebrae of the dogs examined by Bonnet.

Entirely similar phenomena have been observed by Döderlein in the tailless cats which not infrequently occur in Japan. In these cats the rudimentary vertebrae of the tail were reduced to a short, thin, inflexible spiral, which formed a knot densely covered with hair on the posterior part of the animal.

Such a reduction of the tail occurs quite independently of artificial injury, in individuals of which the parents were not injured: it is even found in races, such as the dachshund, which, as far as we know, have never been habitually mutilated.

But the fact is rendered especially interesting because the reduction of the vertebral column in the region of the tail takes place in very various degrees. Sometimes only four vertebrae are absent, sometimes as many as ten. The degree of abnormality in shape and the degree of coalescence between the vertebrae also differ greatly. Hence Bonnet rightly concludes that a slow and gradual process of reduction is going on in these animals, a process which tends, as it were, to shorten the tail. I intentionally say 'as it were,' for of course the statement must not be taken literally, and we must not conclude that the process of reduction is a consequence of some hypothetical developmental force seated in the organism, of which the purpose is to remove the tail. On the contrary, this instance shows very clearly that the appearance of a development guided in a certain direction may be produced without the existence of any motive developmental force.

The disposition of the tail to become rudimentary, in cats and dogs, may be explained in the simplest way, by the process which I have formerly called panmixia. The tail is now of hardly any use to these animals; and neither dog nor cat would perish because they possessed only an incomplete tail. Hence natural selection does not now exercise any influence over these parts, and an occasional reduction is no longer eliminated by the early destruction of its possessor: therefore such reduction may be transmitted to the offspring.

The race of tailless foxes which, according to Settegast, existed during the present century on the hunting-grounds of Prince Wilhelm zu Solms-Braunfels, very soon disappeared; while cats and dogs with rudimentary tails have been preserved in many cases. Such results are to be expected, because in these domesticated animals the absence of the tail did not cause any inferiority in the struggle for existence.

But these facts appear to me to be remarkable in another direction. I previously mentioned the tailless race of Manx cats. Tradition does not tell us how it happened that the descendants of the first tailless cat in the Isle of Man were able to increase and spread in such a manner as to form the dominant race in the island. But we can easily imagine how it happened, when we learn that tailless cats are especially prized[300] in Japan, because people think that they are better mousers. Every one in Japan wishes to possess a tailless cat, and people even cut off the tails of normal cats when they cannot obtain those with congenital rudimentary tails, because they believe that cats become better mousers in consequence of taillessness. In Waldkirch the same account of the superiority of tailless cats is curiously enough also found. We thus see how a slight but striking variation may at once cause an energetic process of artificial selection, which helps this variation to predominance: a hint for us to be careful in passing judgment upon sexual selection, for the latter also works upon such functionally indifferent but striking variations. In the case of the cats, man has favoured a particular variation, because the novelty rather than the beauty of the character surprised and attracted him. He has attached an imaginary value to the new character, and by artificial selection has helped it to predominate over the normal form. I see no reason why the same process should not take place in animals by the operation of sexual selection.

But now, after this little digression, let us return to the transmission of mutilations.

We have seen that the rudimentary tails of cats and dogs, as far as they can be submitted to scientific investigation, do not depend upon the transmission of artificial mutilation, but upon the spontaneous appearance of degeneration in the vertebral column of the tail. The opinion may, however, be still held that the customary artificial mutilation of the tail, in many races of dogs and cats, has at least produced a number of rudimentary tails, although, perhaps, not all of them. It might be maintained that the fact of the spontaneous appearance of rudimentary tails does not disprove the supposition that the character may also depend upon the transmission of artificial mutilation.

Obviously, such a question can only be decided by experiment: not, of course, experiments upon dogs and cats, as Bonnet rightly remarks, but experiments upon animals the tails of which are not already in a process of reduction. Bonnet proposes that the question should be investigated in white rats or mice, in which the length of the tail is very uniform, and the occurrence of rudimentary tails is unknown.

Before this suggestion was made, I had already attacked the problem experimentally. Such a course might, perhaps, have been more natural to those who maintain the transmission of mutilations, to which I am opposed. Although I undertook the experiments expecting to obtain purely negative results, I thought that the latter would not be entirely valueless; and since the numerous supporters of the transmission of acquired characters do not seem to be willing to test their opinion experimentally, I have undertaken the not very large amount of trouble which is necessary in order to conduct such an experimental test.

The experiments were conducted upon white mice, and were begun in October of last year (1887), with seven females and five males. On October 17 all their tails were cut off, and on November 16 the two first families were born. Inasmuch as the period of pregnancy is only 22-24 days, these first offspring began to develope at a time when both parents were without tails. These two families were together eighteen in number, and every individual possessed a perfectly normal tail, with a length of 11-12 mm. These young mice, like all those born at later periods, were removed from the cage, and either killed and preserved, or made use of for the continuance of the breeding experiments. In the first cage, containing the twelve mice of the first generation, 333 young were born in fourteen months, viz. until January 16, 1889, and no one of these had a rudimentary tail or even a tail but slightly shorter than that of the offspring of unmutilated parents.

It might be urged that the effects of mutilation do not exercise any influence until after several generations. I therefore removed fifteen young, born on December 2, 1887, to a second cage, just after they were able to see, and were covered with hair; their tails were cut off. These mice produced 237 young from December 2, 1887 to January 16, 1889, every one of which possessed a normal tail.

In the same manner fourteen of the offspring of this second generation were put in cage No. 3 on May 1, 1888, and their tails were also cut off. Among their young, 152 in number, which had been produced by January 16, there was not a single one with an abnormal tail. Precisely the same result occurred in the fourth generation, which were bred in a fourth cage and treated in exactly the same manner. This generation produced 138 young with normal tails from April 23 to January 16.

The experiment was not concluded with the fourth generation; thirteen mice of the fifth generation were again isolated and their tails were amputated; by January 16, 1889 they had produced 41 young.

Thus 901 young were produced by five generations of artificially mutilated parents, and yet there was not a single example of a rudimentary tail or of any other abnormity in this organ. Exact measurement proved that there was not even a slight diminution in length. The tail of a newly-born mouse varies from 10.5 to 12 mm. in length, and not one of the offspring possessed a tail shorter than 10.5 mm. Furthermore there was no difference in this respect between the young of the earlier and later generations.

What do these experiments prove? Do they disprove once for all the opinion that mutilations cannot be transmitted? Certainly not, when taken alone. If this conclusion were drawn from these experiments alone and without considering other facts, it might be rightly objected that the number of generations had been far too small. It might be urged that it was probable that the hereditary effects of mutilation would only appear after a greater number of generations had elapsed. They might not appear by the fifth generation, but perhaps by the sixth, tenth, twentieth, or hundredth generation.

We cannot say much against this objection, for there are actual phenomena of variation which must depend upon such a gradual and at first imperceptible change in the germ-plasm, a change which does not become visible in the descendants until after the lapse of generations. The wild pansy does not change at once when planted in garden soil: at first it remains apparently unchanged, but sooner or later in the course of generations variations, chiefly in the colour and size of the flowers, begin to appear: these are propagated by seed and are therefore the consequence of variations in the germ. The fact that such variations *never* occur in the first generation proves that they must be prepared for by a gradual transformation of the germ-plasm.

It is therefore possible to imagine that the modifying effects of external influences upon the germ-plasm may be gradual and may increase in the course of generations, so that visible changes in the body (*soma*) are not produced until the effects have reached a certain intensity.

Thus no conclusive theoretical objections can be brought forward against the supposition that the hereditary transmission of mutilations requires (e. g.) 1000 generations before it can become visible. We cannot estimate *a priori* the strength of the influences which are capable of changing the germ-plasm, and experience alone can teach us the number of generations through which they must act before visible effects are produced.

If therefore mutilations really act upon the germ-plasm as the causes of variation, the possibility or even probability of the ultimate appearance of hereditary effects could not be denied.

Hence the experiments on mice, when taken alone, do not constitute a complete disproof of such a supposition: they would have to be continued to infinity before we could maintain with certainty that hereditary transmission cannot take place. But it must be remembered that all the so-called proofs which have hitherto been brought forward in favour of the transmission of mutilations assert the transmission of a single mutilation which at once became visible in the following generation. Furthermore the mutilation was only inflicted upon one of the parents, not upon both, as in my experiments with mice. Hence, contrasted with these experiments, all such 'proofs' collapse; they must all depend upon error.

It is for this reason important to consider those cases of habitual mutilation which have been continually repeated for numerous generations of men, and have not produced any hereditary consequences. With regard to the habitually amputated tails of cats and dogs I have already shown that there is only an apparently hereditary effect. Furthermore, the mutilations of certain parts of the human body, as practised by different nations from times immemorial, have, in not a single instance, led to the malformation or reduction of the parts in question. Such hereditary effects have been produced neither by circumcision[301], nor the removal of the front teeth, nor the boring of holes in the lips or nose, nor the extraordinary artificial crushing and crippling of the feet of Chinese women. No child among any of the nations referred to possesses the slightest trace of these mutilations when born: they have to be acquired anew in every generation.

Similar cases can be proved to occur among animals. Professor Kühn of Halle pointed out to me that, for practical reasons, the tail in a certain race of sheep has been cut off, during the last hundred years, but that according to Nathusius, a sheep of this race without a tail or with only a rudimentary tail has never been born. This is all the more important because there are other races of sheep in which the shortness of the tail is a distinguishing peculiarity. Thus the nature of the sheep's tail does not imply that it cannot disappear.

A very good instance is mentioned by Settegast, although perhaps with another object in view. The various species of crows possess stiff bristle-like feathers round the opening of the nostrils and the base of the beak: these are absent only in the rook. The latter, however, possesses them when young, but soon after it has left the nest they are lost and never reappear. The rook digs deep into the earth in searching for food, and in this way the feathers at the base of the beak are rubbed off and can never grow again because of the constant digging. Nevertheless this peculiarity, which has been acquired again and again from times immemorial, has never led to the appearance of a newly hatched individual with a bare face.

Thus there is no reason for the assumption that such a result would occur in the case of the mice even if the experiments had been continued through hundreds or thousands of generations. The supposition of the accumulative effect of mutilation is entirely visionary, and cannot be supported except by the fact that accumulative transformations of the germ-plasm occur; but of course this fact does not imply that mutilations belong to those influences which are capable of changing the germ-plasm. All the ascertained facts point to the conclusion that they have not this effect. The transmission is all the more improbable because of the striking form of the mutilation in all cases which are relied upon as evidence. The only objection which can be raised is to suppose that the absence of the tail is less easily transmitted than other mutilations, or that mice possess smaller hereditary powers than other animals. But there is not the slightest evidence in favour of either of these suggestions; the supporters of the Lamarckian principle have, on the contrary, always pointed to the transmission of mutilated tails as one of their principal lines of evidence.

The opinion has often been expressed that such transmission need not occur in every case, but may happen now and then under quite exceptional conditions with which we are unacquainted: for this reason it might be urged that all negative experiments and every refutation of the 'proofs' of the transmission of mutilations are not conclusive. Only recently, a clever young zoologist said in reference to Kant's statements upon the subject, that perhaps the most decided opponent of the transmission of mutilations would not venture nowadays to maintain his view with such certainty, 'for it must be admitted that the transmission of acquired characters may take place at any rate as a rare exception.' Similar opinions are often expressed, especially in conversation, and yet they can mean nothing except that the transmission of acquired characters has been proved; for if such transmission can take place at all, it exists, and it does not make the least difference theoretically whether it occurs in rare cases or more frequently. Sometimes heredity has been called capricious, and in a certain sense this is true. Heredity appears to be capricious because we cannot penetrate into its depths: we cannot predict whether any peculiar character in the father will reappear in the child, and still less whether it will reappear in the first, second, or one of the later children: we cannot predict whether a child will possess the nose of his father or mother or one of the grandparents. But this certainly does not imply that the results are due to chance: no one has the right to doubt that everything is brought about by the operation of certain laws, and that, with the fertilization of the egg, the shape of the nose of the future child has been determined. The co-operation of the two tendencies of development contained in the two conjugating germ-cells produces of necessity a certain form of nose. The observed facts enable us to know something of the laws under which such events take place. Thus, for instance, among a large number of children of the same parents some will always have the form of the nose of the mother or of the mother's family; others will have the nose of the father's family, and so on.

If we apply this argument to the supposed transmission of mutilations, such transmission, if possible at all, must occur a certain number of times in a certain number of cases: it must occur more readily when both parents are mutilated in the same way, or when the mutilation has been repeated in many generations, etc. It is extremely improbable that it would suddenly occur in a case where it was least expected, while it did not occur in 900 cases of the most favourable kind. Those who recognise in the doubtful cases of transmission of a single mutilation present in only one of the parents, proofs of the existence of the disputed operation of heredity, quite forget that the transmission presupposes a very marvellous and extremely complex apparatus which if present at all ought, under certain conditions, to become manifest regularly, and not only in extremely exceptional cases. Nature does not create complex mechanisms in order to leave them unused: they exist by use and for use. We can readily imagine how complex the apparatus for the transmission of mutilations or acquired characters generally must be, as I have tried to show in another place. The transmission of a scar to the offspring e. g. presupposes first of all that each mechanical alteration of the body (*soma*) produces an alteration in the germ-cells: this alteration cannot consist in mere differences of nutrition, only affecting an increased or decreased growth of the cells: it must be of such a kind that the molecular structure of the germ-plasm would be changed. But such a change could not in the least resemble that which occurred at the periphery of the body in the formation of the scar: for there is neither skin nor the preformed germ of any of the adult organs in the germ-plasm, but only a uniform molecular structure which, in the course of many thousand stages of transformation, must tend to the formation of a soma including a skin. The change in the germ-plasm which would lead to the transmission of the scar, must therefore be of such a kind as to influence the course of ontogeny in one of its later stages, so that an interruption of the normal formation of skin, and the intercalation of the tissue of the scar, would occur at a certain part of the body. I do not maintain that equally minute changes of the germ-plasm could not occur: on the contrary, individual variation shows us that the germ-plasm contains potentially all the minutest peculiarities of the individual; but I have in vain tried to understand how such minute changes of the germ-plasm in the germ-cells could be caused by the appearance of a scar or some other mutilation of the body. In this respect I think that Blumenbach's condition is nearly fulfilled: he was inclined to declare himself against the transmission of mutilations, but only if it were proved that such transmission was *impossible*. Although this cannot be strictly proved, it can nevertheless be shown that the apparatus presupposed by such transmission must be so immensely complex, nay! so altogether inconceivable, that we are quite justified in doubting the possibility of its existence as long as there are no facts which prove that it *must* be present. I therefore do not agree with the recent assertion[302] that Blumenbach's condition cannot be fulfilled to-day, just as it was impossible at the time when it was first brought forward. But if nevertheless such a mysterious mechanism existed between the parts of the body and the germ-cells, by means of which each change in the former could be reproduced in a different manner in the latter, the effects of this marvellous mechanism would certainly be perceptible and could be subjected to experiment.

But at present we have no evidence of the existence of any such effects; and the experiments described above disprove all the cases of the supposed transmission of single mutilations.

Of course, I do not maintain that such cases are to be always explained by want of sufficient observation. In order to make my position clear, I propose to discuss two further classes of observations. First of all, there are very many cases of the apparent transmission of mutilations in which it was not the mutilation or its consequences which was transmitted, but the predisposition of the part in question to become diseased. Richter[303] has recently pointed out that arrests of development, so slight as to be externally invisible, frequently occur, and that such arrests exhibit a tendency to lead to the visible degeneration of parts in which they occur, as the result of slight injuries. Since therefore the predisposition towards such arrest is transmitted by the germ—occasionally even in an increased degree—the appearance of a transmitted injury may arise. In this way Richter explains, for instance, the frequently quoted case of the soldier who lost his left eye by inflammation fifteen years before he was married, and who had two sons with left eyes malformed (microphthalmic). Microphthalmia is an arrest of development. The soldier did not lose his eye simply because it was injured, but because it was predisposed to become diseased from the beginning and readily became inflamed after a slight injury. He did not transmit to his sons the injury or its results, but only microphthalmia, the predisposition towards which was already innate in him, but which led in his sons from the beginning, and without any obvious external injury, to the malformation of the eye. I am inclined to explain the case which Darwin in a similar manner adduced, during the later years of his life, in favour of the transmission of acquired characters, and which seemed to prove that a malformation of the thumb produced by chilblains can be transmitted. The skin of a boy's thumbs had been badly broken by chilblains associated with some skin disease. The thumbs became greatly swollen and remained in this state for a long time; when healed they were malformed, and the nails always remained unusually narrow, short, and thick. When this man married and had a family, two of his children had similarly malformed thumbs, and even in the next generation two daughters had malformed thumbs on both hands. The case is too imperfectly known to admit of adequate criticism; but one may perhaps suggest that the skin of different individuals varies immensely in its susceptibility to the effects of cold, and that many children have chilblains readily and badly, while others are not affected in this way. Sometimes members of the same family vary in this respect, and the greater or less predisposition towards the formation of chilblains corresponds with a different constitution of the skin, in which some children follow the father and others follow the mother. In Darwin's instance a high degree of susceptibility of the skin of the thumb was obviously innate in the father, and this susceptibility was certainly transmitted, and led to the similar malformation of the thumbs of the children, perhaps very early and after the effect of a comparatively slight degree of cold[304].

The last class of cases which I should wish to consider, refer to observations in which the mutilation of the parent was certain, and in which a malformation similar to the mutilation had appeared in the child, but in which exact investigation shows that the malformations in parent and child do not in reality correspond to each other.

In this class I include an instance which has only become known during the present year (1888), and which has been observed as exactly as possible by an anthropologist and physician, who has worked up the history of the case. Dr. Emil Schmidt communicated to this year's meeting of the German Anthropologists' Association at Bonn a case which indeed seems at first sight to prove that mutilations of the human ear can be transmitted. As Dr. Schmidt has been kind enough to place at my disposal all the material which he collected upon the subject, I have been able to examine it more minutely than is generally possible in such cases; and I will discuss it in detail, as it seems to me to be of fundamental importance in the history of human errors upon this subject.

In a most respectable and thoroughly trustworthy family, the mother possesses a cleft ear-lobe upon one side. She quite distinctly remembers that when playing, between the ages of six and ten years, another child tore out her ear-ring, and that the wound healed so that the cleft remained. Later on a new hole was made in the posterior part of the lobe. She had seven children, and the second of these, who is now a full-grown man, has a cleft ear-lobe on the same side as the mother. It is not known whether the mother possessed an innate malformation of the ear before it was mutilated, but, judging from the present appearance of the ear, this is extremely improbable. Furthermore, the existence of an innate cleft in the ear-lobe has never been previously observed. The parents of the mother did not possess any malformation of the ear. The conclusion seemed to be therefore inevitable that the transmission of an artificial cleft in the ear-lobe had really taken place.

But we must not be too hasty in forming an opinion. When we compare the figures I. and II., representing the two ears, we are first of all struck by the fact that the malformation of the ear of the son has an entirely different appearance from that of the mother. The ear-lobe of the latter is quite normally formed; it is broad and well-developed, and only exhibits a median vertical furrow which is the result of the mutilation. The ear-lobe of the son, on the other hand, is extremely minute, one might even maintain that it is completely wanting. In my opinion a cleft is not present at all, but the far higher posterior corner of the ear forms the end of its posterior margin—the so-called helix. But even if another opinion were pronounced with regard to the interpretation of this part, there is one other circumstance to be taken into account, which appears to me to be absolutely conclusive, and which completely excludes the interpretation of this malformation as the transmission of a mutilation.

Fig. I. *H.* Helix. *Cr. Ah.* Crura anthelicis. *Ah.* Anthelix. *Cch.* Concha. *Hl^1.* and *Hl^2.* Holes 1 and 2 for ear-rings. *Lob.* Ear-lobe. *Sp. H.* Spina helicis. *Inc.* Incisura intertragica. *Tr.* Tragus. *Atr.* Antitragus.

If we compare the ears with each other, that of the mother with that of the son, not only the anatomist but every trained observer will at once be struck by the fact that they are totally different in their outlines as well as in every detail. The upper margin of the ear is very broad in the mother, in the son it is quite pointed: the so-called *crura anthelicis* are normally developed in the mother, in the son they can hardly be distinguished and open in an anterior direction, while in the mother they are directed upwards. The concha itself, the *incisura intertragica*, in short everything in the two ears, is as different as it can possibly be in the ears of two individuals.

But this fact obviously proves that the son does not possess the ear of his mother, but probably that of his father or grandfather. Unfortunately the father and grandfather have been now dead for a long time, so that we cannot obtain certain evidence upon this point. At all events, the son does not possess the ear of his mother, and it would be very rash to suppose that he has inherited the ear from the father, but the malformation of the ear-lobe from the mother—a malformation which, as it seems to me, is certainly quite different from that of his mother's ear. I said that this case was of fundamental importance chiefly because it shows very distinctly, on the one hand, how difficult it is to bring together the material which is absolutely necessary for the correct understanding of a single case, and on the other hand, how carefully the abnormalities must be compared and examined if we wish to escape wrong conclusions. Such precautions have hitherto been rarely taken with the necessary accuracy; people are in most cases satisfied with the knowledge that an abnormality is present in the child on the same part which had been malformed by mutilation in the parent.

But if we are to speak of the transmission of a mutilation, it must be shown, before everything else, that the malformation of the child corresponds precisely to the mutilation of the parent.

For this reason the older observations upon this subject are, in most cases, entirely valueless.

The readiness with which we may be deceived is shown by the fact that I myself nearly became a victim during the past year (1888). A friend of mine, in order to convince me of the transmission of mutilations, called my attention to a linear scar on his left ear, which extended from the upper margin of the helix for some distance upon the posterior part of the anthelix, giving it the appearance of a small, rather sharp ridge. The scar had been caused by a cut from a duelling sword, which the gentleman had received during his residence at the University. Strangely enough, the left ear of his daughter, who is five years old, exhibits a similar peculiarity. The posterior part of the anthelix forms a rather sharp and narrow ridge like that of the father, although the scar itself is wanting.

I must admit that I was at first rather puzzled by this fact, but the mystery was soon solved in a very simple manner. I asked the father to show me his right ear, and I then saw that this ear possessed a similar ridge on the posterior part of the anthelix. Only the scar was absent, which in the left ear brought the crest of the ridge into still greater prominence. The ridge was therefore only an individual peculiarity in the formation of the ear of the father,—a peculiarity which had been transmitted to one ear of the child. No transmission of the mutilation had taken place.

In the same manner, many of the so-called proofs of the transmission of mutilations would be shown, by a careful examination, to be deceptive. We must not expect to succeed in all of them, for in most cases the investigation cannot be completed, chiefly because the condition of the part in question in the ancestors is not known or is only known in an insufficient manner. This is the reason why fresh examples of such so-called proofs continue to appear from time to time,— proofs which do not admit of a searching criticism because something, and in most cases very much, is invariably wanting. But it will be admitted that even a very large number of incomplete proofs do not make a single complete one. On the other hand, it may be asserted that a single instance of coincidence between a mutilation in the parent and a malformation in the offspring, even if well established, would not constitute a proof of the transmission of mutilations. Not every *post hoc* is also a *propter hoc*. Nothing illustrates this better than a comparison between the 'proofs' which are even now brought forward in favour of the transmission of mutilations and the 'proofs' which supported the belief in the efficacy of so-called 'maternal impressions' during pregnancy, a belief which was universally maintained up to the middle of the present century. Many of those 'proofs' were simply old wives' fables, and were based upon all kinds of subsequent inventions and alterations. But it cannot be denied that there are a few undoubtedly genuine observations upon cases in which some character in the child reminds us in a striking manner of a deep psychical impression by which the mother was strongly affected during pregnancy.

Thus a trustworthy person told me of the following case. A well-known medical authority cut his leg above the ankle with a knife: his wife was present at the time and was much frightened. She was then in the third month of pregnancy: the child when born was found to have an unusual mark upon the same place above the ankle. People almost forget nowadays the tenacity with which the idea of maternal impressions was kept up until the middle of this century; but it is only necessary to read the received German text-book on physiology of fifty years ago, viz. that of Burdach, in order to be convinced of the accuracy of this statement. Not only does Burdach give a number of 'conclusive' cases in man and even in animals (cows and deer), but he also attempts to construct a theoretical explanation of the supposed process. This is undertaken in the following manner,—'Imagination influences the function of organs;' but the function of the embryo is the 'tendency towards development, and hence the influence [of maternal imagination] can make itself felt only as variations in the mode of development.' Thus by exchanging the conception of function for that of the development of organs, Burdach comes to the conclusion that 'homologous organs of the mother and the embryo are in such connexion' that when the former are disturbed a corresponding 'change in the formation of the latter may arise.'

It seems to be not without value for the appreciation of the questions with which we are dealing to remember that the idea of 'maternal impressions' was only comparatively recently believed to be a scientific theory, and that the proofs in support of it were brought forward in form and language as scientific proofs. In Burdach's book we even meet with detailed 'proofs' that violent mental shocks produced by maternal impressions may not only exercise their influence upon one but even upon several children born successively, although with diminishing strength. 'A young wife received a shock during her first pregnancy upon seeing a child with a hare-lip, and she was constantly haunted with the idea that her child might have the same malformation. She was delivered of a child with a typical hare-lip: her next child had an upper lip with a less-marked cleft; while the third possessed a red mark instead of a cleft.'

Now what can be said about such 'proofs'? We may probably rightly conjecture that Burdach, who was in other respects a clever physiologist, was in this subject somewhat credulous: but there are also instances about which there is not the slightest doubt. I may remind the reader of a case which has been told by no other than the celebrated embryologist, Carl Ernst von Baer[305].

'A lady was very much upset by a fire, which was visible at a distance, because she believed that it was in her native place. As the latter was seven German miles distant, the impression had lasted a long time before it was possible to receive any certain intelligence, and this long delay affected the mind of the lady so greatly, that for some time afterwards she said that she constantly saw the flames before her eyes. Two or three months afterwards she was delivered of a daughter who had a red patch on the forehead in the form of a flame. This patch did not disappear until the child was seven years old.' Von Baer added, 'I mention this case because I am well acquainted with it, for the lady was my own sister, and because she complained of seeing flames before her eyes before the birth of the child, and did not invent it afterwards as the "cause" of the strange appearance.'

Here then we have a case which is absolutely certain. Von Baer's name is a guarantee for absolute accuracy. Why then has science, in spite of this, rejected the whole idea of the efficacy of 'maternal impressions' ever since the appearance of the treatises by Bergmann and Leuckart[306]?

Science has rejected this idea for many and conclusive reasons, all of which I am not going to repeat here. In the first place, because our maturer knowledge of the physiology of the body shows that such a causal connexion between the peculiar characters of the child and, if I may say so, the corresponding psychical impressions of the mother, is a supposition which cannot be admitted; but also and chiefly because a single coincidence of an idea of the mother with an abnormality in the child does not form the proof of a causal connexion between the two phenomena.

I do not doubt that among the many thousands of present and past students in German Universities, whose faces are covered with scars, there may be one with a son who exhibits a birth-mark on the spot where the father possesses a scar. All sorts of birth-marks occur, and why should they not sometimes have the appearance of a scar? Such a case, if it occurred, would be acceptable to the adherents of the theory of the transmission of acquired characters; it would in their opinion completely upset the views of their opponents.

But how could such a case, if it were really established, be capable of proving the supposed form of hereditary transmission, any more than von Baer's case could prove the theory of the efficacy of 'maternal impressions'?

I am of opinion that the extraordinary rarity of such cases strongly enforces the fact that we have to do with an accidental and not a causal coincidence. If scars could be really transmitted, we should expect very frequently to find birth-marks which correspond to scars upon the face of the father,—viz. in almost all cases in which the son had inherited the type of face possessed by the father. If this were so we should have to be seriously concerned about the beauty of the next generation in Germany, as so many of our undergraduates follow the fashion of decorating their faces with as many of these 'honourable scars' as possible.

I have spoken of 'maternal impressions' because I wished to show that, until quite recently, distinguished and acute scientific men have adhered to an idea, and believed that they possessed the proof of an idea, which has now been completely and for ever abandoned by science. But in addition to this, there is a very close connexion between the theory of the efficacy of maternal impressions and that of the transmission of acquired characters, and sometimes they are even confounded together.

Last year a popular scientific journal quoted the following case as a proof of the transmission of mutilations. I do not, however, wish to imply that the editor must be held responsible for the errors of a correspondent. 'In November, 1864, a pregnant merino sheep broke its right fore-leg, about two inches above the knee-joint; the limb was put in splints and healed a long time before the following March, when the animal produced young. The lamb possessed a ring of black wool from two to three inches in breadth round the place at which the mother's leg had been broken, and upon the same leg.' Now if we even admitted that a ring of black wool could be looked upon as a character which corresponds to the fracture of the mother's leg, the case could not possibly be interpreted as the transmission of a mutilation, but as an instance of the efficacy of maternal impressions; for the ewe was already pregnant when she fractured her leg. The present state of biological science teaches us that, with the fusion of egg and sperm-cell, potential heredity is determined[307]. Such fusion determines the future fate of the egg-cell and the individual with all its various tendencies.

Such tales, when quoted as 'remarkable facts which prove the transmission of mutilations,' thoroughly deserve the contempt with which they have been received by Kant and His. When the above-mentioned instance was told me, I replied, 'It is a pity that the black wool was not arranged in the form of the inscription "To the memory of the fractured leg of my dear mother."'

The tales of the efficacy of 'maternal impressions' and of the transmission of mutilations are closely connected, and break down before the present state of biological science. No one can be prevented from believing such things, but they have no right to be looked upon as scientific facts or even as scientific questions. The first was abandoned in the middle of the present century, and the second may be given up now; when once discarded we need not fear that it will ever again be resuscitated.

It is hardly necessary to say that the question as to the transmission of acquired characters is not completely decided by the unconditional rejection of the transmission of mutilations. Although I am of opinion that such transmission does not take place, and that we can explain the phenomena presented by the transformation of species without this supposition, I am far from believing that the question is settled, simply because the transmission of mutilations may be dismissed into the domain of fable. But at all events we have gained this much,—that the only facts which appear to directly prove a transmission of acquired characters have been refuted, and that the only firm foundation on which this hypothesis has been hitherto based has been destroyed. We shall not be obliged, in future, to trouble about every single so-called proof of the transmission of mutilations, and investigation may be concentrated upon the domain in which lies the true decision as to the Lamarckian principle, it may be concerned with the explanation of the observed phenomena of transformation.

If, as I believe, these phenomena can be explained without the Lamarckian principle, we have no right to assume a form of transmission of which we cannot prove the existence. Only if it could be shown that we cannot now or ever dispense with the principle, should we be justified in accepting it. I do not think that I can represent the state of the subject better than by again referring to the metaphor of the ship. We see it moving along with all sails set, we can discern the presence of neither paddles nor screw, and as far as we can judge there is no funnel, nor any other sign of an engine. In such a case we shall not be justified in concluding that an engine is present and has some share in the movement of the vessel, unless the movement is of such a kind that it is impossible to explain it as due to the unaided action of the wind, the current, and the rudder. Only if the phenomena presented by the progress of organic evolution are proved to be inexplicable without the hypothesis of the transmission of acquired characters, shall we be justified in retaining such an hypothesis.

Footnotes for Essay VIII.

293. It is true that he based his opinions upon entirely erroneous theories as to the constancy of species. Compare Brock, 'Einige ältere Autoren über die Vererbung erworbener Eigenschaften' in 'Biolog. Centralblatt,' Bd. VIII, p. 491 (1888): see also Hugo Spitzer, 'Beiträge zur Descendenz-theorie und zur Methodologie der Naturwissenschaft,' Leipzig, 1886, pp. 515 et seq.

294. W. His, 'Unsere Körperform,' Leipzig, 1875.

295. See Essay II in the present volume.

296. [One of the most remarkable forms of this revival of Lamarckism is the establishment in America of a 'Neo-Lamarckian School,' which includes among its members many of the most distinguished American biologists. One of the arguments upon which the founders of the school have chiefly relied is derived from the comparative morphology of mammalian teeth. The evolution of the various types are believed to be due to modifications in shape, produced by the action of mechanical forces (pressure and friction) during the life of the individual. The accumulation of such modifications by means of heredity explains the forms of existing teeth.

It may be reasonably objected that the most elementary facts concerning the development of teeth prove that their shapes cannot be altered during the lifetime of the individual, except by being worn away. The shape is predetermined before the tooth has cut the gum. Hence the Neo-Lamarckian School assumes, not the transmission of acquired characters, but the transmission of characters which the parent is unable to acquire!—E. B. P.]

297. See p. 412 of the preceding Essay (VII).

298. [See 'Nature,' vol. xxix. p. 20, and vol. xxxv. p. 38. In the latter article nine generations are recorded, and in both articles figures of the normal and abnormal feet are given. Additional generations and many more families have been since observed, and an account of these observations will shortly be published in the same paper. The breed originally came from Bristol. In the observations recorded, the abnormality of the offspring is an indication of the hereditary strength of the female parents, while the degree of normality is a similar test of heredity through the male parents; for the female parents were always abnormal, the male parents always normal.

The most abnormal kitten observed possessed seven toes on each forefoot, seven toes on the right hind foot (three more than the normal number), and six on the left hind foot. Kittens with seven toes on the forefeet and six on the hind were comparatively common, and all intermediate conditions between this and the normal were of frequent occurrence. Cats with extra toes are, I think, not uncommon in most countries, and the fact that the peculiarity is transmitted is also well known. The object of the investigation alluded to was to observe the transmission systematically through many generations.—E. B. P.]

299. Bonnet, 'Die stummelschwänzigen Hunde im Hinblick auf die Vererbung erworbener Eigenschaften,' Anat. Anzeiger, Bd. III, 1888, p. 584; see also 'Beiträge zur patholog. Anatomie und allgem. Pathologie' by Ziegler and Nauwerck, Bd. IV, 1888.

300. See the interesting remarks by Döderlein on this point, which Dr. Ischikawa of Japan tells me are quite correct. Döderlein, 'Ueber schwanzlose Katzen,' Zool. Anzeiger, vol. x. Nov. 1887, No. 265.

301. It is certainly true that among nations which practise circumcision as a ritual, children are sometimes born with a rudimentary prepuce, but this does not occur more frequently than in other nations in which circumcision is not performed. Rather extensive statistical investigations have led to this result.

302. See Brock, 'Biolog. Centralblatt,' Bd. VIII. p. 497, 1888.

303. W. Richter, 'Zur Vererbung erworbener Charaktere,' Biolog. Centralblatt, Bd. VIII. 1888, p. 289.

304. This case was not observed by Darwin himself, but was communicated to him by J. P. Bishop of Perry, in North America (see 'Kosmos,' vol. ix. p. 458). Quite apart from the fact that it is by no means certain whether the father did not already possess an innate malformation of the thumb, exact data are wanting as to the time during which the thumb was diseased, and as to the time when the malformation of the thumb was first observed in the children and the grandchildren; whether at birth or at a later period. For a thorough criticism it would also be necessary to have figures of the thumbs. I should not have alluded to this case, because of its incomplete history, if it had not appeared to me to illustrate the ideas mentioned above. Of course I do not maintain that I have suggested the right explanation in this particular case. It is possible that the father possessed an inherent malformation of the thumb which he had forgotten by the time that he came to have children and grandchildren, and was struck by the abnormality of their thumbs.

305. See Burdach, 'Lehrbuch der Physiologie,' Bd. II, 1835-40, p. 128.

306. See Handwörterbuch der Physiologie von Rud. Wagner, Artikel 'Zeugung,' von Rud. Leuckart.

307. See V. Hensen, 'Physiologie der Zeugung.' Leipzig, 1881.

INDEX.

polar bodies of, <u>218</u>.

Molluscs, length of life of, <u>55</u>;

determined by markings on shell, <u>14</u>;

enemies of, <u>58</u>.

Monoplastides, <u>115</u>, <u>122</u>, <u>125</u>, <u>146</u>, <u>159</u>;

definition of, <u>120</u>;

reproduction of, <u>149</u>.

Mouse, length of life of, <u>6</u>; gestation of, <u>7</u>.

Müller, F., on heredity of acquired characters, <u>320</u>, <u>322</u>.

Müller, H., on colours of flowers, <u>259</u>;

on nectaries, <u>307</u>.

Multicellular organisms, division of labour in, <u>27</u>.

Musca domestica, length of life of, <u>43</u>.

Musca vomitoria, polar bodies of, <u>353</u>.

Myopia, <u>89</u>.

Nägeli, <u>167</u>, <u>171</u>, <u>175</u>;

on idioplasm, <u>174</u>, <u>182</u>, <u>190</u>, <u>201</u>, <u>340</u>, <u>414</u>;

on inherent tendency to vary, <u>256</u>, <u>298</u>;

on Alpine plants, <u>269</u>;

on adaptation, <u>300</u>;

on medium of heredity, <u>318</u>, <u>355</u>.

Najadae, length of life of, <u>56</u>.

Natica heros, length of life of, <u>56</u>.

Nautilus, persistence of, <u>300</u>.

Nematodes, polar bodies of, <u>188</u>;

Vespa, <u>53</u>.

Viola calcarata, fertilization of, <u>310</u>;

tricolor, <u>414</u>.

Vitrinae, length of life of, <u>55</u>.

Volvocineae, <u>202</u>, <u>248</u>.

Volvox, <u>204</u>, <u>248</u>.

Vorticellidae, conjugation of buds of, <u>287</u>.

Vultures, length of life of, <u>11</u>, <u>37</u>.

Wallace, on constancy of number of individuals in successive generations, <u>12</u>;

on production of death by natural selection, <u>23</u>.

Wasps, duration of life of male and female, <u>18</u>, <u>53</u>;

loss of embryonic limbs in development of larva of, <u>89</u>.

Westphal, on epilepsy in guinea-pigs, <u>314</u>.

Whales, length of life of, <u>6</u>; adaptation in, <u>261</u>.

White mice, experiments in the transmission of mutilations on, <u>431</u>.

Will, on origin of ova, <u>222</u>.

Wolff, on theory of epigenesis, <u>316</u>.

Zacharias, on the inheritance of mutilations, <u>426</u>.

THE END.

Transcriber's notes:

Footnotes were collected at the end of each essay and of the appendices for each essay.

On page <u>198</u> the formula $(1/p)$ should probably be $(1/pn)$.

Obvious typographical errors have been corrected. Any inconsistencies in spelling have been left.

"Notes" as used in this volume often means the same as footnotes. The Appendix to Essay I contains the "Notes" as used in that essay. In the footnotes to the Appendix to Essay I, one note refers to a note on another page which has been collected with all the other notes. The note referred to is indicated by a local note in braces.

In the footnotes for Essay VII this

"See 'Nature,' vol. 36, pp. 491-407.—E. B. P."

was changed to

"See 'Nature,' vol. 36, pp. 491-507.—E. B. P."

on supposition of a typo.

www.ingramcontent.com/pod-product-compliance
Lightning Source LLC
Chambersburg PA
CBHW051853170526
45168CB00001B/84